The Ecological Basis of Conservation

The Ecological Basis of Conservation

Heterogeneity, Ecosystems, and Biodiversity

Edited by

S.T.A. Pickett[1]   R.S. Ostfeld[1]
M. Shachak[1,2]   G.E. Likens[1]

[1]Institute of Ecosystem Studies
Mary Flagler Cary Arboretum
Millbrook, New York

[2]Mitrani Center for Desert Ecology
The Jacob Blaustein Institute for Desert Research
Ben-Gurion University of the Negev, Sede Boker, Israel

 CHAPMAN & HALL

 **International Thomson Publishing**
Thomson Science

New York • Albany • Bonn • Boston • Cincinnati • Detroit • London • Madrid • Melbourne
Mexico City • Pacific Grove • Paris • San Francisco • Singapore • Tokyo • Toronto • Washington

Cover photo courtesy of: Jerry Franklin

Cover design: Carolyn Walaitis

Printed in the United States of America

Chapman & Hall
115 Fifth Avenue
New York, NY 10003

Thomas Nelson Australia
102 Dodds Street
South Melbourne, 3205
Victoria, Australia

International Thomson Editores
Campos Eliseos 385, Piso 7
Col. Polanco
11560 Mexico D.F
Mexico

International Thomson Publishing Asia
221 Henderson Road #05-10
Henderson Building
Singapore 0315

Chapman & Hall
2-6 Boundary Row
London SE1 8HN
England

Chapman & Hall GmbH
Postfach 100 263
D-69442 Weinheim
Germany

International Thomson Publishing-Japan
Hirakawacho-cho Kyowa Building, 3F
1-2-1 Hirakawacho-cho
Chiyoda-ku, 102 Tokyo
Japan

3 4 5 6 7 8 9 10 XXX 01 00 99 98 97

Library of Congress Cataloging-in-Publication Data

The ecological basis of conservation : heterogeneity, ecosystems, and
    biodiversity / edited by S.T. A. Pickett.
        p.   cm.
    Originated with the Sixth Cary Conference held at the Institute of
Ecosystem Studies, May 1995.
    Includes bibliographical references and index.
    ISBN 0-412-09851-2 (HB : alk. paper)
    1. Conservation biology--Congresses.   2. Ecology--Congresses.
I. Pickett, Steward T.   II. Cary Conference (6th : 1995 :
Institute of Ecosystem Studies)
QH75.E25   1996
333.95 ' 16--dc20   96-18325
                                                                          CIP
Chapters 3 and 33 © 1997 by Joel E. Cohen

British Library Cataloguing in Publication Data available

To order this or any other Chapman & Hall book, please contact **International
Thomson Publishing, 7625 Empire Drive, Florence, KY 41042.**
Phone: (606) 525-6600 or 1-800-842-3636.
Fax: (606) 525-7778. e-mail: order@chaphall.com.

For a complete listing of Chapman & Hall titles, send your request to
Chapman & Hall, Dept. BC, 115 Fifth Avenue, New York, NY 10003.

Contents

Foreword: *Bruce Babbitt*   ix

Preface   xi

Contributors   xv

Participants   xix

I. Introduction: The Needs for a Comprehensive Conservation Theory 1

  1. Defining the Scientific Issues   3
    R. S. Ostfeld, S. T. A. Pickett, M. Shachak, and G. E. Likens

  2. Part 1. *Gretchen Long Glickman*—Science, Conservation,
    Policy, and the Public   11
    Part 2. *H. Ronald Pulliam*—Providing the Scientific Information
    that Conservation Practitioners Need   16
    Part 3. *Michael J. Bean*—A Policy Perspective on Biodiversity
    Protection and Ecosystem Management   23

  3. Conservation and Human Population Growth: What are
    the Linkages?   29
    Joel E. Cohen

  4. Developing an Analytical Context for Multispecies
    Conservation Planning   43
    Barry Noon, Kevin McKelvey, and Dennis Murphy

  5. Operationalizing Ecology under a New Paradigm:
    An African Perspective   60
    Kevin H. Rogers

II. Foundations for a Comprehensive Conservation Theory　　　**81**
Themes—*S. T. A. Pickett, R. S. Ostfeld, M. Shachak, & G. E. Likens*

　6. The Paradigm Shift in Ecology and Its Implications
　　 for Conservation　　　83
　　 Peggy L. Fiedler, Peter S. White, and Robert A. Leidy

　7. The Emerging Role of Patchiness in Conservation Biology　　　93
　　 John A. Wiens

　8. Linking Ecological Understanding and Application: Patchiness
　　 in a Dryland System　　　108
　　 Moshe Shachak and S. T. A. Pickett

III. Biodiversity and Its Ecological Linkages　　　**123**
Themes—*R. S. Ostfeld, S. T. A. Pickett, M. Shachak, & G. E. Likens*

　9. The Evaluation of Biodiversity as a Target for Conservation　　　125
　　 M. Philip Nott and Stuart L. Pimm

　10. Conserving Ecosystem Function　　　136
　　 Judy L. Meyer

　11. The Relationship between Patchiness and Biodiversity in
　　 Terrestrial Systems　　　146
　　 Lennart Hansson

　12. Reevaluating the Use of Models to Predict the Consequences of
　　 Habitat Loss and Fragmentation　　　156
　　 Peter Kareiva, David Skelly, and Mary Ruckelshaus

　13. Managing for Heterogeneity and Complexity on
　　 Dynamic Landscapes　　　167
　　 Norman L. Christensen, Jr.

　14. Toward a Resolution of Conflicting Paradigms　　　187
　　 S. L. Tartowski, E. B. Allen, N. E. Barrett, A. R. Berkowitz,
　　 R. K. Colwell, P. M. Groffman, J. Harte, H. P. Possingham,
　　 C. M. Pringle, D. L. Strayer, and C. R. Tracy

　15. The Land Ethic of Aldo Leopold　　　193
　　 A. Carl Leopold

IV. Toward a New Conservation Theory　　　**201**
Themes—*R. S. Ostfeld, S. T. A. Pickett, M. Shachak, and*
G. E. Likens

　16. The Future of Conservation Biology: What's a Geneticist to Do?　　202
　　 Kent E. Holsinger and Pati Vitt

17. Habitat Destruction and Metapopulation Dynamics   217
    Ilkka Hanski

18. How Viable is Population Viability Analysis?   228
    Katherine Ralls and Barbara L. Taylor

19. Reserve Design and the New Conservation Theory   236
    Nels E. Barrett and Juliana P. Barrett

20. Ecosystem Processes and the New Conservation Theory   252
    John J. Ewel

21. Measurement Scales and Ecosystem Management   262
    Doria R. Gordon, Louis Provencher, and Jeffrey L. Hardesty

22. Biogeographic Approaches and the New Conservation Biology   274
    Daniel Simberloff

23. Conserving Interaction Biodiversity   285
    John N. Thompson

V. **The Application of Conservation Ecology**   **297**
   Themes—*R. S. Ostfeld, S. T. A. Pickett, M. Shachak and
   G. E. Likens*

24. State-Dependent Decision Analysis for Conservation Biology   298
    H.P. Possingham

25. Expanding Scientific Research Programs to Address
    Conservation Challenges in Freshwater Ecosystems   305
    Catherine M. Pringle

26. Standard Scientific Procedures for Implementing Ecosystem
    Management on Public Lands   320
    *Robert S. Peters, Donald M. Waller, Barry Noon, S. T. A. Pickett, Dennis
    Murphy, Joel Cracraft, Ross Kiester, Walter Kuhlmann,
    Oliver Houck, and William J. Snape III*

27. Whatever It Takes for Conservation: The Case for Alternatives
    Analysis   337
    Mary H. O'Brien

28. Conservation Activism: A Proper Role for Academics?   345
    Joy B. Zedler

29. Getting Ecological Paradigms into the Political Debate:
    Or Will the Messenger Be Shot?   351
    Graeme O'Neill and Peter Attiwill

VI. Synthesis and a Forward Look   **361**
Themes—*S. T. A. Pickett, R. S. Ostfeld, M. Shachak, and
G. E. Likens*

30. A Summary of the Sixth Cary Conference   363
      Thomas E. Lovejoy

31. The Linkages between Ecology and Conservation   368
      Lee M. Talbot

32. The Central Scientific Challenge for Conservation Biology   379
      John Harte

33. Toward a Comprehensive Conservation Theory   384
      S. T. A. Pickett, Moshe Shachak, R. S. Ostfeld, and G. E. Likens

Epilogue: A Vision of the Future—*Joel E. Cohen*   400

Bibliography   405

Index   453

Foreword

From its inception, the U.S. Department of the Interior has been charged with a conflicting mission. One set of statutes demands that the department must develop America's lands, that it get our trees, water, oil, and minerals out into the marketplace. Yet an opposing set of laws orders us to conserve these same resources, to preserve them for the long term and to consider the noncommodity values of our public landscape.

That dichotomy, between rapid exploitation and long-term protection, demands what I see as the most significant policy departure of my tenure in office: the use of science—interdisciplinary science—as the primary basis for land management decisions.

For more than a century, that has not been the case. Instead, we have managed this dichotomy by compartmentalizing the American landscape. Congress and my predecessors handled resource conflicts by drawing enclosures: "We'll create a national park here," they said, "and we'll put a wildlife refuge over there." Simple enough, as far as protection goes. And outside those protected areas, the message was equally simplistic: "Y'all come and get it. Have at it." The nature and the pace of the resource extraction was not at issue; if you could find it, it was yours.

But what we're learning, thanks to modern biology and a more reasoned and thoughtful understanding of the landscape, is that we cannot protect the splendor and biological diversity of the natural world by simply fencing off a few protected areas within an overall landscape of exploitation. It doesn't work for a number of reasons.

The empty open spaces of the American West are filling up; there simply is not much more land to set aside as a means of protecting biological diversity.

The lands we manage do not fit into neat compartments. Everglades National Park is dying because forces outside the park—forces beyond the fence—are affecting the hydrological flows so essential to the Everglades ecosystem. Lines on a map don't always protect lands inside the preserve.

Migratory wildlife doesn't stay within the boundaries. Some birds migrate across the North American continent and some, like the Arctic tern and a few of the plovers, all the way down to the tip of South America. But their flyways are now threatened because the blocks of habitat are being fragmented and pulled out from under them at critical points on their course.

Our new mission, then, is to look beyond the fences. It is a recognition of modern ecosystem biology and of the island biogeography work done by E. O. Wilson and others. These concepts pose a much harder challenge by saying: "We can't satisfy our debt to nature with a few parks, scattered like postage stamps here and there." More and more, we have to come to grips with a much larger and more challenging problem, and that is how we live in equilibrium with our surroundings.

Bruce Babbitt

Preface

This book originated with the Sixth Cary Conference, held in May 1995 at
the Institute of Ecosystem Studies. That conference was stimulated by the recognition of major changes in ecology and in the practice of conservation. Because
it seemed to us that the changes were complementary, it was valuable to link
the two. Indeed, successes and failures in each area helped to change the other.
Therefore, exploring the similarities and needs for change in the two areas might
offer support for the development of each independently, and for better integration
of the science of ecology and the practice of conservation. One particular change
in ecology deserves to be singled out here as a motivation because it is a
sociological change within ecology and is not explored in the body of the book.
The change toward the increasing respect and legitimization of ecological applications is one of the healthiest developments in the discipline over the last few
years. Conservation biology, or perhaps more broadly conservation ecology,
includes both fundamental science, applications of the basic science, and scientific
knowledge specifically motivated by the real-world problems of conservation. If
we can contribute to the growing legitimization of ecological applications, we
are happy to do so. Therefore, the book attends to both science and the use of
the science. And because the comfort of ecologists with applications, including
the involvement in the public and policy sphere is of such recent vintage, we
believe that a volume that emphasizes future directions, rather than past difficulties, will help the science underlying conservation move more effectively into
the future.

These sorts of goals are shared by all the Cary Conferences, which emphasize
a more philosophical and synthetic approach to some important concern or connection of ecological science than is possible in most scientific meetings. This
volume, however, is different from a proceedings because we include several
chapters that fill out the scope beyond that which was possible to include in a
single meeting, and it includes chapters that were stimulated by interactions at

the meeting. Hence the book has evolved beyond the content of the meeting. Furthermore, we have provided linking material and have identified themes that connect the various contributions. Finally, we have attempted a synthesis, which would not have been possible at the time of the meeting due to the subsequent maturation and revision of the individual contributions.

The collection of essays we have assembled is diverse. We have attempted to match the breadth of professional concern with conservation. Hence the scientific scope runs from genetics to large landscapes. The applications are similarly diverse, ranging from contributions from conservative, land-based, nongovernmental organizations to federal agencies to active environmental advocacy organizations. Even among the scientists represented, the degree and mode of involvement in inserting ecological science into the public discourse on conservation is broad. Some of the chapters are rather technical, whereas others are more philosophical, although all are reflective. We have grouped the chapters to highlight common themes and to point out opportunities for integration. However, the book need not be read straight through. One advantage of a collection, even when integration is attempted, is that the contributions represent a discrete discussion by one or a group of authors, and are relatively self-contained. Hence, readers can sample. In essence, readers can assemble their own anthology, or selection of favorite blossoms, from the book.

We have designed the book to engage a variety of audiences. Of course, being professional ecologists ourselves, we have designed the book to help bring new material and perspectives to the attention of our colleagues in the profession. We particularly address graduate students, who will be the people who actually bring to fruition many of the syntheses and integrations suggested in the book. There is an overlapping audience of professional conservation biologists. Many conservation biologists are ecologists, but many are not. Hence, we have tried to alert ecologists to those areas of conservation practice that may not be congruent with ecology, and include material that can help advance conservation science as a pursuit that is in some ways broader than ecology. This breadth addresses the concerns of environmental managers inside and outside government, and addresses the informed decision maker as well.

The chapters are unusual. We charged the authors to take a broad overview that would attack important issues in conservation science and its practical use. We furthermore charged them to look forward, rather than to look backward. Therefore, the chapters do not contain detailed reviews or present new data from a specific research project—such detailed technical information can be found by using the comprehensive bibliography assembled and cited by the contributors. Rather, the chapters are in essay style, and are designed to discuss scientific needs, application strategies and techniques, and the shape and content of a new conservation science that takes into account the changes in the fundamental science of ecology and the real world in which it must be applied. And we gave the authors a target or perspective to stimulate their critical and creative juices.

We asked them to analyze the emerging conservation science in light of a potentially unifying and broad perspective—spatial heterogeneity and patchiness. Although we did not require authors to stick to that path, we did wish everyone to start from at least the same trailhead. They were free to indicate when shortcuts, U-turns, a sharp right, or a new trail needed to be made.

An additional agenda we set is to provoke involvement by scientists in the discourses required by conservation. Conservation ecology is in part a science, and in part a practice. The large proportion of conservation science that is practice requires involvement by the scientists who espouse the discipline. Such involvement can take many forms. Within the science, involvement can be through building bridges between disparate scientific specialties, such as ecosystem and population ecology, or building bridges between theory and empiricism, or even between different modeling or empirical approaches. Perhaps more difficult is the interface of the science with the public arena. The public, including those who are environmentally aware and those who are not, legislators and their staff, agency personnel from local to federal levels, the media, and environmental activist organizations are all examples of the kinds of constituencies with which science must communicate. And in this era of rapid political change worldwide, realignment of nations, and globalization of corporations and economies, the voice of science is at risk of being marginalized. Even the most reticent and theoretical of scientists must be concerned that science keep visible to the public its ability to pose intelligent questions, to solve problems, and to analyze complex connections. An activity that is not visible cannot be valued; and science that is not involved is not visible. Science is a major part of the intellectual backbone that allows modern society to stand erect. Without the involvement of scientists, scientific information, and scientific perspectives in the public dialog, appreciation, support, and use of science will atrophy. Involvement of scientists is crucial to their disciplines, as well as salutary for the policy concerns their science can inform.

Finally, the book provides a framework to show the context of involvement. The framework shows the needs in terms of scientific information and dialogs with potential users. Some dialogs, such as the one between theory and observation, appropriately remain within science. The framework is used to indicate crucial scientific dialogs that must be pursued. Other dialogs occur between scientists and conservation managers and planners who may not work as scientists, although some have scientific training. The framework helps to show the nature and place of such dialogs. Combined with the hypothesized conceptual basis for unification in conservation ecology, a view of how the science and its involvement can improve, emerge from this book.

This effort has had an unusually broad range of financial support, for which we are grateful. The breadth of concerns and uses of conservation required a similar breadth of participants in the meeting from which this book sprang. The Mary Flagler Cary Charitable Trust, which has showed dedicated support of scientifically based conservation through the years, provided support for the 1995 Cary Conference. Indeed, the conference originated as a suggestion of a small workshop held at the Institute of Ecosystem Studies (IES) during the summer of 1992 and sponsored by the trust. The Baldwin Foundation provided early encouragement for this project and supported the effort with a generous contribution. Core support from IES was crucial to the conference. The Rockefeller Brothers Fund stretched us to be better engaged with policy and to make more substantive connections with policy makers. Their contribution enabled us to reach out to the media, as well as to invite a number of key policy makers and persons familiar with the workings of the public discourse on conservation. The National Science Foundation, through its Directorates of Ecosystem Studies, Population Biology, Ecology, and Special Programs, supported many of the participants. The U.S. Department of Agriculture Competitive Grants Program also contributed to the support of participants at the conference. Without the generosity of these groups, it would have been impossible to assemble as large and as diverse a group of contributors.

We thank all the participants in the 1995 Cary Conference for their serious and generous commitment to the intellectual community we constructed over three days in May, and for cheerfully and unstintingly sharing their knowledge and experience during the conference. Not all these people have chapters in the book, but the intellect and generosity of each and every one of them is represented in its pages. We are most grateful to them for their contributions.

Finally, but far from least, we sincerely thank the operations, administrative, education, and research staffs of the Institute of Ecosystem Studies. For several days, and in some cases weeks, their working lives were disrupted to make the Sixth Cary Conference not only possible but enjoyable and stimulating for all of us. We personally wish to thank Jan Mittan for her inestimable contributions to the success of this effort from a germ of an idea to the finished book and media outreach.

S.T.A. Pickett
Richard S. Ostfeld
Moshe Shachak
Gene E. Likens

Contributors

Edith B. Allen, Botany and Plant Sciences, University of California, Riverside, CA 92521-0124.

Peter Attiwill, School of Botany, University of Melbourne, Parkeville, Victoria 3052, Australia.

Bruce Babbitt, U.S. Department of Interior, Washington, DC.

Juliana P. Barrett, The Nature Conservancy, Connecticut Chapter, 55 High Street, Middletown, CT 06457.

Nels E. Barrett, Institute of Ecosystem Studies, Millbrook, NY 12545 (Present address: The Nature Conservancy and Connecticut Geological and Natural History Survey, State of Connecticut, Department of Environmental Protection, 79 Elm Street, Hartford, CT 06106-5127).

Michael J. Bean, Wildlife Program, Environmental Defense Fund, 1875 Connecticut Avenue, NW, Washington, DC 20009.

Alan R. Berkowitz, Institute of Ecosystem Studies, Millbrook, NY 12545.

Norman L. Christensen, Jr., School of the Environment, Duke University, Durham, NC 27708.

Joel E. Cohen, Laboratory of Populations, Rockefeller University, 1230 York Avenue, New York, NY 10021-6399, and Global Systems Initiative, Columbia University, New York.

Robert K. Colwell, Department of Ecology and Evolutionary Biology, U-42, University of Connecticut, Storrs, CT 06269-3042.

Joel Cracraft, American Museum of Natural History, Central Park West and 79th Street, New York, NY 10024.

John J. Ewel, U.S. Department of Agriculture, Forest Service, Pacific Southwest Forest Experiment Station, Institute of Pacific Islands Forestry, 1151 Punchbowl Street, Room 323, Honolulu, HI 96813.

Peggy L. Fiedler, Department of Biology, San Francisco State University, San Francisco, CA 94132.

Gretchen Long Glickman, PO Box 1034, Wilson, WY 83014.

Doria R. Gordon, The Nature Conservancy, Department of Botany, PO Box 118526, University of Florida, Gainesville, FL 32611.

Peter M. Groffman, Institute of Ecosystem Studies, Millbrook, NY 12545.

Ilkka Hanski, Department of Ecology and Systematics, Division of Population Biology, P.O. Box 17, FIN-00014 University of Helsinki, Finland.

Lennart Hansson, Department of Wildlife Ecology, Swedish University of Agricultural Sciences, S-750 07 Uppsala, Sweden.

Jeffrey L. Hardesty, The Nature Conservancy, Department of Botany, P.O. Box 118526, University of Florida, Gainesville, FL 32611.

John Harte, Energy and Resources Group, University of California, Berkeley, CA 94720.

Kent E. Holsinger, Center for Conservation and Biodiversity, Department of Ecology and Evolutionary Biology, U-43, University of Connecticut, Storrs, CT 06269-3043.

Oliver Houck, Tulane Law School, 6801 Freret Street, New Orleans, LA 70118.

Peter Kareiva, Department of Zoology, University of Washington, Seattle, WA 98195.

Ross Kiester, U.S. Department of Agriculture, Forest Service, Pacific Northwest Research Station, 3200 SW Jefferson Way, Corvallis, OR 97331.

Walter Kuhlmann, Boardman, Suhf, Curry & Field, Firstar Plaza, Suite 410, Madison, WI 53701-0927.

Robert A. Leidy, U.S. Environmental Protection Agency, 75 Hawthorne Street, San Francisco, California 94105, and Department of Wildlife and Fisheries Conservation Biology, University of California, Davis, CA 95616.

A. Carl Leopold, Boyce Thompson Institute for Plant Research, Cornell University, Ithaca, NY 14853.

Gene E. Likens, Institute of Ecosystem Studies, Millbrook, NY 12545.

Thomas E. Lovejoy, Smithsonian Institution, Washington, DC 20560.

Kevin McKelvey, U.S. Department of Agriculture, Forest Service, Redwood Sciences Laboratory, 1700 Bayview Drive, Arcata, CA 95521.

Judy L. Meyer, Institute of Ecology, University of Georgia, Athens, GA 30602-2602.

Dennis Murphy, Center for Conservation Biology, Stanford University, Stanford, CA 94305.

Barry Noon, U.S. Department of Agriculture, Forest Service, Redwood Sciences Laboratory and National Biological Service, 1700 Bayview Drive, Arcata, CA 95521.

M. Philip Nott, Department of Ecology and Evolutionary Biology, University of Tennessee, Knoxville, TN 37996-1610.

Mary H. O'Brien, Environmental Research Foundation, Annapolis, MD, and Ecosystem Policy Analyst, Hells Canyon Preservation Council, Joseph, OR, P.O. Box 12056, Eugene, OR 97440.

Graeme O'Neill, Science Ink, 2 Banool Road, Selby, Victoria 3159, Australia.

Richard S. Ostfeld, Institute of Ecosystem Studies, Millbrook, NY 12545.

Robert S. Peters, Defenders of Wildlife, 1101 14th Street, NW, Suite 1200, Washington, DC 20005.

Steward T. A. Pickett, Institute of Ecosystem Studies, Millbrook, NY 12545.

Stuart L. Pimm, Department of Ecology and Evolutionary Biology, University of Tennessee, Knoxville, TN 37996-1610.

Hugh P. Possingham, Department of Environmental Science, University of Adelaide, Roseworthy Campus, Roseworthy SA 5371, Australia.

Catherine M. Pringle, Institute of Ecology, University of Georgia, Athens, GA 30602.

Louis Provencher, The Nature Conservancy, Longleaf Pine Restoration Project, P.O. Box 875, Niceville, FL 32588.

H. Ronald Pulliam, National Biological Service, Washington, DC 20240.

Katherine Ralls, National Zoological Park, Smithsonian Institution, Washington, DC 20008.

Kevin H. Rogers, Centre for Water in the Environment, University of the Witwatersrand, Johannesburg, P. Bag 3 WITS 2050, South Africa.

Mary Ruckelshaus, Department of Zoology, University of Washington, Seattle, WA 98195 (Present address: Department of Biological Sciences, Florida State University, Tallahassee, FL 32306-2043).

Moshe Shachak, Ben Gurion University of the Negev, The Jacob Blaustein Institute for Desert Research, Sede Boker Campus, Israel, and the Institute of Ecosystem Studies, Millbrook, NY 12545.

Daniel Simberloff, Department of Biological Science, Florida State University, Tallahassee, FL 32306.

David Skelly, Greely Lab, Yale School of Forestry and Environmental Studies, 370 Prospect Street, New Haven, CT 06511.

William J. Snape III, Defenders of Wildlife, 1101 14th Street NW, Suite 1400, Washington, DC 20005

David L. Strayer, Institute of Ecosystem Studies, Millbrook, NY 12545.

Lee M. Talbot, Lee Talbot Associates International, 6656 Chilton Court, McLean, VA 22101.

Sandy L. Tartowski, Institute of Ecosytem Studies, Millbrook, NY 12545.

Barbara L. Taylor, National Marine Fisheries Service, Southwest Fisheries Science Center, P.O. Box 271, La Jolla, CA 92038.

John N. Thompson, Departments of Botany and Zoology, Washington State University, Pullman, WA 99164.

C. Richard Tracy, Biological Resources Research Center, University of Nevada, Reno, NV 89557.

Pati Vitt, Center for Conservation and Biodiversity, Department of Ecology and Evolutionary Biology, U-43, University of Connecticut, Storrs, CT 06269-3043

Donald M. Waller, Department of Botany, 430 Lincoln Drive, University of Wisconsin, Madison, WI 53706-1381.

Peter S. White, CB #3280 Biology, University of North Carolina, Chapel Hill, NC 27599.

John A. Wiens, Department of Biology and Graduate Degree Program in Ecology, Colorado State University, Fort Collins, CO 80523.

Joy B. Zedler, Pacific Estuarine Research Lab, Biology Department, San Diego State University, San Diego, CA 92182-0057.

Participants

J. David Allan, School of Natural Resources, University of Michigan, 430 East University Street, Ann Arbor, MI 48109-1115.

Michael F. Allen, National Science Foundation, 4201 Wilson Boulevard, Arlington, VA 22230.

Edward A. Ames, The Mary Flagler Cary Charitable Trust, 122 East 42nd Street, Room 3505, New York, NY 10168.

Torsten W. Berger, Institute of Ecosystem Studies, Box AB, Millbrook, NY 12545.

Patrick J. Bohlen, Institute of Ecosystem Studies, Box AB, Millbrook, NY 12545.

Douglas T. Bolger, Department of Biology, Dartmouth College, Hanover, NH.

Joseph N. Boyer, Florida International University, SE, Environmental Research Program, University Park, Miami, FL 33199.

James T. Callahan, National Science Foundation, 4201 Wilson Boulevard, Arlington, VA 22230.

Charles D. Canham, Institute of Ecosystem Studies, Box AB, Millbrook, NY 12545.

Scott L. Collins, National Science Foundation, 4201 Wilson Boulevard, Arlington, VA 22230.

Diana W. Freckman, Natural Resources Ecology Laboratory, Colorado State University, Fort Collins, CO 80523.

Samuel M. Hamill, Jr., The Baldwin Foundation, 974 Lawrenceville Road, Princeton, NJ 08540.

Gary S. Hartshorn, World Wildlife Fund, 1250 Twenty-fourth Street NW, Washington, DC 20037.

Elaine Hoagland, Association of Systematics Collections, 730 11th Street NW, 2nd Floor, Washington, DC 20001.

Thomas W. Hoekstra, U.S. Department of Agriculture Forest Service, Rocky Mountain Research Station, 240 W. Prospect Road, Fort Collins, CO 80526-2098.

Wm. Robert Irvin, J.D., Center for Marine Conservation, 1725 DeSales Street NW, Suite 500, Washington, DC 20036.

Deborah B. Jensen, The Nature Conservancy, 1815 North Lynn Street, Arlington, VA 22209.

Clive G. Jones, Institute of Ecosystem Studies, Box AB, Millbrook, NY 12545.

L. Katherine Kirkman, Joseph W. Jones Ecological Research Center, Ichauway and University of Georgia, Route 2, Box 2324, Newton, GA 31770.

Nancy Knowlton, Smithsonian Tropical Research Institute, Unit 0948, APO AA, Miami, FL 34002-0948.

Russell Lande, Department of Biology, University of Oregon, Eugene, OR 97403-1210.

John H. Lawton, Centre for Population Biology, Imperial College at Silwood Park, Ascot SL5 7PY, England.

Karin E. Limburg, Institute of Ecosystem Studies, Box AB, Millbrook, NY 12545.

Gary M. Lovett, Institute of Ecosystem Studies, Box AB, Millbrook, NY 12545.

Rosemary H. Lowe-McConnell, Streat near Hassocks, Sussex BN6 8RT, England.

Mark J. McDonnell, Bartlett Arboretum, University of Connecticut, Stamford, CT 06903-4199.

Gary K. Meffe, Savannah River Ecology Laboratory, Drawer E, Aiken, SC 29802.

Eric S. Menges, Archbold Biological Station, P.O. Box 2057, Lake Placid, FL 33852.

Peter B. Moyle, Department of Wildlife, Fish and Conservation Biology, University of California, Davis, CA 95616.

Michael L. Pace, Institute of Ecosystem Studies, Box AB, Millbrook, NY 12545.

Gregory Payne, Chapman and Hall, 115 Fifth Avenue, New York, NY 10003.

Richard V. Pouyat, U.S. Department of Agriculture Forest Service, c/o Institute of Ecosystem Studies, Box AB, Millbrook, NY 12545.

William Robertson IV, Andrew W. Mellon Foundation, 140 East 62nd Street, New York, NY 10021.

Nicholas A. Robinson, Center for Environmental Legal Studies, Pace University, 78 North Broadway, White Plains, NY 10603.

William H. Shaw, Department of Biology, Sullivan County Community College, P.O. Box 4002, Loch Sheldrake, NY 12759.

Thomas A. Spies, PNW Research Station, U.S. Department of Agriculture Forest Servcie, Forestry Sciences Laboratory, 3200 Jefferson Way, Corvallis, OR 97331.

Monica G. Turner, Department of Zoology, University of Wisconsin, Madison, WI 53706.

Peter A. Tyler, Aquatic Science, Deakin University, P.O. Box 432, Warrnambool, Victoria 3280, Australia.

Joseph S. Warner, Institute of Ecosystem Studies, Box AB, Millbrook, NY 12545.

Richard Warwick, Plymouth Marine Laboratory, Prospect Place, The Hoe, Plymouth PL1 3DH, England.

Kathleen C. Weathers, Institute of Ecosystem Studies, Box AB, Millbrook, NY 12545.

Cathleen Wigand, Institute of Ecosystem Studies, Box AB, Millbrook, NY 12545.

Raymond J. Winchcombe, Institute of Ecosystem Studies, Box AB, Millbrook, NY 12545.

SECTION I

Introduction: The Needs for a Comprehensive Conservation Theory

1

Defining the Scientific Issues

R. S. Ostfeld, S. T. A. Pickett, M. Shachak,
and G. E. Likens

Summary

This chapter introduces the problems with which the book deals and lays out the structure of the analysis. We introduce the needs, both scientific and practical, for developing a new theory of conservation. Such a theory will have both empirical and conceptual aspects. The structure and content of this new theory must recognize both the internal needs of the science—objectivity, completeness, integration, and as a guide for future research—and the political reality and variety of motivations for the practice of conservation.

Needs for a New Theory for Conservation

Over the past several years, a major change has occurred in the way biological conservation and management policies are devised at the national level both in the United States and elsewhere (Keiter and Boyce, 1991; Jones, J. R. et al., 1993; Kempf, 1993). Recently, U.S. national policy has begun shifting away from the conservation of single species and their habitats and toward conservation and management of the interactive networks of species and large-scale ecosystems on which species depend. Although this change is supported by many prominent scientists, the underlying justification appears to be based on logistics and practicality more than on clearly articulated scientific principles (e.g., Franklin, 1993b; Orians, 1993; Edwards, May, and Webb, 1994). Moreover, conservation of species and habitats under the U.S. Endangered Species Act has developed into a crisis-oriented undertaking, since remedial action is not triggered until a species or its habitat is seriously imperiled (Fiedler et al., 1993; Orians, 1993). New national policy ostensibly is designed to obviate the need for crisis intervention on a species-by-species basis by managing the ecosystems that sustain interacting networks of species (National Research Council, 1992).

Ideally, conservation policy and practice must be based on scientific principles. In the case of the management of species and their habitats, conservation practitioners have available a rich and well-established theoretical foundation in conservation biology (e.g., recent undergraduate texts by Primack, 1993; Meffe and Carroll, 1994). For example, to protect populations of an endangered large carnivore, practitioners can devise plans that are based on established theories of island biogeography (Shafer, 1990), population and quantitative genetics (Lande, 1988b), population viability analysis (Soulé, 1987), and metapopulation dynamics (Kareiva, 1990). Although theoretical principles in conservation biology are often controversial and inadequately tested in the real world (Abbott, 1980), at least such principles are available as frameworks to guide management plans (e.g., Harrison, 1994). On the other hand, the new approach of ecosystem management does not have a foundation of well-developed theory to guide it (May, 1994b; Jones, C. G. and Lawton, 1995). Nor has the role of traditional conservation biology theory in the development of new theories of ecosystem management been explored thoroughly.

This book, which is an outgrowth of the 1995 Cary Conference at the Institute of Ecosystem Studies, is intended to aid in the establishment of a theoretical framework for ecosystem conservation and management. In the chapters that follow, the authors identify issues for which further development of theory is particularly important and explore the role of case studies in understanding and applying general theory (e.g., Shrader-Frechette and McCoy, 1993). We provide a synthesis that recognizes the parallel changes of paradigms in ecology and conservation. For example, background assumptions and principles of ecology have changed so that the new nonequilibrium paradigm emphasizes that ecological systems are often open to external control, frequently nondeterministic, and rarely at equilibrium (Botkin, 1990; Pickett et al., 1992; Pickett and Ostfeld, 1995). Meanwhile, conservationists increasingly recognize that individual species are embedded in interactive communities and that disturbance and dynamics are fundamental processes in natural systems (e.g., McNaughton, 1989; Reid and Miller, 1989; Society of American Foresters, 1993).

Heterogeneity: A Hypothetical Foundation

The future of the scientific basis for conservation requires that we address the question, What conceptual framework can be used to facilitate a more effective use of ecology in conservation, especially including higher levels of organization and extensive spatial scales? Because heterogeneity (Box 1.1) is an inescapable fact of the natural world (e.g., Bell, McCoy, and Mushinsky, 1991; Milne, 1991; Tilman and Pacala, 1993; Campbell, 1994), we hypothesize that it can serve as an important principle for conservation. Although heterogeneity in the guise of

---

Box 1.1. Key Concepts

Heterogeneity—Any form of variation in environment, including physical and biotic
  components. Such variation may appear as spatial or temporal patterns, and may
  be fixed or dynamic. If dynamic, heterogeneity may appear as a steady- or a
  transient-state, depending on both the nature of the process and the scale of
  observation.

Patchiness—A form of spatial heterogeneity in which boundaries may be discerned.
  Patchy heterogeneity appears as contrasting, discrete states of physical or biotic
  phenomena. An array of patches may be seen as a mosaic at a particular scale.

Graded Patchwork—A form of heterogeneity intermediate between discrete patches
  and unbounded heterogeneity. The gradients may be nonlinear and may be steep
  or gradual. In cases where the gradients are narrow relative to the relatively uniform patches, the gradients may be considered to be ecotones, or tension zones
  between discrete patches.

---

variability is often treated as a problem and inconvenience, it also has crucial
ecological functions (Kolasa and Pickett, 1991). Fundamentally, heterogeneity
or patchiness (Box 1.1) in site, dispersal, performance, or interaction is a root
of biological diversity at all levels of ecological organization and scales (Bormann
and Likens, 1979a; Pulliam, 1988; Naeem and Colwell, 1991; McLaughlin and
Roughgarden, 1993).

We suggest that the concept of ecological heterogeneity or patchiness may
effectively link new views in ecology to new approaches in conservation. Although patchiness as yet has no complete or unified theory (Levin, 1989), it can
serve as a conceptual tool to accommodate several important aspects of ecological
systems. Patchiness calls attention to the spatial matrix of ecological processes
(Forman and Godron, 1986); it highlights fluxes of organisms and materials
in nature (Bormann and Likens, 1979b; Wiens, Crawford, and Gosz, 1985); it
suggests the dynamic role of ecotones and edges (Forman and Moore, 1992); it
encompasses the dynamics of entire mosaics as well as their parts (Bormann and
Likens, 1979b; Remmert, 1991); and it is applicable to various scales, from
microsites to biomes (Milne, 1991). Hence, we view patchiness as a suitable and
exciting candidate framework for new theory in conservation, a notion that is
explored in depth in the chapters that follow. Note that we use the term patchiness
in many instances as a synonym for the more inclusive term of heterogeneity.
We do this in the interest of readability, rather than to suggest that all heterogeneity
that is important for conservation practice and theory will be discretely bounded.
Indeed, heterogeneity can be discreet or gradual (Box 1.1). Hence, "graded
patchiness," discrete patchiness, and other forms of spatial heterogeneity are
important for conservation.

Patchiness can also be a useful tool for expanding the scope of conservation biology to multiple levels and across various scales. Patchiness is not solely a province for landscape ecology. For instance, in addition to its relevance to many levels of organization, patchiness is relevant to such aspects of ecology as population regulation, dispersal, niche partitioning, succession, disturbance, biogeochemical flux and cycling, habitat fragmentation, connectivity and landscape linkages. Furthermore, patchiness can affect the productivity of ecosystems (Shachak and Brand, 1991). Because frequency of disturbance may affect food chain length (Pimm, 1988) and biogeochemistry (Bormann and Likens, 1979b; Likens and Bormann, 1995), patch structure may have ecosystem consequences. Biogeochemical flux and cycling are also affected by the patchiness generated by human land use (e.g., Groffman and Likens, 1994). In spite of its importance, patchiness in ecosystem fluxes on various scales requires much additional research (Vitousek and Denslow, 1986; Reiners, 1988). The examples just outlined suggest that patchiness is relevant to a wide variety of the structures and functions that must be maintained to conserve biodiversity. Indeed, the conservation biology literature has recognized an important role of patchiness and spatial heterogeneity for a long time (Pickett and Thompson, 1978), and attention to the topic is growing (Meffe and Carroll, 1994).

An additional reason for proposing patchiness as a framework for conservation theory is a practical one. Human activities in both industrialized and developing countries have divided and are continuing to divide contiguous natural areas (Turner et al., 1990; Likens, 1991; McDonnell and Pickett, 1993). Conservation is increasingly practiced in a fragmented world, where human land use, ownership, and jurisdictional boundaries impose patchiness on various aquatic and terrestrial systems (Noss and Harris, 1986; Hansson, 1992b; National Research Council, 1993). In fact, classical theories in wildlife biology are sometimes used to increase the fragmentation of landscapes to favor certain game species without consideration of the broader, negative consequences of such actions (Alverson et al., 1994). Conservationists require clear and well-documented conceptual tools to evaluate the de facto patchiness of the world and of conservation and management plans. The understanding of heterogeneity as a foundation for unifying and advancing conservation ecology is explored in depth in this book.

To lay the groundwork for a set of principles and an organizing conceptual framework for conservation, it is important to recognize that the narrow view of theory as idealized mathematical models alone is no longer considered accurate (Richerson, 1977; Haila, 1986; Fagerstrom, 1987; Caswell, 1988), nor appropriate for the needs of conservation biology (May, 1994a; Shrader-Frechette and McCoy, 1993). Therefore, a subsidiary goal of this book is to utilize the broad, modern conception of theory and its emphasis on application in the real world (National Research Council, 1986; Pickett et al., 1994) to better exploit ecological knowledge in conservation. Exposing a theoretical framework and set of principles to the harsh light generated by the need for success in conservation, management, and restoration is a rigorous test (Bradshaw, 1987a; Ewel, 1987; Harper, 1987).

Scope of the Book

This book focuses on the most widely recognized targets of conservation: biodiversity and ecosystem function (Jones, G. E., 1987; National Research Council, 1992; Franklin, 1993b; Orians, 1993). Focusing on both individual biotic entities and on fluxes of matter and energy incorporates the two major topic areas in ecology, the organismal and the ecosystem paradigms (Jones, C. G., and Lawton, 1994). Not only is there a need to integrate these two paradigms for progress in the basic science of ecology (Likens, 1992; Pickett, Kolasa, and Jones, 1994), but conservationists also call for the use of the two paradigms jointly (Franklin, 1993b). In parallel, the concept of ecosystem management as a conservation strategy explicitly calls for consideration of populations, species, communities, ecosystems and landscapes (Jones, G. E., 1987; Walker, 1989; Grumbine, 1990; Hansson, 1992a), along with the relevant economic and social systems (Likens, 1992; Johnson, 1993; Kempf, 1993; Keystone Center, 1993). The fact that both nongovernmental organizations and federal agencies (Keiter and Boyce, 1991; Kempf, 1993; Morrisey et al., 1994) have called for employing such a broad scope in conservation activities, and for adopting a diversity of tools, makes this exploration all the more compelling.

In addition to the ecological topics that must be integrated to support effective conservation, there are a number of management approaches, for example, adaptive management, ecosystem management, and restoration, that are required to ensure success (Hunter, 1990; Myers, 1993; Slocombe, 1993; Morrisey et al., 1994). Therefore, conservation efforts should also embrace the broadest possible interpretation of biological conservation to include relevant management. To this end, this book explores situations where there are common ecological principles that support biodiversity conservation, ecological restoration, and ecosystem management (Meffe and Carroll, 1994; Pickett and Parker, 1994).

Ecology, Conservation, and the Real World

Conservation ecology is a complex pursuit because it brings science into contact with the real world. But because conservation aims to preserve, reserve, maintain, or restore some part of the natural world, it is fraught with value and value judgments. The science may be as objective as ever, but the application is squarely in the realm of value. This is an uncomfortable place for most scientists to be. Some analysis of this difficult terrain is worthwhile so that scientists can better understand just why it is so difficult.

Why do people wish to conserve, in all the senses just outlined, some part of the natural world? Two kinds of values bear on this decision (Proctor, 1995). Conservation can be pursued because of the intrinsic value of nature, or because of its instrumental value. Intrinsic values attach to nature for its own sake. In contrast, instrumental values derive from the practical or inspirational use of nature by humans. Instrumental values are complex, because they may center on species (biocentrism), ecosystems (ecocentrism), or resources. Ethics is the combination of values and conduct. Hence, each of the kinds of values will have an ethical stance associated with it. Ecology interacts with these values and ethics in several ways. Ecological knowledge can bring natural phenomena and processes to people's attention, and so make those phenomena the subject of valuation. Ecological knowledge can expose aspects of natural phenomena formerly observed from one ethical perspective to the attention of other or contrasting ethics. Or ecological knowledge can be used to manipulate or maintain natural systems to sustain the values recognized in those systems. Clearly, these uses of ecological knowledge put ecologists and evolutionary biologists into "hot water" because of the variety of values and ethical stances that can be placed on the same part or parcel of the natural world by different people or institutions.

The second major kind of complexity in conservation ecology is the concept of nature itself. What is natural? How does our conception of nature inform what we seek to conserve? Nature has at least three aspects that are relevant to conservation ecology and conservation biology. Nature first of all refers to the world as a collection of interacting nonhuman organisms and environments—of ecosystems and their components and contexts, and ultimately of the entire biosphere and its connected biotic and abiotic constituents. This would be the only face of nature were it not for people. Of course, humankind is real and present, and that introduces two other ways to look at nature. Nature in the simple sense of the biosphere alone does not exist. The biosphere and people exist together. Therefore, nature includes people, their institutions and effects, and is modified from some primordial, prehuman state. In fact, such a state can now only be imagined or extracted from the deepest paleoecological records, and people and their lighter or more weighty, subtle or direct effects, are ubiquitous on the Earth. Humans are components of ecosystems and of the biosphere and are partners and influences on the rest of the Earth's biota. Nature can no longer be conceived of without humans.

This recognition introduces the third face of nature. Nature is, in part, a social construct—a human conceptual invention (Cronon, 1995). Of course there is that part of nature that would exist without humans, and whose processes cannot be manipulated by humans. In other words, there is always some "other" in nature. But how much "other" we recognize, as individuals, institutions, societies, and nations is determined jointly by our involvement in nature and the model we make of nature. Such models can take the form of stories, myths, advertisements, art, literature, and so on. The cultural models of nature are also informed by values and conduct, and hence ethics. Nature, as conceived by farmers, industrialists, urbanites, loggers, and the like, usually means very different things, and will be approached and valued very differently. The problem of values attached to nature therefore appears again. The science of conservation treads in this complex, value-driven, ethical maelstrom. Whether the step is bold or timid, effective or bumbling, depends in part on the quality of our objective science. But success also depends on how well we recognize the complexity of the dance of values and ethics itself.

It is probably fair to say that most ecologists were drawn to their discipline in part by a fascination with, and strong aesthetic or ethical connection to, the natural world. As scientists, our job is to attempt to understand nature as rigorously and generally as possible, but as human beings, most of us strive to experience the manifold gifts that nature offers us, with or without involvement of our intellects. Indeed, human societies survive and prosper on the basis of the resources provided by "nature." Thus, ecologists face the dilemma of balancing the pitting of our intellectual energy and creativity against the mysteries of nature, while we dwell with wonderment, perhaps asking entirely different, nonintellectual questions of that same mystery. This dilemma exists only because ecologists confront scientifically the same entities—ecological systems—with which they feel strong ethical connections. As is abundantly clear in the chapters that follow, ecologists differ strongly in their need to separate versus meld together their scientific and ethical/aesthetic interactions with the natural world.

Conservation ecology is a useless intellectual exercise unless it can help society to discover, prevent, and correct environmental degradation and other problems. Thus, our discipline is firmly grounded in the real world of policy and management, and is increasingly relevant as the degradation of environmental quality accelerates. However, the uses and abuses of our science exist in political forums that traditionally have been outside the interests and expertise of ecologists (Meyers, 1991). For the science of ecology to be used most effectively for conservation, this withdrawal of scientists from political decision-making processes must end. One purpose of this book is to remind ecologists about the real, political world in which their science is used, abused, and ignored. Another is to provide a forum for scientists to reflect on the advantages and pitfalls of application of ecological knowledge to the political debate on environmental issues.

Conclusion

The goals and complexities outlined in this introductory chapter motivate the structure of this book. Section I continues with policy perspectives on the science of conservation and its application and couches the entire effort in the largest context of human presence and expansion on the Earth. We then present two cases in which interesting and difficult conservation problems invite the development of new integration and frameworks. Section II highlights some of the key conceptual motivations for development of a new way to conceive of and integrate the science underwriting conservation. The basics of heterogeneity are laid out in preparation for assessing its integrating role between biodiversity and ecosystem function. Section III examines the meaning and approaches to biodiversity and ecosystem function as targets for conservation and explores how they might be integrated. The chapters in Section IV analyze how the recognized specialties within current conservation biology can contribute to the more integrated and anticipatory conservation science of the future. Section V explores exactly how conservation biology is or can be applied to real cases and alerts scientists to the costs and benefits of involvement in the public discourse about conservation. Section VI presents a multifaceted synthesis of the volume, returning to policy implications and motivations and inviting us to consider a vision of the future as a motivation and guide for how to apply the science of ecology.

2

Part 1: Science, Conservation, Policy, and the Public

Gretchen Long Glickman

Months ago when Steward Pickett and Rick Ostfeld recommended the subject for the Sixth Cary Conference of "Enhancing the Ecological Basis of Conservation," I heartily welcomed the idea. Having been associated with the conservation community for many years, I knew how important, really basic, the contribution of science is to conservation goals. But I didn't recognize how especially valuable it would be to have this discussion now—in this era of political hotheadedness and rampant misinformation—in this period of negation of environmental progress that we have made.

It seems a bit ironic that we should be gathered together, in an atmosphere of rational, thoughtful inquiry and discourse, when the Congress of the United States was entertaining what I would term irresponsible proposals to dismantle many of the environmental/ecological gains of the past 25 years. While we at IES discussed heterogeneity and ecosystem functioning and biodiversity, our Congress seriously considered a bill—Senator Gorton's (R-WA) bill to amend the Endangered Species Act, which would eliminate habitat protection and greatly weaken species protection. Simultaneously another bill, HRS 961, passed overwhelmingly in the House, arbitrarily removed 60-80 percent of current wetlands from its definition, according to Environmental Protection Agency (EPA) officials. And Senator Dole's proposal, S 343, drastically opened up all environmental regulations to new reviews and demanded cost/benefit analysis that clearly focused on values that are traded in the marketplace versus those intrinsic values that are not, and required scientific proof of risks that may exceed the limits of scientific knowledge in assessing risks. These three potentially devastating bills are a cynical assault on ecological values. They are an effort to stamp out what some term rampant, intrusive, costly environmentalism. Biodiversity has become a dirty word.

So if our goal in those three days was to better determine how we can enhance the ecological basis of conservation, how science can be more influential in protecting ecosystems, there is no time like the present. How can science convey its knowledge of ecological systems and the consequences of human disturbances in a way that will support the conservation ethic and assist the conservation activist?

It is noteworthy that IES, so rigorous in the integrity of its basic research, would take on this subject. It is consistent with the institute's mission statement, which states IES is dedicated to the creation, dissemination, and application of knowledge about ecological systems. This statement says further that the scientific research will be applied through participation in decisionmaking regarding the ecological management of natural resources and through promotion of a broader awareness about the importance of ecological relationships to human welfare.

How do we best do this? From my experience as a practitioner, there are three basic dilemmas that limit the success of science in meeting these objectives.

I raise these dilemmas from the vantage point of a practitioner, as I have been involved with the conservation community for many years. In addition to the really extraordinary honor to associate with Gene Likens and the IES staff as a member of the IES Board of Trustees, I also have been closely involved with the Environmental Defense Fund (EDF) as a board member since 1981, and with the Greater Yellowstone Coalition (GYC), serving as its president for the last three years.

The GYC experience has been particularly relevant to today's discussion. The GYC mission statement reads "to preserve and protect the Greater Yellowstone Ecosystem and the unique quality of life it sustains." From its inception 12 years ago, brought about at a meeting at the Teton Science School, this organization, very much an advocacy organization, has based its efforts on a large-scale ecosystem approach to protecting the magnificent wildlands and wildlife that is Greater Yellowstone. It is the first time to my knowledge an activist group had focused on ecosystem management as its primary goal. Over the years, GYC has published extensive, scientifically descriptive material including "An Environmental Profile of the Greater Yellowstone Ecosystem," in 1991, and "Sustaining Greater Yellowstone, Blueprint for the Future" in 1994. *Conservation Biology* for the first time dedicated an entire issue to excerpts from the Profile.

GYC coined the expression "Greater Yellowstone," now a familiar reference depicting the region we view as an intact ecosystem. But it was the fires of 1988 that established the concept more firmly in people's minds. It was the fires that demonstrated the truth that natural processes disregard arbitrary jurisdictional boundaries—that is, the park, the forest regions, private land, and so on. The fires made evident the need for coordinated ecosystem management. The fires demonstrated the impact of natural and human-induced disturbances. And the fires served as an excellent platform for public education, as the message regarding the fires changed from one of "Yellowstone National Park is destroyed" to true wonder at the regenerative results of this event. The scientific community and the National Park Service can take pride in that. And the concept Greater Yellow-

stone is now a familiar one. But even as science has been the bedrock of GYC—
we have a Science Council, in addition to our Board of Directors—we see the
limits in the effectiveness of ecological knowledge to influence conservation
goals.

From a practitioner's point of view, the three areas of concern about the
effectiveness of science in enhancing conservation goals are

1. How can the knowledge of science be conveyed in a way that is meaningful and relevant?

2. Who's science is sound science?

3. When is there enough scientific knowledge, although not complete, to
   be predictive, to enable scientists to speak out on a controversial issue?

Regarding the first question, for science to contribute optimally to conservation
goals, science needs to better get the message across. It needs to translate its
findings and convert its language in a text that is meaningful to the different
audiences that determine conservation goals—grass roots, legislators, agency
managers, and the like. Translating science to sound bites is a repugnant task,
but some simplification and common language is needed if scientists are to be
most effective in directly communicating, which I believe will be increasingly
important for scientists to do.

Ecological findings will often be most helpful when coordinated with the
knowledge of other disciplines, an integration, if you will, of the data of the
social sciences to the hard sciences. Correlation with other ecological studies
will enlarge comprehensive understanding.

The voice of science needs to be more assertive. When there is misinformation,
science needs to take the leadership in getting the facts out on the table. The
recent discussion of the accuracy of the assertions of the Gregg Easterbrook
(1995) book, *A Moment on the Earth*, is an interesting example of when the
scientific community might speak up. In this case, the scientists of EDF asserted
their "corrections." But it would be helpful if scientists independent of an advocacy group would speak out.

Science is a validation. The more concerted the voice, coordinated rather than
disparate, the more consensus takes place. We've seen this in the tobacco issue.
Yet this style flies in the face of the scientific legacy to respect differences in
scientific conclusions, to protect the integrity and independence of each research
project, and to focus on one's own work, rather than to influence others.

Which takes us to the second dilemma: Who's science is accurate anyway?
In ecology, just as in public health, the public and the policymakers are confused
by multiple claims. Differences of data collection, interpretation, and synthesis
among the agencies and independent studies add to the confusion.

The sponsor of the research tends to influence the design of the research, thus
bringing different results. Industry, public land agencies, and advocacy groups

all are driven by different goals. These institutional limitations focus research efforts differently. Biodiversity conservation is not always a high priority. The U.S. Forest Service, for example, has a multiple use mandate that emphasizes timber production rather than forest health. The service employs 12,000 productive foresters and only 100 biologists. Among the public agencies, it has been impossible to get a comprehensive, integrated database—which is why we all hoped for a national biological survey.

Taking from my GYC experience a current very dramatic example of conflicting science, of who's science is accurate, anyway, is the debate centering on the proposal to delist the grizzly bear from the Endangered Species list. The Wyoming Game and Fish Department wants to delist the bear so they can have more "flexibility" in managing the bear's interface with humans. With population growth and consequent habitat encroachment, coupled with this past season's reduced supply of food, the grizzly has been much in evidence in the backyards of ranchers, and they are up in arms. Political pressure being what it is, the public agencies say the population is restored. Private scientists long involved in bear studies scorn federal and state estimates of the bears' population and its status. Among scientists, differences abound in what constitutes a viable population, stabilization versus long-time viability, and so on. The newspapers are full of contradicting scientific comments every day. For instance, a long article in the *Casper Star Tribune* (April 6, 1995) quoted various scientists in clear dispute. Chris Serveen, USFWS Grizzly Bear Recovery Coordinator stated, "We have evidence that the population is going up. We don't make these statements casually. We wouldn't be saying that unless we had confidence that's occurring." And John Craighead, scientist with the Wildlife and Wildlands Institute, retorted, "The indices set for recovery are not valid and the methods used for measuring those indices are not statistically valid." Who's science is credible? How can the nonscientific community differentiate good science from bad, or at least understand the design differences that contribute to differing results.

The third dilemma confronting science in its efforts to enhance conservation goals is the issue of: when is there sufficient knowledge to speak out? When can science speak out on a possible trend without having all the data? How can science play a role as an early warning system, rather than being reactive? The earlier there are indicators, the sooner practitioners and others can come up with constructive alternatives. Aren't there some issues where the severity of the potential results suggests raising questions sooner rather than later? Doesn't it depend on the nature of the resources at risk, what can be lost and not regained, including human health? To me issues of high, irremediable risk auger the importance of science speaking out, even if all the data is not in. If the trends are sufficiently broad, the data essentially there, and the risks of failure to act high, then I would suggest it is not inappropriate for the scientific community to express its concerns, even if in a conditional manner. Sharing the scientific basis for concern contributes to an informed public. I would cite the EDF's position on global warming as an example.

In this period of highly charged, political rhetoric and simplistic anecdotal examples, it is particularly imperative that scientists speak out and reinforce what they do know about how ecosystems function and what's at stake. Providing this information to the public and to practitioners will make a difference. Selecting projects in consultation with the needs of practitioners could also be helpful.

Science has all the ingredients to be a strategic force in the conservation movement. Ecology is the intellectual framework of much of the conservation community's efforts. I look forward to learning from others at the conference how they view the prospects and problems of enhancing the ecological basis of conservation.

Part 2: Providing the Scientific Information that Conservation Practitioners Need

H. Ronald Pulliam

To understand the impact of our science on the practice of conservation, ecologists and other environmental scientists need to realize that the information we provide is ultimately used, or ignored, by people with a variety of backgrounds and perspectives. I am pleased that this volume, and the conference from which it was derived, incorporates the views and opinions not just of scientists, but also of conservation practitioners. Conservation practitioners include both people who make regulations and laws and practicing land and natural resource managers in the private and in the public arenas.

In this brief essay, I address three interrelated questions all focused on what kinds of advice, information, and tools need to be provided to conservation practitioners by the scientific community. These questions are

1. What do landowners need from us, particularly private landowners?

2. What do land and natural resource managers need from us, particularly federal and state land managers?

3. What do policy makers need from us?

The needs of these three communities overlap very broadly, but there are differences that need to be recognized, including the kinds of information and experience that each group brings to the table with them and the kinds of biases that each group might have or, alternatively, the kinds of biases that these conservation practitioners might perceive that we scientists have. A good starting point for working with any of these conservation practitioners is the recognition that the information, experiences, and biases that we, as scientists, bring to the table may be substantially different from those of the practitioners who will use the information we provide.

Private Landowners

Many, perhaps most, decisions that impact natural resources are made by private landowners. When we discuss natural resource management issues and the relevance of our science with private landowners, we need to bring with us a sense of respect for landowners as individuals, and we need to realize that their management goals and desires, although often different from ours, may nonetheless be reasonable. Unfortunately, many scientists tend to begin by mistrusting the goals of landowners, whether the owners are corporations or individual citizens. I believe that progress will be encouraged by the realization that, although there are no doubt many exceptions, most landowners want to do the "right thing" and, although they are looking for ways to accomplish their own goals, most do not want to "destroy the environment."

This does not mean that we should not educate people about the possible deleterious impacts of their proposed activities. Indeed, I believe that we have an obligation to inform people so that they can make better decisions. I also believe that we have an obligation to advise them, when possible, about how their goals might be achieved with the minimum damage to the environment. Within the bounds of existing law, it is landowners who will usually make the final decisions regarding land and resource management. I believe that it is our obligation to provide them with the best scientific information available so that they can make their decisions fully aware of their options and of the probable impacts of those options. I also believe that most landowners want to have good scientific information and will welcome our attempts to work with them, so long as we respect their rights as landowners.

As an aside, I think one of the real problems with environmental policy right now is that too often good land stewards are not rewarded, but instead are penalized. By the very way that we approach these problems, very often it is the people who have been the best land stewards who feel they have the most to be concerned about. For example, a landowner will probably not have endangered species on his or her land unless he or she has been a good land steward, and many owners of property occupied by threatened or endangered species fear that they will be penalized for their good stewardship in the past. I am pleased that the Department of the Interior recognizes this problem and has proposed changes in the way in which the Endangered Species Act is administered that will provide more incentives for good stewardship. For example, under a proposed new Department of the Interior rule for the enforcement of the Endangered Species Act, landowners who purposefully make improvements to their land that create habitat for endangered species will not later be penalized if they convert their land back to its former use.

Communication between scientists and landowners must be improved. Scientists must realize that most landowners are intelligent, well-educated people whose backgrounds and intellects should be respected. Most landowners, however, have not had courses in ecology or resource management, and thus they do not know our science, or even what it is we might bring to dialogues about conservation issues. We have to explain things in a language that they can understand. I think this is perhaps best illustrated in the current, very politicized, and sometimes emotional, debate over ecosystem management.

I have spent a lot of time in the last year talking to various "stakeholder" groups who feel they have a particular interest in what the National Biological Service is doing or might do. These are groups like the Cattlemen's Association, the Home Builders Association, and the American Farm Bureau. I have attempted to find simple ways of explaining what it is that ecologists study and know and to describe our work in a manner understandable and relevant to issues the typical landowner has to deal with. For example, with regard to ecosystem management, ecologists often argue that we need to expand the scope and scale of management decisions. These terms are not necessarily in the vocabulary of the landowners and other citizens. However, the notion of expanding scale is clearly illustrated by the statement that you can't manage the ducks on a single pond. Clearly, people who want to have more ducks on their ponds understand that the dynamics of duck populations are not controlled by what happens on their one pond; duck population dynamics are controlled by what happens in a very large area including the ponds and other wetlands of many neighbors. From this perspective, it could be said that one of the groups that has been among the best implementers of ecosystem management is Ducks Unlimited, a group that has understood that it is necessary to make cooperative arrangements with large numbers of people to achieve common goals.

Resource Managers

Resource managers, particularly at the federal and state levels, need concrete management goals and practical tools for achieving those goals. Scientists often assume that their latest research results will be quickly incorporated into changes in management practices. In reality, it often takes decades for new scientific paradigms to be translated into useful management tools. Many new paradigms although relevant to resource management are unfamiliar to resource managers trained years before the paradigm shift. Even if managers are familiar with the recent literature, they may not see how new paradigms will help them manage resources. Managers are rewarded for achieving results, not for testing new theories; accordingly, they are often unwilling to give up a proven method to try a new unproven one.

I was trained in ecology at a time when most ecologists believed not only that there was a "balance of nature," but also that there was what I call a "hand-in-glove" relationship between habitat and the species distribution. A Gleasonian view of the world predominated in which all niches were assumed to be full and all species were thought to be in their proper places (Gleason, 1926). In the last decade or two, a fundamental paradigm shift has taken place that greatly impacts the way ecologists view the world we study. Ideas about metapopulations have taught us that very frequently we might observe empty but perfectly suitable habitat (Hanski, 1985). Furthermore, ideas about source-sink dynamics have taught us that we may often have species in an unsuitable habitat (Pulliam, 1988). This leads to a paradigm shift in the way that scientists think about the relationship between habitat and species distribution and abundance.

Unfortunately, this paradigm shift could lead to more confusion rather than to more enlightened management when presented to resource managers. If ecologists say there is little relationship between habitat occurrence and species distribution, we clearly are opening a door that can lead to misinterpretation. We have to be very careful to package our information in a way that can be understood and used by people who need the information. In terms of the hand-in-glove analogy, there clearly is a relationship between the hand and the glove, that is, between the occurrence of habitat and the presence of species, but that relationship is more of a statistical one (Hanski et al., 1995b) than a determinate one-for-one law of nature. I think of it as a glove that has a few extra finger pockets where there are no fingers, and perhaps two fingers sticking into a single pocket, but still, by-in-large, most fingers are in pockets and most pockets are occupied.

Despite the lack of a perfect fit, there is a clear statistical relationship between the amount of habitat and the viability of populations. This relationship relates directly to a central issue surrounding the reauthorization of the Endangered Species Act, that is whether or not the taking of habitat is related to the viability of populations. Ecologists need to speak in a clear voice. I would dare say that in many of these issues there is scientific consensus. No ecologist would argue that there is no relationship between the amount of habitat and the viability of a population. It may not be as clear and clean a relationship as we once thought, but the relationship is real.

In fact, metapopulation theory can be used to argue that it is even more important to understand this relationship between habitat and population abundance than we previously thought. According to current theory, only when habitat is very abundant is there a fairly close linear relationship between the amount of habitat and population size. Overall, the relationship between available habitat and population viability is nonlinear, and there are thresholds of habitat availability below which populations are very likely to go locally extinct. As habitat dwindles, populations decline in a more or less linear manner until a threshold is reached, after which extinction can be very rapid. This is an important message to get across to resource managers and one that argues for the need for actively managing and monitoring landscapes dominated by human activities.

Resource managers need concrete goals to guide their management decisions. It is not helpful to tell a manager "just go out there and manage for biodiversity." It is also unhelpful, in many cases, to tell managers only what not to do. Many ecologists have criticized land managers for "narrow" goals such as maximizing timber production or deer numbers. These goals may be far too narrow, but they are at least concrete enough to guide management decisions. Management goals cannot be as vague as "manage all species." Ecologists can be helpful by working with land managers to identify species that are particularly sensitive indicators of undesirable change or that provide information about ecosystem function. It also helps if indicators are responsive to known management practices and are relatively easy to monitor. It does little good to know that a species is a good indicator of a particular ecosystem function if that function does not respond to any known management option, or if the species indicating the system's response to management activities cannot be monitored effectively.

Once indicators and other resource management goals have been chosen and the baseline status of the indicators has been established, the next step is to analyze all the options that are available to managers. This of course depends on the system. In the Everglades, the management options might include alterations to the hydrology of the system or changes in the regulations affecting pesticide runoff from nearby agricultural areas. In other areas, the management options available might include alterations of the fire regime, control of exotic pests, or changes in hunting or fishing regulations.

Scientists and managers can work together to identify the primary driving forces impacting an ecosystem, determine which of those driving forces are most amenable to management control, and develop predictive tools that project future trends, including the future status of the indicators chosen. Depending on the degree of understanding of the system, these predictive tools might be complex mathematical models of ecosystem dynamics or much less formal conceptual models of how the system works. Also, depending on the knowledge of the system, the models might be ready to use as management tools or might be better suited to guide research until the system is better understood.

Whatever the available management tools are, or whatever the degree of understanding of how the system works, monitoring the future trends of the ecosystem indicators is essential to determining the efficacy of the management options chosen (Christensen, this volume). Managers often lament that they do not know how well their management schemes are working because they do not have adequate monitoring of the resources they manage. A well-conceived monitoring program can simultaneously provide a measure of management success and provide for the parameterization and verification of ecosystem models. In turn, as a model of the system is refined and more accurately reflects the observed trends in the system, the more likely it will reflect future trends under alternative management regimes.

We must continually ask how well we were able to predict changes and continually update the models we use for making predictions. One of the tools that clearly is relevant for managing landscapes, is the use of vegetation maps that can be translated into maps of habitat distribution. Together with information on the habitat requirements of organisms, remote sensing and Geographic Information Systems (GIS) can be used to make predictions about where future habitat will be. These spatial analyses can be combined with modeling efforts to make multispecies and multiresource projections of the consequences of alternative ways of managing land resources.

Policy Makers

Policy makers and decision makers of all sorts need to know not only what the current thinking is, but they need to know how certain the scientific community is of the information we are providing. Accordingly, ecologists need a reasoned approach to balancing what we consider to be facts and consensus in the community with estimates of the uncertainty that is associated with that consensus.

Documents like National Research Council reports have a strong impact because they are consensus statements about what scientists know, and they usually point out areas of scientific disagreement where substantial uncertainties remain. We will not always speak with consensus, but it is important to point out where there is agreement and where the areas of uncertainty remain.

Should ecologists wait until we have all the answers before we provide information to the policy makers? From my perspective the answer is a resounding no. We will never have all the answers. Policy makers need to understand that our answers are always tentative, that they are always being updated, and that they are always subject to revision. That's the nature of science. Scientists need to be ready and willing to communicate what our current understanding is, to update that understanding, and to admit when we are wrong. We must be frank and open about the status of our science and the uncertainty that is associated with the information we bring to the table. Scientists should, however, be equally vocal in communicating that, as uncertain as the information is, we base our opinions on scientific observations and experiments and the collective information we provide represents the best information and understanding currently available from the scientific community.

In dealing with policy makers or with the public at large, it is essential that we continually be aware that, despite our best efforts, information can be distorted and misused. We are living in the age of electronic communications. Facts can be communicated and disseminated very rapidly. But misinformation can be disseminated just as rapidly as information. Whether or not this is the age of information or the age of misinformation remains to be seen. If we aren't careful, it will be the age of confusion. It will be the age in which people are being bombarded with so much conflicting information that they will not know what to make of it. We have all seen indications that some people with political agendas will provide misinformation to policy makers or to the public when it suits their purposes. Scientists must provide the facts as they see them to counter such misinformation.

Let me end by providing one example of how a scientific theory that I have contributed to has been twisted and misused to serve a political agenda. The following is a quote from a newspaper article (*Washington Times*, March 21, 1994) on the president's Northwest Forest Plan, written by syndicated columnist Alston Chase:

> While the administration proposal is designed to preserve all so-called "late successional" species (those living in mature forest), concern for northern spotted owls drives the process. Yet this bird may not even need old growth. This possibility was raised to me in March, when I was invited to give a lecture at the Yale School of Forestry and Environmental Studies. During my visit, a wildlife biologist put his reservations to me succinctly: Everyone, he explained, assumes mature forests are essential for owl survival, but the truth may be exactly the opposite: that these creatures depend on younger tree communities. The professor's argument derives from a scholarly field known as "source-sink dynamics." Whereas all creatures require a source habitat where births outnumber deaths, writes biologist H. Ronald Pulliam, a leader in this specialty, a large fraction . . . may occur regularly in "sink" habitats, where within habitat reproduction is sufficient to balance local mortality. "Given that a species may commonly occur and successfully breed in sink habitats," Mr. Pulliam continues, "an investigator could easily be mislead about the habitat requirements of a species. . . . Population management decisions based on studies in sink habitats could lead to undesirable results." In other words, it's easy to confuse sources—where creatures multiply— with sinks—where they dwindle. The Yale professor was suggesting there may be just such a confusion about the owl. A mistake is entirely possible because the thesis that owls need old growth habitat has never been tested.

This quote is a blatant example of how scientific facts can be twisted and quoted out of context. In fact, the thesis that northern spotted owls require old growth forest has been tested and the results strongly support the notion that old growth forest in Oregon and Washington are source habitats and that excess production in this source results in immigration of surplus individuals into secondary growth forests (Ripple et al., 1991; Lehmkuhl and Raphael, 1993), just the opposite of Mr. Chase's contention. To complicate matters, northern spotted owls do breed in secondary growth areas, although their reproductive success in these sinks is not sufficient to balance mortality there. To make matters even more complex, recent studies in California have suggested that secondary forests there may indeed be source habitat for the California subspecies of the spotted owl (Blakesley et al., 1992).

The preceding example points out not only how complicated the natural world is but also how difficult our role as information providers can be. Scientific facts are frequently distorted to serve a political agenda. As if that were not bad enough, what is false in one place may be true elsewhere, and we have to explain all of this to contentious policy makers and a confused public.

Part 3:   A Policy Perspective on Biodiversity Protection and Ecosystem Management

Michael J. Bean

Secretary of Labor Robert Reich gave a speech this past fall on the subject of "competitiveness." Here is part of what he said: "What do we mean by 'competitiveness' anyway? Rarely has a term of public discourse gone so directly from obscurity to meaninglessness without any intervening period of coherence" (Reich, 1994). Much the same, I fear, can be said about "ecosystem management."

Just as everyone embraces the notion that our country must be "competitive," so too everyone—or nearly everyone—now thinks that our natural resources must be managed using an "ecosystem approach." Unfortunately, not everyone shares a very clear idea of just what that means. This conference intended to provide some clarity to that idea. The challenge is more than just that of assembling a cook book of "how-to" steps for practical managers. There is a deeper problem, as evidenced by the invitation each of us received to the conference. The letter of invitation noted that "the ongoing change in conservation policy, from the focus on single species and individual land parcels to functioning communities, ecosystems, and landscapes, is taking place in the near absence of a firm foundation in ecological theory." That statement ought to give us all considerable pause. More recently, I took part in a gathering of timber industry biologists and environmentalists where the view of at least some was that although they could not clearly define ecosystem management, they knew it when they saw it—rather like Supreme Court Justice Potter Stewart once said of pornography.

Despite the uncertainties being voiced by some, I am pleased to report that at least a few people have a clear fix on what ecosystem management means. Take Ike Sugg of the Competitive Enterprise Institute, for example. According to Sugg (1993),

> this so-called "ecosystem approach" will be the effective end of the right to private property in rural America. .... In short, the ecosystem approach is nothing more than a pretext for shattering what few fragile limits remain on government's

ability to regulate land use. . . . Defenders of property rights be forewarned, ecosystem management is the new rhetoric for regulating everything.

If that sounds a little alarmist, keep in mind that the same issue of the Competitive Enterprise Institute's newsletter in which Sugg's statement appeared also carried a story reporting that "Organic farming looms as the largest current threat to both humanity and wildlife" (Avery, 1993).

Another who has figured out what ecosystem management means is Alexander Cockburn, a columnist who leans nearly as far to the left as Sugg leans to the right. According to a column Cockburn wrote for *The Washington Post*, "as now being employed by Babbitt and by looters of the public domain, an ecosystem approach is just a piece of conceptual flim-flam disguising dismemberment of existing environmental protections" (Cockburn, 1993).

These are, of course, quite diametrically opposite views, from two government outsiders. What is the inside view? I found fascinating the definition given by Assistant Interior Secretary George Frampton in a recent published interview. Here is what he said:

> Ecosystem management is going to be the hot phrase of the 1990s. I'm not sure the Bush administration ever knew what it meant. . . . What we mean by ecosystem management is good science, looking at a range of species in communities and trying to plan for some optimization of resources. Not just one resource at a time. (Frampton, 1993)

I think what Frampton said could be paraphrased as follows: "the management of all the various renewable resources of the [land] so that they are utilized in the combination that will best meet the needs of the American people." And that, sadly, comes straight from the definition of "multiple-use" in the Multiple-Use, Sustained-Yield Act of 1960. Perhaps everyone can now relax: ecosystem management is nothing more than the familiar old multiple-use management in new rhetorical garb.

I cannot resolve this debate, or even contribute much of value to it. Neither will lawmakers, whose grasp of what the new terminology really means in terms of on-the-ground decisions about resource management will be superficial at best. With consensus elusive among managers and ecological theorists about what ecosystem management is or should be, don't expect the increasingly polarized U.S. Congress to provide meaningful guidance. Don't even expect it to show much interest. Consider what Wayne Gilchrest, a Republican Congressman from Maryland, was quoted in the press recently as saying about the new Congress's attitude toward environmental issues. Gilchrest declared: "I have never seen so many people afraid of information in my life [or] so extravagantly funded by interest groups that stand to make a lot of money from misinformation" (Gilchrest, 1995).

This Congress, quite clearly, is not going to blaze any new policy trails in the area of ecosystem management. The task, therefore, is to put some flesh on the bones of this new concept through the development of policy within existing legal authorities. I will try to illustrate what I mean by reference to the Endangered Species Act and one of the species that it has long sought to conserve.

The Endangered Species Act never uses the term "biodiversity" and uses the word "ecosystem" only once. The act's statement of purposes begins with the declaration that "[t]he purposes of this Act are to provide a means whereby the ecosystems upon which endangered species and threatened species depend may be conserved." Given that this is the first stated purpose of the act, one might reasonably expect that the operative provisions of the act would specify how that purpose is to be achieved. They don't, at least not explicitly and directly. No mention of conserving ecosystems appears in the provisions of the act governing the development of recovery plans for listed species, the delegation of conservation authority to the states, the detailed duties of federal agencies, or the general prohibitions applicable to both public and private parties. The task thus falls to managers and policy makers within the Fish and Wildlife Service to design the strategies that achieve the act's purposes without the benefit of detailed congressional guidance.

When Congress enacted the Endangered Species Act in 1973, it "grandfathered" onto the endangered list about 200 species that had been designated as endangered under predecessor legislation. Among these was the red-cockaded woodpecker, which was listed as an endangered species in 1970. The woodpecker once occurred throughout the pine forests of the Southeast. Alhough it could be found in a number of different forest types, the type with which it was most closely associated and within which it was most abundant, was the longleaf pine forest type of the Southeastern coastal plain.

The original range of the longleaf pine stretched from extreme southern Virginia to east Texas and may have included some 60 million acres (Croker, 1990). It supported a variety of species long since gone. For example, "[p]ractically all southern records of the historic and prehistoric bison are within the longleaf pine range," including droves of bison observed by the early American explorer Mark Catesby (Landers et al., 1990). By 1955, 80 percent of the original longleaf pine acreage had been lost; only 12.2 million acres remained. By 1985, two thirds of that was gone—only 3.8 million acres remained; 94 percent of the original acreage was gone. Between 1975 and 1985, despite the Endangered Species Act, nearly a million acres of longleaf pine forest—some 20 percent of the total remaining in 1975—was lost throughout the Southeast. Since 1985, the loss of longleaf pine forest has continued apace.

Not surprisingly, just as the ecosystem on which it principally depends shrank, the red-cockaded woodpecker also declined. In the 1980s the number of red-cockaded woodpeckers on private land in the Southeast declined by an estimated 20 percent. On public land, the trend was also downward, but less dramatic.

Two conclusions seem inescapable from these figures. First, the red-cockaded woodpecker, despite nearly a quarter century of protection as an endangered species, is much closer to the brink of extinction today than it was when protection was first bestowed on it. Second, the ecosystem on which it principally depends has not been conserved, but has steadily and dramatically declined.

Let me add to those fairly obvious conclusions three more controversial conclusions of my own. The first is that the strategies pursued to conserve the woodpecker have contributed to the decline of its ecosystem, at least on private land; second, that the policies that have guided those strategies have implicitly abandoned what many regard as one of the central tenets of ecosystem management; and third, that the conservation of the woodpecker and its supporting ecosystem can yet be achieved through new but untried approaches that the Endangered Species Act allows.

On the first point, most of the ecosystem on which the red-cockaded woodpecker depends is found on private land. Thus, any effective strategy to conserve any more than isolated fragments of that ecosystem must seek to preserve the bird and its ecosystem on private land. In fact, however, there is increasing evidence that at least some private landowners are actively managing their land to avoid potential endangered species problems—that is, restrictions on land use stemming from the act's prohibition of "taking" endangered species—by avoiding endangered species. Because the woodpecker principally uses older trees for nesting and foraging, some landowners are harvesting their trees before they reach sufficient age to be attractive to woodpeckers. Because the woodpecker prefers open forest conditions, with minimal hardwood understory—a condition maintained historically by regular and recurrent fires—landowners can eliminate or avoid endangered species complications by refraining from understory management. Because the woodpecker prefers longleaf pine over other species, landowners can reduce their likelihood of having woodpeckers by planting other species.

It is important to acknowledge that actions like these are not necessarily the result of malice toward the woodpecker or the environment. Rather, they can be rational decisions motivated by a desire to avoid potentially significant economic constraints. In short, they are nothing more than the predictable responses to the familiar "perverse incentives" that sometimes accompany regulatory programs.

That is point 1: the strategies pursued to date to conserve the red-cockaded woodpecker on private land have contributed to the decline of its ecosystem. Point 2 is that one of the central tenets of ecosystem management has been essentially abandoned in those strategies. Specifically, the one thing that nearly everyone writing about ecosystem management agrees on is that it implies looking beyond the property boundary to some larger watershed, ecosystem, or landscape unit, and that it seeks to influence management decisions within that broader unit. By contrast, the recovery plan for the red-cockaded woodpecker essentially writes off the larger, private landscape. With only a single exception, the areas where recovery is to be pursued are major federal landownings. More recently, the Fish and Wildlife Service has encouraged the development of habitat conservation plans under which private landowners with existing breeding birds can make the progeny of those birds available for relocation onto public lands for a period of years, after which the birds and their habitat on the private land can be eliminated.

These strategies may secure the survival of the red-cockaded woodpecker for many more decades. They will not, however, promote the restoration and maintenance of the ecosystem on which it depends, except on widely scattered federal landholdings where recovery efforts are to be focused. Are there recovery strategies that could do so? I believe there are, and that they include significant opportunities on private lands.

What are they? First, from my conversations with landowners in North Carolina, I believe there are some private landowners with land that, if properly managed, could provide suitable habitat to support red-cockaded woodpeckers. In most instances, this is forest land that once supported red-cockaded woodpeckers, but no longer does because it has not been actively managed for some time, and the resulting encroachment of the hardwood understory precludes its current use by that species. A "hold harmless" agreement with such landowners would be sufficient to cause some landowners to undertake the needed management measures voluntarily. Under such an agreement, in return for carrying out the management actions that will create the habitat conditions the woodpecker requires, the government simply gives assurances that future incompatible land use decisions will not be regarded as violations of the act. The mechanism for providing such assurances exists, I believe, in the provisions relating to incidental taking under both Sections 7 and 10 of the act.

In short, creative regulatory relief will be a sufficient incentive for some private landowners to begin the sort of management that will benefit both the woodpecker and its ecosystem. For other landowners, stronger economic incentives may be needed. How can they be provided? Unfortunately, the automatic response is often to think solely in terms of tax relief or governmental payments. These certainly can be important incentives, but the likelihood of securing either in the current fiscal and political climate is probably not great.

Other, more innovative approaches are probably needed as well. Such ideas might include mitigation banking, transferrable development rights, and even major investments in forest preservation and management by utilities seeking credit for offsetting increased emission of carbon dioxide. For individual private landowners, the revenue stream from sources such as these, when added to revenue from the sale of timber and pine straw, may make longleaf pine restoration and management an economically attractive alternative. Suddenly, halting the decades long decline of longleaf pine forest acreage can begin to look less like a pipe dream and more like a potential reality.

The lesson in this, I believe, is that one of the beneficial effects of the recent interest in ecosystem management—whatever it is—is that it has encouraged all of us to think in ways that break down some of the traditional molds and barriers. It has encouraged us to recognize the shortcomings of past ways of business and think more broadly about possibilities. We need to extend this way of thinking and this willingness to reexamine old approaches. Just as ecosystems don't stop at the property line, opportunities to conserve them do not lie solely in the province of traditional resource management policies. By creatively aligning conservation interests with the economic impulses that drive private behavior, some of the goals that have seemed beyond our grasp may be brought within it.

3

Conservation and Human Population Growth: What Are the Linkages?

Joel E. Cohen[1]

Summary

This chapter offers an overview of the demographic, economic, environmental, and cultural situation of the human species today, then takes a closer look at the distribution of human population density in relation to farming systems. Current views of the relation between conservation and human population growth range from a denial or neglect of any connection whatsoever to an assertion that population growth is the main cause of conservation problems. Although both extreme views may be appropriate sometimes, human interactions with the environment are in general strongly influenced by economics and culture, and rapid population growth makes it more difficult to preserve many aspects of environmental quality. Conservationists need an external agenda that shapes the positions they will support in other fields that affect conservation, such as family planning, agriculture, education, and environmental law. To promote discussion of the external agenda of conservation, I put forward a few proposals.

Introduction

At least six listed participants in this meeting have dealt in print with conservation and human population growth. In 1975, Gene Likens published with Robert Whittaker a classic article on "The Biosphere and Man." They concluded that the biosphere could feed generously a human population stabilized at a low enough level, but that if people failed to stabilize their population, the poor countries could become trapped in poverty, the rich countries in a degraded environment, and the whole world in trouble (Whittaker and Likens, 1975). In 1991, Michael L. Pace and colleagues showed that "on a global scale, human population within a river's watershed is strongly related to the concentration of nitrate in rivers that discharge to coastal ecosystems" (Peierls et al., 1991; see also Cole et al., 1993). In 1993, Gary Meffe, together with Anne Ehrlich and David Ehrenfeld, lamented that not one paper that dealt directly with human population growth had been submitted for publication in the journal *Conservation Biology* (Meffe et al., 1993). Also in 1993, Steward Pickett observed that "American ecology has largely ignored humans. . . . But human effects, both subtle and conspicuous are being increasingly documented at all spatial scales" (in Jolly and Torrey, 1993:37–38). In his presidential address to the Ecological Society of America in 1994, Ronald Pulliam focused on "human population growth and the carrying capacity concept." He estimated that "a Brazilian Amazon population of . . . 200 million would result in 100 percent deforestation and the loss of all endemic forest interior species" (Pulliam and Haddad, 1994:154). In this volume, Judy L. Meyer observes that "Humans are a part of all ecosystems."

These works and others by other participants show that I am following a trail blazed by giants. Yet it is worth revisiting this trail in the hope of seeing something new.

In a nutshell, my message is this. The effects of human population growth on conservation depend strongly on economic and social factors as well as on human numbers, density, and growth rates. Although human population growth can directly intensify the three major threats to the survival of nonhuman species—which are habitat conversion, hunting and species introductions—rapid population growth also makes it harder for a society to solve many of its political, social, environmental, and economic problems, including but not only its problems of conserving biological diversity. Conservationists need to support efforts to slow human population growth for reasons of conservation and of human well-being.

This chapter offers an overview of the demographic, economic, environmental, and cultural situation of the human species today, then takes a closer look at the distribution of human population density in relation to farming systems. Current views of the relation between conservation and human population growth range from a denial or neglect of any connection whatsoever (e.g., DeWalt et al., in Jolly and Torrey, 1993) to an assertion that population growth is the main cause of conservation problems (e.g., Popline, 1992). Although both extreme views may be appropriate sometimes, the reality in general is more complex. Conservationists need an external agenda that shapes the positions they support in fields that affect conservation. To promote the development of an external agenda for conservationists, I put forward several proposals.

Context: Population, Environment, Economics, Culture

I begin with a global overview. Further details appear in Cohen (1995). Compared to history before World War II, the human situation is unprecedented in four respects that are relevant to conservation:

- human population size and growth;
- human impact on the environment, and vice versa;
- enormous wealth and disparities between the rich and the poor; and
- a cultural implosion of diverse traditions.

Population Growth

Shortly after the last Ice Age ended about 12,000 years ago, the human population of the Earth first exceeded 5 million people. By A.D. 1650, the population grew to about 500 million. This 100-fold increase represented a doubling about once every 1,650 years, on the average. After A.D. 1650, population growth accelerated tremendously. The human population increased from roughly half a billion to roughly 5.7 billion today—about three and a half doublings in three and a half centuries, or one doubling per century. Since 1955, the population has doubled in 40 years—more than a 40-fold acceleration over the average population growth rate before 1650. Never before the second half of the twentieth century had any person lived through a doubling of global population—and now some have lived through a tripling of human numbers. Not only is the population bigger than ever before—it is growing much faster. As of 1995, the world's population would double in 45 years if it continued to grow at its present 1.5 percent per year, though that is not likely. Although this growth rate is less than the all-time peak of 2.1 percent per year in the period 1965–70, it greatly exceeds any global population growth rate before World War II. An absolute increase in population by 1 billion people, which took from the beginning of time until about 1830, now requires about 12 years.

These global totals and averages remind me of a story. The story is related to one of the main themes of this book, namely, heterogeneity. An ecologist, an economist, and a statistician went on a deer hunt with bow and arrow. Creeping through the undergrowth, they came on a deer. The ecologist took careful aim and shot. His arrow landed 5 meters to the left of the deer. The economist then took careful aim and shot. Her arrow landed 5 meters to the right of the deer. The statistician looked at the arrow to the left, the arrow to the right, and the deer, and jumped up and down shouting, "We got it! We got it!"

Global statistics conceal vastly different stories in different parts of the world. About 1.2 billion people live in the economically more developed regions: Europe, Northern America, Australia, New Zealand, and Japan. The remaining 4.5 billion live in the economically less developed regions. The population of the more developed regions grows at perhaps 0.2 percent per year, with an implied doubling time of more than 400 years. The population of the less developed regions grows at 1.9 percent per year, a rate sufficient to double in 36 years if continued. The least developed regions with the world's poorest half billion people increase at 2.8 percent per year, with a doubling time of less than 25 years. At current birth rates, the worldwide average number of children born to a woman during her lifetime (the global total fertility rate) is around 3.1. The number ranges from more than 6 in sub-Saharan Africa to 1.5 in Europe.

The populations of some domestic animals have grown even faster than human numbers. For example, the number of chickens, 17 billion, more than doubled from 1981 to 1991. In 1992, domestic animals were fed 37 percent of all grain consumed (World Resources Institute, 1994:296). Some domestic animals have major environmental impacts because they produce methane, liquid and solid wastes, overgraze fragile grasslands, and prevent forest regeneration. The human species lacks any prior experience with such rapid growth and large numbers of its own or of its domestic species. A few of the many complex connections between these growing numbers, the environment and economics are spelled out next.

Environment

Energy use is one simple index of both economic power and human influence on the environment. Energy use per person and population growth have interacted multiplicatively. Between 1860 and 1991, while the human population more than quadrupled from about 1.3 billion to about 5.4 billion, inanimate energy used per person grew from about 0.9 megawatt-hours per year to about 17.6 megawatt-hours per year. Global inanimate energy use (the product of population size and average energy use per person) grew nearly 100-fold from 1 billion megawatt-hours per year in 1860 to 95 billion megawatt-hours per year in 1991.

Vulnerability to environmental impacts is also increasing. For example, the impact of a projected rise in sea levels increases with the tide of urbanization, as the number of people who live in coastal cities rapidly approaches 1 billion (World Resources Institute, 1994:354). With increasing frequency, people make contact with the viruses and other pathogens of previously remote forests. Cities of unprecedented population density and increased global travel provide novel opportunities for transmission, and new diseases are emerging.

Between 1973 and 1988, while world population rose by 1.2 billion, developing countries transformed around 400,000 square kilometers of forest to farms and around 856,000 square kilometers of forest to houses, roads and factories. Deforestation in developing countries in this 15-year period totaled about 1,450,000 square kilometers, almost the area of Alaska (Harrison, 1993:9). For each additional person, the area equal to roughly one quarter of an American football field was deforested. Did the population increase during this time cause the deforestation? We'll return to that question later.

Economic Growth

In the aggregate production of material wealth, the half century since World War II has been a golden era of technological and economic wonders. For example, in constant prices with the price in 1990 set equal to 100, the price of petroleum fell from 113 in 1975, to 76 in 1992. The price of a basket of 33 nonfuel commodities fell from 159 in 1975, to 86 in 1992. Total food commodity prices fell from 196 in 1975, to 85 in 1992 (World Resources Institute, 1994:262).

Remember the ecologist's and economist's arrows that averaged on target? As the world's average economic well-being rose, economic disparities between the rich and the poor increased. In 1960, the richest countries with 20 percent of world population earned 70.2 percent of global income, while the poorest countries with 20 percent of world population earned 2.3 percent of global income. Thus the ratio of income per person between the top fifth and the bottom fifth was 31 to 1 in 1960. In 1970, that ratio was 32 to 1; in 1980, 45 to 1; in 1991, 61 to 1 (United Nations Development Programme 1992:36; 1994:63).

In 1992, the 830 million people in the world's richest countries enjoyed an average annual income of $22,000—a truly astounding achievement. The almost 2.6 billion people in the middle-income countries received only $1,600 each on average. The more than 2 billion people in the poorest countries lived on an average annual income of $400, or a dollar a day (Demeny, 1994:17). If you remember only two numbers from this chapter, please remember these two, based on 1992 statistics: 15 percent of the world's population in the richest countries enjoyed 79 percent of the world's income.

Dollars are not the full measure of human well-being. In 1990–95, while Europe enjoyed a life expectancy above 75 years, Africa still had a life expectancy of 53 years—below the world average 20 years earlier. In developing regions, the absolute numbers and the fraction of people who were chronically undernourished fell from 941 million, and 36 percent, around 1970 to 786 million, and 20 percent, around 1990. In Africa, contrary to the world trend, the absolute number of chronically undernourished increased by two thirds (World Resources Institute, 1994:108). Africa also had the highest population growth rates during this period, and still does.

Food commodity prices dropped by half, as I showed earlier, while nearly a billion people in developing countries chronically did not eat enough. The bottom billion have no money to buy food, so they cannot drive up its price. The asset they are able to produce most easily—an asset that they hope will help them wrest a living from often declining natural resources—is children. In developing countries, high fertility is both a cause and a consequence of poverty.

Culture

The cultural implosion of recent decades is the change that is potentially most explosive. Migrations within and between countries, business travel, tourism, media and telecommunications have shrunk the world stage. In 1800, roughly 2 percent of people lived in cities; today the fraction is about 45 percent. The absolute number of city dwellers rose more than 140-fold from perhaps 18 million in 1800 to some 2.6 billion today. In every continent, in giant city-systems, people who vary in culture, language, religion, values, ethnicity, and socially defined race increasingly share the same space for social, political, and economic activities. The resulting frictions are evident in all parts of the world.

This completes my overview of the current situation. Now I want to zoom in for a closer look at population density.

The Distribution of Human Population Density: A Closer Look

In 1994, the world had an average population density on ice-free land of 0.42 people per hectare. A hectare is a square 100 meters on a side—approximately the area of two American football fields placed side by side. One square kilometer equals 100 hectares, so a population density of 0.42 people per hectare means 42 people per square kilometer, or roughly one person for every 2½ hectares.

The global average, like the arrows of the economist and the ecologist, misses the mark. In the more developed regions, the population density is 0.22 people per hectare, half the global average. In the less developed regions, the population density is 0.54 people per hectare. The countries with less wealth have a higher population density to support.

To analyze population densities in more detail, I examined 1989 data on the population and area of 148 countries (World Resources Institute, 1992). In 1989, the U.S.S.R. still existed. The combined areas and populations of these 148 countries covered almost the entire ice-free land area and human population of the Earth. I divided each country's population by its land area to get its population density, then ranked the countries from the least to the most densely populated.

In the top left panel of Figure 3.1, I added the area of all countries in which the population density was less than or equal to the population density shown. For example, nearly 13 billion hectares had an average population density of ten or fewer people per hectare. Only a tiny area had a population density greater than ten people per hectare.

To see the cumulative distribution of area at low population densities, I replotted the same data with population density on a logarithmic scale in the second row of the first column of Figure 3.1. More than 11 billion hectares had one person per hectare or fewer, and more than 10 billion hectares had on average less than half a person per hectare.

To emphasize the distribution of population density among the countries with the least dense populations, I replotted the same data a third time with both population density and cumulative area on logarithmic scales in the bottom left panel of Figure 3.1. Together these three figures give the global pattern of population density in relation to cumulative area on the scale of all nations.

What is the distribution of population density within a nation? I applied the identical treatment to the populations and areas of the states of the United States plus Washington D.C. in 1990. The three plots in the middle column of Figure 3.1 are remarkably similar to the corresponding plots for the countries of the world.

I then applied the identical treatment to the populations and areas of the 62

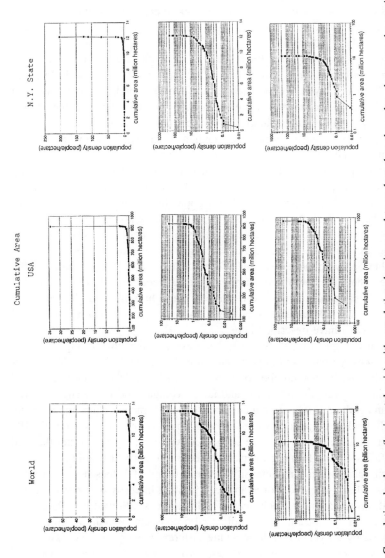

Figure 3.1. Cumulative land area (horizontal axis) with population density equal to or less than the value shown on the vertical axis. Left column: countries of the world in 1989 (data: World Resources Institute, 1992); middle column: 50 states of the United States plus Washington, D.C. (data: U.S. 1990 Census); right column: 62 counties of New York State (data: U.S. 1990 Census). First row: both axes are on a linear scale; second row: vertical axis is logarithmic, horizontal axis is linear; third row: both axes are on a logarithmic scale.

counties of New York State. The plots of county population density as a function of cumulative area in the right column of Figure 3.1 are similar to the plots for states and countries. The total area of New York State, a bit less than 13 million hectares, is about one-thousandth of the total ice-free land area of the world, roughly 13 billion hectares.

Based on this obviously very limited analysis, I conjecture that the distribution of human population density by area is self-similar over a thousandfold range of areas. That is, if you take the labeling off the tic marks on the axes, you cannot tell the size of units you are examining from the shape of the plotted curves. Does this observation hold for other countries and other states? for the human population at earlier times? for nonhuman species? I don't know, but I am curious to find out.

How many people lived at each level of population density in 1989? Using the same data, and replacing cumulative area by cumulative population, Figure 3.2 shows that 2 billion people lived at a population density of one person or fewer per hectare, and about 4 billion people lived at a population density of 2 people per hectare or lower. The distribution of population density within the United States including Washington, D.C., in 1990 was remarkably similar to the distribution by countries. However, the counties of New York State offered a surprise. The large 1990 populations of the five counties of New York City moved the high end of the reverse-L curve to the right, relative to the pattern of countries and states. More data on other regions are required to learn how general are the patterns I have described here.

Historically, population density has been associated with farming intensity. Farming intensity (expressed in percentages) is defined as 100 times the number of crops divided by the number of years in which the land is cultivated and fallowed in one cycle of cultivation and fallow (Pingali and Binswanger, 1987). Hunters and gatherers, who do not cultivate the land, practice a farming intensity of zero. If land is cropped once every year, the farming intensity equals 100 percent. If multiple crop cycles are completed within a single year and the land is never fallowed, the farming intensity exceeds 100 percent (Table 3.1).

A given value of farming intensity between 0 and 100 does not specify the duration of a farming cycle. For example, a farming intensity of 5 percent could mean that, on the average, each year of cultivation is followed by 19 years of fallow. It could also mean, in principle, that 5 consecutive years of cultivation are followed by 95 years of fallow. Quite different amounts of succession and forest recovery can take place under these two regimes. Thus a given value of farming intensity is consistent with very different effects on biological diversity.

In practice, forest fallow consists of one to two annual crops and 15–25 years of fallow; bush fallow of two or more crops and 8–10 years of fallow; short or grass fallow of one to two crops and 1–2 years of fallow; annual cropping of one crop per year, with fallow for only part of a year; and multicropping of two or more crops on the same land each year with no fallow (Boserup, 1981:19). Bush fallow and more intense farming systems prevent forest regeneration.

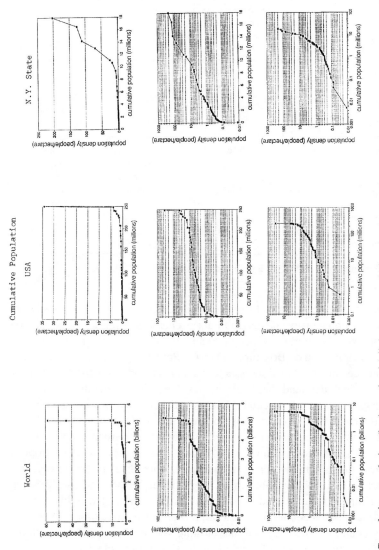

Figure 3.2. Cumulative population (horizontal axis) living at a population density equal to or less than the value shown on the vertical axis, by countries of the world (left column), states of the United States plus Washington, D.C. (middle column), and counties of New York State (right column). Data and linear or logarithmic scaling of the axes are the same as in Figure 3.1.

Table 3.1. *Population density, farming intensity and farming systems in low-technology countries.*

Farming System	Farming Intensity (percent)	Population Density (people/hectare of potentially arable land)	Climate	Tools Used
Hunter/gatherer	0	0–0.04		
Pastoralism	0	0–0.04		
Forest fallow	0–10	0–0.04	humid	axe, matchet, digging stick
Bush fallow	10–40	0.04–0.64	humid or semihumid	above tools plus hoe
Short fallow	40–80	0.16–0.64	semihumid, semiarid, high altitude	hoes and animal traction
Annual cropping with intensive animal husbandry	80–100	0.64–2.56	semihumid, semiarid, high altitude	animal traction and tractors
Multicropping with little animal food	200–300	2.56 and up		

Sources: Boserup (1981: 9, 19, 23); Pingali and Binswanger (1987: 29).

The Danish economist Ester Boserup (1981:23) investigated the association between population density and agricultural systems. Her definition of population density is not simply the ratio of people to all land, but rather the ratio of people to potentially arable land (Boserup, 1981:16). Potentially arable land excludes areas under ice, unirrigable deserts, and mountains too steep for terracing or pasturing, but includes land that could be developed into agricultural land with suitable investments in infrastructure, land now covered by forests that could be cleared and then farmed, grazing lands that are arable, and long-term fallow lands. This definition of "potentially arable land" is difficult, or perhaps impossible, to measure in practice. For example, who knows whether never-cleared tropical forest land will be suitable for agriculture for more than a very few years? Compromising with the available international statistics on land use, Boserup (1981:16) simply excluded land classified as "other" "only if it is likely to be arctic or desert and accounts for . . . a large share of total territory." For low-technology countries, she proposed that farming systems are associated with the population densities per area of arable land as shown in Table 3.1.

The global average population density of 0.42 people per hectare would be compatible with bush fallow or short fallow farming systems if all ice-free land were arable (an unlikely possibility, especially with low levels of technology). Domesticated land (cropland plus permanent pasture) approximated 37 percent of all land (excluding Antarctica) during 1986–89 (World Resources Institute, 1994:284). If all domesticated land is potentially arable using3tic low technology, then the global population density per arable land would be 0.42/0.37, that is, 1.14 people per hectare of arable land. Table 3.1 suggests that annual cropping is required when the population density exceeds 0.64 people per hectare of arable land. It follows that nearly all arable land (defined here as domesticated land) should be cropped at least annually if farmers respond to global population densities rather than to local population densities only, and if farmers use low technology. Because farmers sell food to remote dense populations, domestic and international trade and transport spread the ecological effects of locally dense populations to less populated regions. Thus the effect of global population growth on land use for agriculture, and hence on conservation, depends in part on domestic and international politics, economics and transport, and in part on the level of technology farmers use.

Human Population Growth and Conservation

What is the connection between human population growth and conservation problems? One extreme view is that human population growth has nothing to do with conservation. This point of view is implicit in a conservation agenda that pays no attention to human population growth. Sometimes scholars explicitly deny that human population growth is responsible for conservation problems. For example, from 1970 to 1989, the population of Honduras nearly doubled from 2.63 million to 4.98 million people, while soil eroded, watersheds deteriorated, and forests and coastal resources were destroyed on a massive scale. A 1993 case study of Honduras investigated whether the rapid population increase was directly linked to the natural resource destruction (DeWalt et al., in Jolly and Torrey, 1993:106–123). The report concluded:

> In southern Honduras, environmental degradation and social problems often attributed to population pressure arise from glaring inequalities in the distribution of land, the lack of decent employment opportunities, and the stark poverty of many of the inhabitants. It is not the carrying capacity of the land that has failed to keep pace with population growth. Neither is population growth the primary cause of the impoverishment of the Honduran ecology and its human inhabitants. While the destruction caused by the poor in their desperate search for survival is alarming, it pales in comparison with the destruction wrought by large landowners through their reckless search for profit.

The authors of this study saw no connection between Honduras's extremely rapid population growth and "the lack of decent employment opportunities, and the stark poverty of many of the inhabitants."

At the opposite extreme from this study, some see human population growth as the root cause of conservation problems. For example, in a newsletter devoted to slowing population growth, the caption of a photograph of clear-cut forests read (Popline, 1992:4): "Central America's forests have diminished by more than two-thirds in the past 500 years and are expected to shrink even more as the region's population continues to grow faster than anywhere outside of Africa."

In 60 tropical countries in 1980 (excluding eight arid African countries), the larger the number of people per square kilometer, the smaller the percentage of land covered by forest. The higher the deforestation in these countries, the higher food production also; forests were cleared to open land for agriculture (Pearce and Warford, 1993:166). When 50 countries of unstated geographic distribution were ranked from high to low percentage of habitat loss in the mid-1980s, the amount of habitat loss decreased with decreasing population density, from 85 percent habitat loss and 1.89 people per hectare in the top 10 countries to 41 percent habitat loss and 0.29 people per hectare in the bottom 10 countries (Harrison, 1992:323). Statistical associations such as these suggest, but do not prove, that human population density is responsible for deforestation and the loss of tropical species.

Where relatively small areas of rain forest are surrounded by cleared land, as in Central America, the Philippines, Rwanda, and Burundi, peasants in the cleared areas expand their areas of cultivation, little by little, by nibbling away at the forests. In these cases, variations in rates of deforestation may be explained by variations in local rates of population increase (Rudel, 1991:56).

However, where there are large blocks of rainforest, population growth is not enough to explain deforestation. In addition to rapid population growth, substantial capital investment, for example, in access roads, plus an absence of enforced property rights are also necessary for rapid deforestation. Rates of deforestation were far higher during the 1970s in Brazil, which was relatively capital-rich, than in capital-poor Bolivia and Zaire. In times of economic hardship, if capital becomes scarce, fewer roads may be built in regions with large extents of rainforest. As these large tracts then remain inaccessible to most migrants from other regions, many potential migrants may stay home and pursue the nibbling form of deforestation. Hence capital scarcity may shift the location and nature of deforestation (Rudel, 1991).

When forests are cleared so the land can be farmed to feed an increasing population, the rate of cutting depends in part on how much land is required to produce food for one more person. That requirement depends on yields, farmer education, credit for agricultural investments in land and equipment, culturally acceptable crop varieties, soil types, water resources both natural and human-built, and so on through every aspect of culture and economics and the environment. Forests are sometimes cut because governments give land tenure or tax advantages to those who clear trees, and sometimes because domestic and international markets demand wood in quantities determined more by wealth and population density in cities than by human numbers in forested regions. A one-directional causal model like "human population growth causes the extinction of species" is far too simple (as emphasized by Marquette and Bilsborrow, 1994).

Of the animal extinctions since 1600, it is estimated that hunting caused 23 percent, the destruction of habitat 36 percent, the introduction of alien species 39 percent, and other factors about 2 percent (World Resources Institute, 1994:149). These proximal causes of biodiversity loss are driven in part by population growth and in part by many other factors: culturally determined demands for rhinoceros horn, ivory, and tiger bones; waste disposal in wetlands and water bodies; international trade that pushes developing countries to grow cash crops for export; faulty or insufficient scientific information about the consequence of introducing species; distorted governmental policies regarding land ownership and agricultural prices; inequities in land ownership and management; market failures in valuing unpriced ecosystem services; and inadequate legal definition and enforcement of property rights.

A 1993 report of the National Academy of Sciences on population and land use in developing countries offered the following major conclusions (Jolly and Torrey, 1993:9–11):

> In the long run, population growth almost certainly affects land use patterns. The effects of population growth occur mainly through the extensification and intensification of agricultural production. ... Most of the changes in land use associated with very rapid population growth are likely to be disadvantageous for human beings. ... Population growth is not the only, or in many cases, the most important influence on land use. Other influences include technological change and changes in production techniques ... inequality itself, however, is in part influenced by rates of population growth ... with clear property rights, robust soils, and efficient markets, population growth is less likely to result in land degradation. ... Rapid population growth is likely to make the survival of other members of the animal and plant kingdom more difficult. Accompanying rapid population growth in the past has been greater species loss and a higher attrition within species than would have occurred in the absence of human expansion.

The Agendas of Conservation

Conservationists need two agendas: an internal agenda and an external agenda. The internal agenda of conservation contains answers to questions like: What research is needed? Given what is known about particular conservation problems, what needs to be done? A principal aim of this book is to define and promote conservation's internal agenda.

The external agenda concerns the positions conservationists support in other fields that affect conservation. Such fields include family planning, agriculture, education, and environmental law. How conservationists define their external agenda could influence their internal agenda for research, and vice versa. To promote discussion of the external agenda of conservation, I put forward a few proposals.

A major item on the external agenda of conservation ought to be slowing human population growth voluntarily by means that simultaneously contribute to other goals. For example, the education of women in developing countries—who now get about half the education of men, on the average—could increase their productivity as workers, improve their child-rearing, and defer the age of first marriage, thereby raising the quality of their lives and also slowing population growth. Improved health facilities could simultaneously reduce infant and child mortality, removing one incentive for high fertility. Better health facilities could also lower scandalous rates of maternal mortality, improve the productivity of agricultural workers, and provide a framework for the distribution of family planning products and services.

As a second item on their external agenda, conservationists should promote the selective intensification of agriculture through nonpolluting and nondestructive means. Intensification means extracting more yield from the same area of land. Intensification would make it possible to support additional billions of people, who appear to be almost inevitable, and to improve the lot of those already born, while taking less additional land away from natural habitats. How yields are raised is all important. Methods of cultivation that erode soil and produce polluting effluents are counterproductive. Better understanding of the food webs of agroecosystems would identify the natural enemies of agricultural pests and might suggest improved nontoxic strategies for pest control (Cohen et al., 1994; Schoenly et al., 1995). Farmer education in the developing world should include natural history and practical systematics related to biological pest control as well as training in more efficient use of water.

Institutional innovation will be required. The agricultural economist Vernon W. Ruttan wrote (in Jolly and Torrey, 1993:150): "The challenge to institutional innovation in the next century will be to design institutions that can ameliorate the negative spillover into the soil, the water, and the atmosphere of the residuals from agricultural and industrial intensification."

Third, appropriate property law could make many renewable natural resources less vulnerable to open access (Hardin, 1968). When a peasant asks, "Why should I plant trees if someone else may harvest them?" it is a good question. Unless legal and institutional guarantees promote the conservation of natural resources, these resources may be mined to exhaustion like the fisheries off the Atlantic coast of the United States and Canada. Property law should enhance rather than oppose social equity, and should regulate private actions, such as clearcutting privately owned forests in a watershed, that have adverse external effects on common goods.

I hope these proposals will stimulate thinking about the external agenda of conservationists.

Acknowledgments

I am grateful for the support of U.S. National Science Foundation grant BSR92-07293 and the hospitality of Mr. and Mrs. William T. Golden.

4

Developing an Analytical Context for Multispecies Conservation Planning

Barry Noon, Kevin McKelvey, and Dennis Murphy

Summary

This chapter attempts to apply population modeling techniques that have been applied to single species, and use them to plan conservation strategies for multiple species when species are well known and have common life history features divisible into prereproductive and reproductive phases. Using three exemplary endangered species, we were able to ordinate them on a common response surface to analyze their sensitivity to habitat loss and fragmentation. The modeling strategy enables managers to assess which species are least sensitive to habitat loss, and therefore reduce costs in monitoring such species. Although at present, the strategy of protecting biological diversity through protecting areas large enough to allow the persistence of habitat mosaics and dynamic processes of change within them is the most effective, the modeling strategy presented here is a promising addition to the tools available to managers, and merits further development. The method should be of most utility for sympatric species in habitats threatened by reduction and fragmentation.

Introduction

We have been involved in practical conservation planning for more than a decade—which is to say that we have applied the principles and theories of ecology in attempts to solve real-world conservation problems. One recurring constraint that we have encountered when attempting to incorporate ecological principles into conservation solutions is the desperate lack of time and money available to those who take on the challenge of balancing environmental protection with the rest of the human endeavor. We know that the number of species at risk of extinction is extensive and that their individual ecologies are diverse. The spatial and temporal scales at which those species respond to environmental degradation vary over several orders of magnitude, and imperiled species differ extensively in their life history attributes. We contend that the diversity of life history attributes, and the multitude of physical and biotic forces that affect landscapes and the ecosystems that they support, interact in such complex ways that they currently defy direct application of existing ecological principles in a conservation context. To a large extent, what we hope to accomplish in this chapter is to prompt ecologists to critically examine whether any general ecological principles are emerging that may prove to be applicable to the conservation of communities of species occupying diverse ecosystems.

Here we advance one method that might enable us to move from single-species to multispecies conservation planning. At the outset, we make it clear that we currently cannot offer a definitive algorithm for multispecies analysis. Rather, we are making an initial attempt toward that goal using a life history model applicable to species with the following generalized characteristics: their life histories can reasonably be represented as having two discrete stages (prereproductive and reproductive), a seasonal breeding pulse exists that is short relative to the rest of the annual cycle, the concept of territory or home range is meaningful, and prereproductive individuals search (to varying degrees) for breeding territories and mates. Using this model, we simultaneously explore the relative sensitivities of three species to habitat loss and fragmentation. The three species, all protected under the Endangered Species Act, are the California gnatcatcher (*Polioptila californica californica*), the Bay checkerspot butterfly (*Euphydryas editha bayensis*), and the northern spotted owl (*Strix occidentalis caurina*). All are found in California. Each, however, resides in very different habitat associations, has widely differing area requirements, and extremely distinctive life histories.

In the process of examining the population dynamics of these species in the face of habitat loss and fragmentation, we address the following questions: First, how do we as conservation scientists move from single-species conservation planning to a multispecies approach? Second, is there any possibility that the umbrella or keystone species concept might reemerge to ease the burden of multispecies conservation planning? Finally, is there a common analytical framework for evaluating multispecies sensitivity to habitat loss and fragmentation? As a starting point, we focus here primarily on the third question. We revisit the first two questions relative to how well we have answered the third.

A Common Framework for Population Viability Analysis

Levins (1970) described the dynamics of populations occupying a system of habitat patches of identical size and even spacing. Although this structure may be rare in natural populations (Harrison, 1993), it can be common in human-dominated systems. In a classic example, the U.S. government gave railroad companies every other section (1 mi^2) of land along railroad right-of-ways. As a result, "checkerboard" patterns of ownership and subsequent habitat disturbance and conversion are common in the western states today. In this context, Levins' metapopulation paradigm therefore can provide a useful starting point for analyzing the impacts of human-induced habitat fragmentation.

The dynamics of the original Levins' model are described by the following differential equation:

$$\frac{dp}{dt} = mp(1 - p) - ep, \tag{4.1}$$

where p is the proportion of habitat patches occupied, m is the colonization rate, and e is the extinction rate of patches. The equilibrium proportion of occupied patches is

$$\hat{p} = 1 - \frac{e}{m}, \tag{4.2}$$

hence the population $\rightarrow$ for all $e \geq m$.

The linear density dependence in Levins' model assumes that potential colonizers are only able to search a single habitat patch, and that all patches are suitable for colonization. We can relax these conditions, allowing for multiple searches and unsuitable habitat patches (Noon and McKelvey, 1996):

$$\frac{dp}{dt} = mp\{1 - [(1 - h) + ph]^n\} - ep, \tag{4.3}$$

where h is the proportion of habitat patches that are suitable and n is the number of patches that a colonizer can search. The associated equilibrium occupancy proportion of patches, in discrete form, is

$$\hat{p} = 1 - \left\{ \frac{\left[1 - \left(1 - \frac{e}{m}\right)^{1/n}\right]}{h} \right\}. \tag{4.4}$$

If $h = 1$ and $n = 1$, equation (4.3) collapses to equation (4.1) and the equilibrium collapses to equation (4.2).

Although these changes in Levins' model are relatively minor, they increase the generality of the model and make it algebraically identical to Lande's (1987) individual territory model (Noon and McKelvey, 1996). This crosswalk enables us to exploit the best features of both models: Levins' model has simple equilibrium criteria, but *e* and *m* are not clearly defined; whereas Lande's model is parameterized on the basis of clear biological understanding and measurable criteria. In Lande's model *e* and *m* are functions of birth, death, and the mobility of individual organisms. Hence we are able to rewrite Levins' equilibrium solution in terms of Lande's (1987) model as

$$\hat{p} = 1 - \left\{ \frac{\left[1 - \left(1 - \frac{1-s}{b} \right)^{1/n} \right]}{h} \right\}, \tag{4.5}$$

where *s* is the adult survival rate, *b* is the average fecundity of adults, and *n* is the number of potential territories (habitat patches) that a dispersing juvenile is capable of searching (Noon and McKelvey, 1996). The parallel between equations (4.4) and (4.5) assumes that $e \propto (1 - s)$ and $m \propto b$.

Given these understandings, if we have estimates of an organism's life history parameters and some knowledge of its dispersal behavior, we can estimate its sensitivity to changes in the amount and fragmentation of habitat (*h*). Because both dispersal (*n*) and habitat fragmentation (h) scale to the territory size of the organism, species of different sizes, from different taxonomic groups, and that exhibit different vagility can be directly compared within the same parameter space. This rescaling enables a comparative assessment of species that respond to environmental variation at distinct spatial scales in a common analytical framework.

This parameter space can be presented as a stability response surface, combining habitat proportion, dispersal ability, and growth potential (Fig. 4.1). The stability condition for population equilibrium (eq. [4.5]), $|f'(p)| < 1$, is

$$\frac{b[1 - (1 - h)^n]}{1 - s} > 1. \tag{4.6}$$

The surface represents the threshold between population persistence and extinction. Combinations of points that lie above this surface have an equilibrium occupancy proportion greater than zero. For the sake of generality we refer to the ratio of fecundity to mortality as growth potential (note that this variable is not equivalent to Lande's [1987] definition of demographic potential). A priori, we expect organisms with low fecundity and low vagility to be sensitive to fragmentation.

Three Exemplary Threatened Species

If this common modeling framework has merit, species that are believed to be threatened by habitat loss or fragmentation should exhibit combinations of life

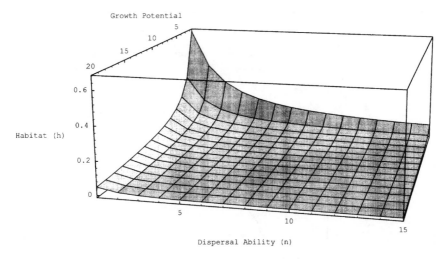

Figure 4.1. Stability response surface for the equilibrium solution to the general metapopulation model. Axes are habitat proportion (h), growth potential [b /(1 - s)] , and dispersal ability (n). The surface is the equilibrium extinction threshold. Note that $\hat{p} >$ 0 for all values above the surface and $\hat{p} < 0$ below it.

history traits (fecundity, adult survival, and vagility) that in the model are associated with a sensitivity to declines in *h*. To test this premise, we have chosen three well-studied species currently listed as threatened under the Endangered Species Act: the California gnatcatcher, the Bay checkerspot butterfly, and the northern spotted owl.

The California gnatcatcher is a small bird, weighing about 6 grams, that is narrowly restricted to coastal sage scrub habitat in coastal southern California. The coastal sage scrub habitat is naturally patchy in distribution, occurring in a mosaic with other plant communities. Habitat quality for the bird apparently is best where sage scrub is interdigitated with grasslands at low elevations close to the coast (Mock, 1992; Mock and Bolger, 1992). The distribution of the California gnatcatcher ranges from Los Angeles south into Baja California. The coastal sage scrub habitat is highly disrupted near to the coast, and the bird now is found largely inland and at higher elevations (Mock, 1992; Mock and Bolger, 1992; Davis et al., 1994; Fig. 4.2). For example, in the southwestern part of San Diego County, urban sprawl restricts the remaining coastal sage scrub habitat for the California gnatcatcher to the eastern part of its historical distribution (Fig. 4.2). Presumably the highest quality or source habitat occurred within sage scrub at lower elevations and on moderate slopes. Whether residual higher elevation patches of coastal sage scrub act as "sink" habitat for the gnatcatcher is really not known (Mock, 1992; Mock and Bolger, 1992), but a proposed regional conservation plan for the California gnatcatcher delimits reserves that are cur-

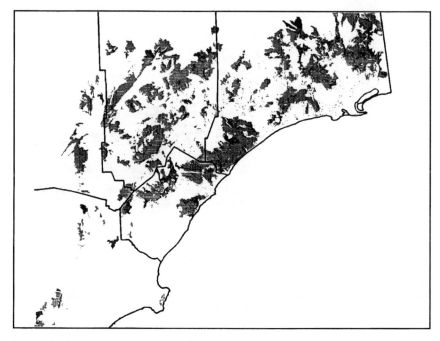

Figure 4.2. Distribution of California gnatcatcher habitat (coastal sage scrub) north of Mexico. The figure includes the entire Natural Community Conservation Planning Region. Lower-elevation habitat has been disproportionally destroyed or disturbed due to agriculture and urbanization.

rently suitable habitat or have the potential to be restored to suitable habitat, and designates specific areas to develop as corridors. Corridors would be restored to enhance connectivity between the reserves (Reid and Murphy, 1995).

The Bay checkerspot butterfly occurs in residual patches of native annual grasses and forbs that are restricted to serpentine-based soils (Harrison, 1989; Murphy et al., 1990; Launer and Murphy, 1994). In most other grasslands that fall within the historic range of the species, the required larval hostplants and adult nectar sources have been replaced by exotic species, but serpentine soils with their unique physical and chemical characteristics have allowed remnant natural habitat to persist.

The remaining amount of serpentine soil habitat in California is quite small, the patches disjunct (Fig. 4.3), and local populations of checkerspot butterflies are therefore prone to local extinction—at this time the butterfly populations are restricted to two large population centers. A reserve has been set aside for the Bay checkerspot butterfly within habitat that supports the more southern of the two population centers. The reserve consists of a portion of one large "mainland"

Figure 4.3. Distribution of serpentine-soil-based habitat in south San Francisco Bay area in central California. Habitat designated MH (Morgan Hill) supports a source population of butterflies that sustains apparent sink populations on habitat patches designated with other initials. Habitat patches not designated with letters have been unoccupied during the past decade.

or source area, whereas a number of smaller "satellite" patches exist nearby but are unprotected.

Spatial scale plays an important role in the dynamics of the Bay checkerspot butterfly population. Habitat quality varies on a microscale, and reproductive success is tightly linked to fine-scale microclimatic and topographic variation (Singer, 1972; Dobkin et al., 1987; Murphy et al., 1990). A map of the thermal environment for the butterfly (Fig. 4.4) indicates the extent of microclimate variation, and suggests that habitat quality varies as a function of the environmental conditions unique to each given year, combined with habitat area, slope, and aspect (Weiss et al., 1993). In simplified terms, the butterflies tend to experience greater reproductive success on warmer, drier slopes during cooler, wetter years, and the converse during warmer, drier years.

The northern spotted owl occupies primarily closed-canopy, late seral stage coniferous forests with nest sites characterized by particularly large diameter trees (Thomas et al., 1990). During the past 50 years, the number and distribution of spotted owls may have been reduced by as much as 50 percent from pre-

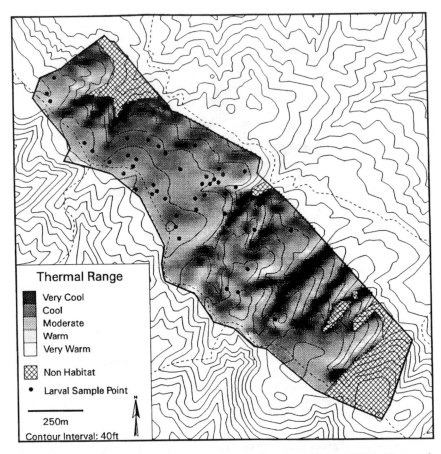

Figure 4.4. The thermal environment within 100 ha of the Morgan Hill habitat patch butterfly reserve in central California. Vital rates for the Bay checkerspot butterfly vary across the topographic gradient from year to year, with substantially higher survival of larvae on cooler slopes in warmer, drought years, and slighly higher survival on warmer slopes in cooler, wetter years.

twentieth century levels (Thomas et al., 1990:20). Threats to the subspecies are primarily a consequence of loss and fragmentation of habitat as a consequence of timber harvest (Fig. 4.5; Murphy and Noon, 1992). The primary silvicultural method in the Pacific Northwest, west of the Cascade crest, was clear-cutting. Because these cutting methods have dominated in the Pacific Northwest, particularly over the last 50 years, habitat within the range of the owl is either undisturbed and suitable, or cut within the last 50 years and unsuitable. The result has changed the forested landscape to an island-like distribution of late seral stage forests imbedded in a matrix of young forest (Ripple et al., 1991).

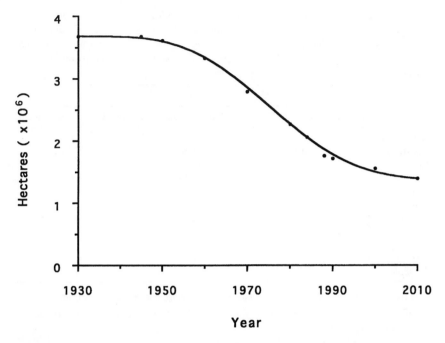

Figure 4.5. Estimated trend in the areal extent of suitable northern spotted owl habitat on National Forest lands in Oregon and Washington for 1930–2010. Estimates for 1990–2010 are projections based on Forest Plans approved prior to 1989.

Developing a Common Modeling Framework

Parameterizing the model requires estimates of vital rates, dispersal ability, and area requirements for breeding. Such data are available for only a handful of species—extensive field work is required to obtain the necessary parameter estimates for any wild population. For these three species, published estimates of fecundity based primarily on field studies and direct enumeration of progeny, estimates of adult survival rates based on mark-recapture studies, and estimates of home range size and mobility derived from following the movement of marked individuals, were available (Table 4.1).

Primary sources for parameter estimates for the Bay checkerspot were Harrison et al. (1988), Cushman et al. (1994), and Carol L. Boggs (personal communication); for the California gnatcatcher, Mock (1992) and Mock and Bolger (1992); and for the northern spotted owl, Burnham et al. (1996). The essential parameters were (1) fecundity, expressed as the number of female offspring per adult female; (2) survival rate during the period of spatial fidelity (the territory period for gnatcatchers and owls, and the period from diapause until metamorphosis for butterflies); and (3) dispersal ability, computed as the average distance moved

Table 4.1. Estimates of the mean annual vital rates (birth and death) and dispersal ability of three species listed under the Endangered Species Act. Dispersal ability is scaled by mean breeding season home range size: Maximum parameter values provide insights into biological potential. These parameter estimates were used as input for model analyses.

Parameter	Species		
	Northern Spotted Owl	California Gnatcatcher	Bay Checkerspot Butterfly
Fecundity[1](b)	0.35	1.05	85
Maximum	0.70	1.50	120
Adult survival[2](s)	0.87	0.55	~ 0.01
Maximum	0.92	0.61	—
Growth potential[3]	2.69	2.33	86
Maximum	8.75	3.85	121
Dispersal ability[4]	10	4	6
Maximum	>20	>20	>>100
Home range	100 ha	14 ha	4 m^2

[1] Fecundity for the bay checkerspot was based on the average number of eggs reaching diapause.

[2] Adult survival for the bay checkerspot was computed as the mean probability of survival from postdiapause to egg laying.

[3] Computed as $b/(1 - s)$.

[4] Average distance moved from birth to breeding, expressed in home-range-sized units.

from birth to breeding and expressed in home-range-sized units. Survival rate during dispersal was implicit in the model, and arises as a function of search ability (n) and the proportion of the landscape that was suitable habitat (h) (Lande, 1987; Lamberson et al., 1992).

The three threatened species have been subject to extensive habitat loss and fragmentation from human activities since the turn of the century, but particularly during the past several decades, either by urbanization or by deforestation. As a consequence, their populations have discrete spatial structure and the paradigm of the metapopulation, to varying degrees, is an appropriate one for these species (Harrison et al., 1988; Mock, 1992; Lamberson et al., 1992). Hanski (1991b) and Hanski and Thomas (1994) have described the different sorts of metapopulation paradigms that one might bring to bear in the development of a theoretical foundation for conservation planning for imperiled species. What we have done is to view these species dynamics in a combined, comparative analysis to contrast their sensitivities to habitat loss and fragmentation. As noted earlier, the same model form can be utilized to look at reserve-level extinction and colonization (Levins' model) or within reserve dynamics at the level of the individual territory (Lande's model). In applying these concepts to these three species, we look first at a within-reserve and then at among-reserve dynamics.

Placing the population dynamics of these three species within a common analytical framework requires some simplification and abstraction. In modeling within-reserve population dynamics, we assumed that the number of suitable sites was static, and that all suitable sites were equally available. There was no within-reserve spatial structure other than the proportion of habitat available. In addition, environmental stochasticity was not modeled, a simplification that is of particular concern in the context of the population dynamics of the Bay checkerspot butterfly (Weiss et al., 1988, 1993).

Population Dynamics Within a Local Population

Both the gnatcatcher and the spotted owl are sensitive to decreases in suitable habitat, with equilibrium populations becoming extinct in the presence of suitable habitat (Fig. 4.6). That is, there is a steep threshold to their population persistence (cf. Lande, 1987; Lamberson et al., 1992). Of the three species, the gnatcatcher is the most sensitive to habitat fragmentation, and the Bay checkerspot, primarily because of its comparatively high reproductive potential, the least sensitive.

Based on available parameter estimates (Table 4.1), we plotted the position of the spotted owl, gnatcatcher, and Bay checkerspot butterfly on the stability

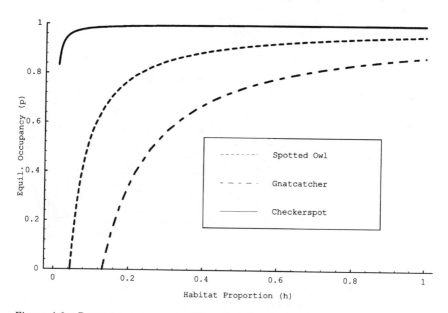

Figure 4.6. Percentage occupancy of breeding sites (home ranges or territories) for the Bay checkerspot butterfly, the northern spotted owl, and the California gnatcatcher against the proportion of the landscape, within a local population, that is suitable habitat.

response surface (Fig. 4.7). Note that the spotted owl and gnatcatcher are located in the portion of the surface that is sensitive to changes in habitat proportion, whereas the Bay checkerspot lies in an insensitive region (Fig. 4.7).

One influence on population persistence that this figure fails to capture, and that is of pronounced significance particularly for the Bay checkerspot (Weiss et al., 1988, 1993), is the annual variation around these points. For example, plotting the range in population growth potential observed in field studies of the Bay checkerspot, even for populations of the size of several hundred individuals, extends beyond the range of that axis (Table 4.1). Thus, in this deterministic analysis, which assumes a global availability of a fixed amount of suitable habitat, the butterfly falsely appears extinction-proof.

To further understand the comparative sensitivities of these three species, the next logical analysis is to take "slices" through each of the points representing the three species on this surface (Fig. 4.7). By keeping either dispersal or growth potential constant, we can determine the proportion of "critical habitat" required for population persistence. These functions can enable us to evaluate the allocation of management efforts. For example, to estimate the changes in population stability associated with facilitating dispersal (e.g., establishing corridors or practicing translocation) versus increasing growth potential (e.g., adding nest sites or supplemental feeding sites).

We first keep dispersal constant and look at the stability requirements, in terms

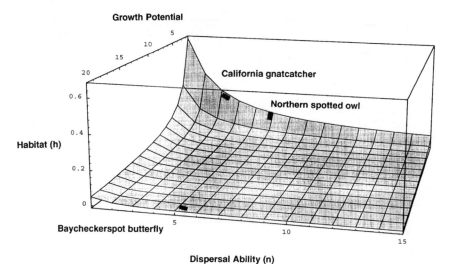

Figure 4.7. Stability response surface for the equilibrium solution to the general metapopulation model. Axes are habitat proportion (h), growth potential [b /(1 - s)], and dispersal ability (n). The California gnatcatcher, Bay checkerspot butterfly, and northern spotted owl are plotted on the stability response surface (Fig. 4.1) based on measured demographic parameters.

of h, for the three species (Fig. 4.8). Stable equilibrium points (combinations of *h* and growth potential) lie above a species' curve (Fig. 4.8). The steeper the slope in the region of the biological potential of a species, the larger the marginal gain (greater freedom from risk) associated with an incremental increase in growth potential. Plotting, along the *x* axis, the interval from the mean to the maximum growth potential (Table 4.1) provides additional insights. The maximum value for a parameter represents the upper limit to a management action. By examining the slope of a species' function over this interval, we can estimate the gain in population stability from management actions that increase b or s.

The trade-off between habitat proportion and growth potential required for population stability is similar in all three species (Fig. 4.8). For the California gnatcatcher, growth potential is relatively invariant (Table 4.1). The gnatcatcher therefore cannot compensate for habitat loss by possible increases in growth potential. For the Bay checkerspot, within-population stability is largely independent of increases to its potential fecundity (Table 4.1 and Fig. 4.8). However, reproductive success for this species can range from zero in some years, to well beyond the end of the scale. Successive years of low recruitment therefore could be disastrous. Relative to the other two species, the spotted owl has the greatest potential decline in risk to extinction from habitat loss given its possible increases in growth potential. However, the gain here is also marginal.

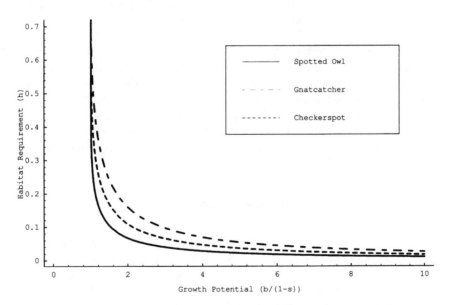

Figure 4.8. Changes in the habitat proportion required for population stability against increases in growth potential, showing the relative sensitivities of the California gnatcatcher, Bay checkerspot butterfly, and northern spotted owl. This figure is based on an orthogonal slice through Figure 4.7, holding dispersal ability constant.

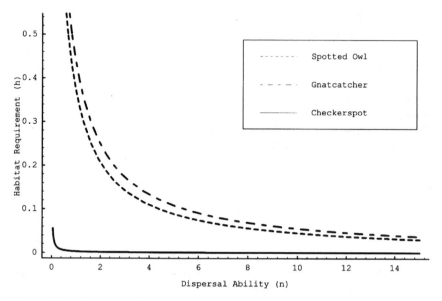

Figure 4.9. Changes in the habitat proportion required for population stability against increases in dispersal ability, showing the relative sensitivities of the California gnatcatcher, Bay checkerspot butterfly, and northern spotted owl. This figure is based on an orthogonal slice through Figure 4.7, holding growth potential constant.

Figure 4.9 shows the trade-off between habitat proportion and dispersal ability required for population stability. At the within-population level, the Bay checkerspot is largely unaffected by changes in dispersal ability. The owl and particularly the gnatcatcher show a greater sensitivity. For these two species, there exists an interval of response in which it is possible to affect population stability by increasing dispersal (Table 4.1). This analysis suggests that specific management actions that increase dispersal success for the two bird species could lead to an important decrease in extinction risk.

Incorporating Between-Population Dynamics

No mathematical differences exist between Lande's (1987) model and our expanded version of Levins' (1970) model, but the biological understanding associated with the model parameters are quite different. The models operate in a strict hierarchical sense both in space and time. Within each local population or reserve, home range (territory) occupancy changes quickly, and vacant sites are recolonized through local recruitment and less frequently by dispersers from other local populations. This is the scale at which Lande's (1987) model operates. The likelihood that a given reserve will experience local extinctions is a function of both the proportion of suitable habitat within the reserve and reserve carrying capacity. The appropriate scales for these dynamics are local and fast.

The proximity and size of other reserves will control recolonization rates and regional population stability. Relative to Lande's model, the appropriate scales for Levins' (1970) are regional and slow. The two temporal and spatial scales—local and fast and regional and slow—and the results of their interactions, collectively determine the overall metapopulation dynamics of the species.

To clarify the trade-offs between within- and among-population dynamics, we used the simulation model of Lamberson et al. (1994). In this model, the metapopulation is composed of a regular array of circular reserves, equal in size and equally spaced, conforming to the original Levins' model. Each reserve, in turn, is composed of a number of breeding sites of which a proportion h are suitable. Individuals first search their local reserve for a suitable breeding site, but if unsuccessful, they leave the reserve and potentially colonize other reserves (Lamberson et al., 1994). To determine $\hat{p}$ for the reserves (eq. [4.2]), we simulated large metapopulations (> 250 reserves) over a long time period (1000 years), counted the reserve transitions: occupied → extinct (e) and extinct → occupied (m), and computed the ratio (Noon and McKelvey, 1996). We simulated reserves with 10, 20, 30, and 40 sites per reserve and allowed the number of suitable sites per reserve and the distance between the reserves to vary.

On the basis of the Lamberson et al. (1994) model we can examine the trade-offs between the proportion of the landscape that is suitable habitat (landscape level h), the proportion of a given reserve that is suitable habitat (local level h), and local population (reserve) size. In this model, for a fixed reserve size, as the proportion of landscape that is suitable declines, the spacing among reserves increases. The analyses discussed below are for the spotted owl; however, the conceptual insights are applicable to the other species as well.

The curves in Figure 4.10 represent stability boundaries for the spotted owl. For a given reserve size, populations represented by points above the curves are unstable and go to extinction; locations below the curves are stable. None of the metapopulations persists if the proportion of suitable sites within each reserve falls below 30 percent. This is analogous to the steep persistence threshold shown in Fig. 4.6, but with additional uncertainty associated with mate finding (see Lamberson et al., 1994; Lande, 1987). In addition, if all sites are suitable, then reserves containing more than 20 sites are stable regardless of their density on the landscape. Between these two extremes lie diagonal boundaries representing the relative sensitivities of metapopulations to dynamics at the within- and among-population scales.

As the size of individual reserves declines, metapopulation dynamics become sensitive both to changes in internal habitat quality and in reserve spacing. That is, as the amount of suitable habitat within a local reserve is reduced, individuals spend more time moving among reserves and are impacted by among-reserve survival costs. These dynamics may be particularly relevant to the Bay checker-

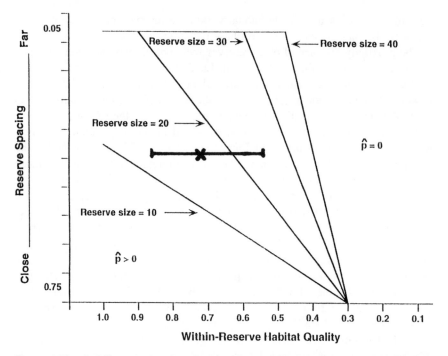

Figure 4.10. Stability response functions based on solutions to the combined individual territory and metapopulation models (Lamberson et al., 1994). Functions illustrate the trade-offs between reserve spacing (proportion of the landscape within reserve boundaries) and within-reserve habitat quality (proportion of sites suitable for colonization) as a function of reserve size. The X in the center of the horizontal bar represents the mean locus of a hypothetical metapopulation. The width of the horizontal bar represents the magnitude of environmental variability in habitat quality. In the deterministic case, if the reserves have 20 or more breeding sites, the metapopulation is stable, but environmental variation can destabilize it.

spot butterfly. The annual effects of weather on larval hostplants across microclimatic gradients result in highly variable habitat quality (Weiss et al., 1988, 1993), as is reflected in the demographic values (Table 4.1). The general insight from this result is that even though the boundaries of a reserve are fixed, the habitat quality (h) of the reserve may remain extremely dynamic. Changes in habitat quality and quantity due to environmental stochasticity and catastrophic loss can be portrayed in Fig. 4.10 as a line running parallel to the x axis. For a fixed amount of the landscape in suitable habitat, if the distribution of within-reserve habitat quality is highly variable, a metapopulation which appears to be stable in the deterministic case may follow extinction trajectories a large proportion of the time.

Insights to Multispecies Conservation Planning

The models presented here have been used in the development of single-species conservation plans, most notably for the northern spotted owl. Because of the obvious limitations associated with using single-species plans to protect biological diversity, however, the framework for evaluation needs to be expanded. The first step in multispecies planning is to develop a common analytical framework in which multiple species can be compared. The approach presented here offers a preliminary tool to accomplish this, at least for species with compatible ecologies and life histories. We were able to ordinate different species on a common stability response surface and explore their comparative sensitivities to habitat loss and fragmentation. For example, we were able to take slices through that surface and look specifically at marginal gains in reduced extinction risk resulting from incremental increases in fecundity, adult survivorship, or dispersal ability.

Given the impossibility of exploring the threats to persistence for all species, the analyses presented here may help identify those species that are most sensitive to habitat loss and fragmentation, and thus would serve as the best indicators of this type of environmental change. Conversely, it may be possible to determine which species are less likely to be of concern, and thereby potentially reduce the costs incurred by monitoring such species. Furthermore, if a number of species show similar sensitivities to habitat modification, then we could choose to monitor those that are least expensive to study.

The most effective way to protect biological diversity is to protect areas that are large enough to allow the existence of habitat mosaics and the dynamic processes of change within these areas (National Research Council, 1995). For most species, however, details on birth and death rates and dispersal abilities are not and will not be available. So how does one apply the data-intensive methods discussed in this paper to the challenge of multispecies conservation planning? One way is to focus research and monitoring on those species that are both good indicators of habitat change and have large area requirements. Managing habitat in a manner consistent with their persistence may indirectly ensure the persistence of numerous other species with overlapping habitat needs and smaller area requirements.

Our inclusion of the Bay checkerspot butterfly in our multispecies view pointed out a major weakness of our modeling framework. Because of the deterministic structure of the models, the stability of species for which vital rates and habitat quality are both dominated by stochastic environmental factors may be greatly overestimated. Despite this limitation, these models provide a valuable tool to explore the sensitivities to habitat change of multispecies communities and to rank member species by their sensitivities to change. We believe that the greatest value of the approaches we have outlined will be found in the future in comparative analyses of sympatric species sharing habitats threatened by reduction and fragmentation.

5

Operationalizing Ecology under a New Paradigm: An African Perspective

Kevin H. Rogers

Summary

Under the balance-of-nature paradigm, conservation managers adopted contrasting approaches to management of large national parks in Kenya, South Africa, and the United States. Whereas ecologists assumed omnipotence of the science as the driving force behind conservation, managers responded uniquely to multiple pressures from the societies they served. The managers' need for pragmatic, adaptive action contrasts starkly with the scientists' search for detailed understanding under a rather self-serving peer review system.

If ecology is to better serve conservation under a new paradigm it must discard the "strategy of hope" that good science will inevitably lead to informed management and develop an explicit interface for technology development and transfer. A review system that judges the utility of ecological research on a par with its quality, would promote technology transfer by turning the focus from the producer to the consumer. The emerging field of technology transfer has much to offer ecologists in their efforts to become more pragmatic in enhancing the ecological basis of conservation.

Patch dynamics provides a useful framework for dealing with heterogeneity in ecology and can be pragmatically reformulated to provide management with a set of forces to engineer the landscape mosaic to achieve conservation goals.

A program of research on the rivers of the Kruger National Park, South Africa, operates around a consensus-building management process facilitated by a decision support system (DSS). The DSS integrates an *operational framework* for setting and evaluating attainable and acceptable goals; a *predictive modeling framework* for assessing the nature, rate, and direction of system change; and a *system response framework* for monitoring response to management action and natural disturbances. A hierarchical model of patch structure and dynamics provides the basis for incorporating heterogeneity into research and management.

Introduction

There is much debate about the usefulness of the science of ecology in managing natural systems or the resources they provide (cf. Botkin, 1990; Peters, 1991; Ludwig et al., 1993; Shrader-Frechette and McCoy, 1993). Much of this argument is by ecologists, for ecologists, and it seldom effectively includes the management arena. This is largely because few conservation practitioners write in scientific journals and ecologists generally perceive nonscientific literature to be "gray" and of limited value. Ecologists and practitioners, therefore, operate from very different world views. This chapter contrasts these different operational environments and develops a proposal for increasing the influence of the science of ecology on conservation practice under a new ecological paradigm.

Three main issues are discussed:

1. In what way did the old balance-of-nature paradigm in ecology influence conservation practice and what are the implications for future interaction between the science and practice?

2. What form should a future operational interface between ecology and conservation take and why?

3. The basis of the new paradigm of the flux of nature is ecological heterogeneity in space and time. How can the concept of patch dynamics be used to move conservation beyond the current species focus?

A final section shows how some of the principles discussed are being incorporated into a program to integrate research and management of the rivers of the Kruger National Park, South Africa.

Influences of Ecology on Conservation

There is little doubt that most ecologists have a deep conviction that by unraveling the complexity of nature they will contribute to the conservation of natural systems. It is not surprising then that ecologists have a sense of responsibility and "ownership" of conservation and tend to assume the omnipotence of their science as the driving force behind conservation philosophy and practice. Thus, as ecologists have found fault with the balance-of-nature paradigm, there has been the tendency to assume that conservation has fallen short under its tenure (Pickett and Ostfeld, 1995). Is this so? What are the implications for ensuring a constructive relationship between ecology under the new paradigm and the evolving technology (Harper, 1992) of conservation?

A historical contrast of two of Africa's "great" national parks, Tsavo in Kenya and Kruger in South Africa, helps answer these questions. Comparison with

Yellowstone National Park (United States) reinforces the answers in an international context.

Most people, and especially ecologists, stand in awe of the size of these great parks and their potential to represent nature as it was in the absence of human interference. These "unspoiled" areas are seen as the ultimate sites in which ecology and conservation can be practiced. If this is so, then they should also provide a test of the influence of the dominant ecological paradigm on the practice of conservation.

Tsavo: Letting Nature Take Its Course

When the 28,000-km^2 Tsavo National Park was proclaimed in 1948, colonial authorities evicted all those who had traditionally hunted or grazed stock in the area and the only legal land use was motorized game viewing. The dominant vegetation was dense *Comiphera* woodland, with a visibility of less than 50 m. The park was noted for high densities of elephants and rhinoceroses, although these had been low in the 1890s (Parker, 1983). In 1957 the first signs that elephants were noticeably thinning areas of woodland were recorded. By 1961 this was so extensive that hundreds of rhino died of starvation in a severe drought. In 1971 thousands more elephants and rhinos died in another drought.

As this saga unfolded, it was accompanied by acrimonious debate between scientists, managers, politicians, and others on the correct course for management. Some assumed that this was part of a "natural cycle" and should thus be allowed to proceed naturally. Others argued that elephants were being driven into the reserve by increased human pressure in surrounding areas. Thus, the trend was "unnatural" and should be halted by large-scale reduction in elephant numbers. A third group maintained that no action could be judged "right" or "wrong" until goals were clearly defined. In the end, the voice of the enigmatic chief warden won the day, and "nature" was allowed to take its course.

By 1973 the entire woodland had been changed to open grassland and seasonal fire had become the dominant disturbance. Over the 20-year period from 1963 to 1982 the population size of lesser kudus (*Tragelaphus inbergis*) decreased by 90 percent, gerenuk (*Litocramius walleri*) by 80 percent, giraffes (*Giraffa camelopardis*) by 40 percent, and black rhinoceroses (*Diceros bicornis*) declined to very low numbers (Parker, 1983). The elephant population was decimated by tribesmen who, driven by drought, "poached" on previously traditional hunting grounds and by bandits who found hunting in the open terrain highly profitable. Populations of grazers such as oryx, Grant's gazelles, and zebras, however, more than doubled in size. The old Tsavo was no more, a new Tsavo was born!

Kruger: Managing a Heritage

In the 19,455-km^2 Kruger National Park, managers have been explicit in its acceptance of the balance-of-nature paradigm since proclamation in 1898. Following the initial years of little active management, signs of change drove managers to control the perceived balance by "pragmatic intervention" (Pienaar, 1983). Lions were culled to reduce their impact on certain herbivore populations, but the herd size of other herbivores (elephants and ungulates) was maintained by culling. Boreholes and dams were constructed to prevent water shortages, strict burning regimes were initiated to control vegetation composition, and active measures were taken to control diseases (anthrax).

It is difficult to ascertain the ecological impacts of these actions because they could, at least in part, have been effective in preventing the obvious Tsavo-style "boom-and-bust" process. In typical manner, the "Afrikaner" heritage was protected from all that threatened to disrupt it.

Under the tenure of the balance-of-nature paradigm diametrically opposed management regimes were imposed on two of the worlds largest conservation areas. There was no consensus from managers as to how to put the overriding paradigm into practice, and ecology was unable to help. It may be argued that these were special cases since there was little ecological research to support management decisions and that most ecological expertise was of temperate origin and of little use in African savannas. The history of Yellowstone National Park in the United States, however, illustrates similar conflicts, despite a relatively large research base.

The Yellowstone Confirmation

For close to 100 years since its proclamation, management of Yellowstone expressly suppressed fire in the park and, during the early part of this century culled predators and fed ungulates (Wagner and Kay, 1993). Much of this policy can be attributed to political influences that stemmed from the need to protect neighboring forestry and livestock (Keiter and Boyce, 1991). In the mid-sixties the role of the balance-of-nature paradigm was entrenched in management philosophy by acceptance of the Leopold Committee recommendations to "maintain biotic associations" as they occurred in "primitive America" (Christensen, this volume).

Since then management practice has changed to accept natural fires, reintroduce predators, and allow "natural regulation" of elk herds, all of which will increase variability and change in the system. Ecology cannot, however, assume all the credit for these changes, at least some have been attributed to political, rather than ecological, pressure (Keiter and Boyce, 1991).

Increasing the potential for ecosystem change will be in conflict with the mandate to "maintain" primitive associations (Christensen et al., 1989). A conflict not helped, particularly in the public and political eye, by debate between ecologists and conservationists over the scientific justification of the actions proposed for their achievement (Wagner and Kay, 1993).

It seems that because it has had to satisfy the ever changing persuasions of a democratic, litigation conscious society, management of Yellowstone has been unable to achieve political and ecological consensus on a long-term, integrated path of conservation practice.

Implications for Ecology and Conservation

These examples demonstrate three main points about the past relationship between ecology and conservation.

1. There has been poor, or at least ineffective, dialogue between the science of ecology and conservation. Ecology has not consistently been a major driving force at the higher levels of conservation decision making.

2. Conservation operates in a multidimensional decision-making environment influenced as much by economics, real-estate transactions, land use, taxation, cultural traditions, political expediency, and public opinion as it is by ecology (Falk, 1992; Bean, this volume; Barrett and Barrett, this volume; Christensen, this volume).

3. The focus of dialogue has been almost exclusively in the context of controlling species populations. Broader ecosystem functioning has been the implicit consequence of species management and has received little explicit recognition as a management goal.

Unfortunately, the science of ecology still seems unable to generate achievable goals and to make them competitive in the multidimensional decision-making environment of conservation (Christensen et al., 1989; Noss and Murphy, 1995). This task has become all the more complex with the recognition, by ecologists and conservationists alike, that nature does not tend toward balance but is in a continual state of flux (Pickett et al., 1992).

If ecology is to become an effective participant in modern conservation beyond the species level, it must develop an efficient and recognizable interface between the science and its practitioners.

An Operational Contrast of Ecology and Conservation

If ecology and conservation are to develop a more explicit and efficient interface, it is essential that they reach mutual understanding and respect of their individual roles and operating procedures. If they do not they will remain isolated in their own worlds. A contrast between the way scientists operate in ecology and managers operate in conservation (Fig. 5.1) is instructive in achieving this understanding.

Ecology

The dominant mode of operation for ecologists is the hypothetico-deductive methodology. Armed with a body of theory that has been well honed by peer

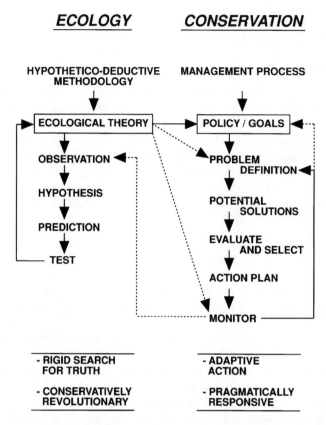

Figure 5.1 A contrast of the modes of operation in the science of ecology and the practice of conservation. Dotted lines represent information exchanges that need to be formalized to promote effective dialogue and technology transfer.

review, the ecologist follows an established process from observations about nature to hypothesis to prediction and test.

The process is designed to ensure a highly structured and repeatable search for the truth about nature, which periodically results in parcels of theory being overthrown and replaced. Although the methodology demands conformity in approach from scientists, the philosophy behind it is revolutionary.

The driving force behind the ecologist's search for complete understanding of natural systems has become a continual but inspirational discontent with the status quo. It has also become somewhat self-serving in that the peer review system rewards ability to select problems of intellectual difficulty rather than immediate usefulness (Cullen, 1990).

This is why it is so rare for an ecological study to tell us both how an ecological system works and how to maintain or achieve a particular management goal (Alpert and Maron, 1994).

Conservation

In contrast, socioeconomic and political circumstances are often more prominent in the formulation of the policies under which conservation managers operate than is peer participation or review.

The mode of translating policy into action is presently in a state of flux from a bureaucratic to a managerial system (Cullen, 1990). The conventional adherence to rules and procedure is therefore gradually being replaced by a system that rewards goal achievement and problem solving (Figure 5-1).

Much of the managers' day-to-day work is adaptive action, be it strategic or remedial, and requires highly pragmatic responses to the vast array of events and processes they must manage.

The Interface between Ecology and Conservation

The pragmatism of the conservation manager is based on the need for action dictated by the problem at hand, not the completeness of understanding to which ecologists aspire.

The practitioner faces a complex and demanding task in implementing actions to ensure that the system under his management "behaves" as policy dictates. It is sobering to contrast the many components and complex interactive processes that ecology has unraveled with the small number and simplicity of the tools (fire, a gun/trap, food, water, earth moving equipment, occasionally money, etc.) available to the manager. We face an enormous challenge to convert the complex products of science into achievable goals and implementable solutions for practical conservation. An important component of this challenge is to ensure that more of the headlong rush of ecologists to understand nature is explicitly diverted to ensuring the applicability of their results (Harte, this volume).

Ecology has played many important roles in conservation and many products of ecology have been used to set policy, define problems, select solutions, and design monitoring programs. However, the lines of communication are generally informal, fuzzy, and distinguished by the lack of a recognizable feedback loop between science and practice (Fig. 5.1).

Many scientists work under the assumption that making information available as reports or journal publications also makes it useful. It may be useful to their peers but just as the instructions to authors in the *Journal of Ecology* warns that Ph.D. theses need complete rewriting to make them publishable, the results of most peer-reviewed publications need major transformation before they can be applied in conservation.

The system of scientific peer review has encouraged ecologists and funding agencies to indulge in what the technology transfer field terms the "strategy of hope" (Hamel and Prahalad, 1989)—the "hope" that good science will inevitably lead to information that someone will find useful!

Good research and development (R&D) organizations know that to avoid the strategy of hope and remain competitive they must institute vigorous and structured technology transfer systems (van Vliet and Greber, 1992). If ecology is to move beyond the strategy of hope, perhaps it too must have an explicit avenue for "product development" and transfer to conservation.

Development of an interface to ensure appropriate product development from ecological research and successful technology transfer within a multidimensional environment will be crucial to the future impact of ecology on conservation. We will not be able to measure the impact of ecology under a new paradigm unless there are clearly identifiable products and steps in the transfer process.

Elements of a Technological Interface for Ecology and Conservation

It is unlikely that a single universally applicable technological interface could be devised to suit all conservation problems under any socioeconomic circumstances, but there are five basic sets of principles that should guide development of systems appropriate to particular circumstances.

Product Development

Many products that conservation needs have already been identified and some have been developed, at least in prototype (Possingham, this volume). Management under the new paradigm will, however, need new methods for appropriately describing system heterogeneity and the way it changes (Pickett and Rogers, in preparation). Dealing with change puts special emphasis on the need for predictive models, monitoring procedures, the means to deal with uncertainty and the situational ambiguity that arises when transferring models between systems and problems (Peters et al., this volume).

Conservation is also in dire need of clear and measurable policy and goals (Christensen et al., 1989; Holsinger and Vitt, this volume; O'Brien, this volume; Ralls and Taylor, this volume). We must establish transferable protocols for developing policy and setting goals, for measuring their success, and for evaluating their acceptability among relevant parties.

Recognition of the multidimensional nature of the decision-making environment also demands techniques and currencies for consensus building (Costanza et al., 1995). Environmental impact assessment, for example, is a highly adversarial process, but adaptations such as the integrated environmental management procedure used in South Africa (Fuggle and Rabie, 1992) include mechanisms that encourage inclusivity and consensus in dealing with a wide range of environmental issues.

Transfer Processes

Two common misconceptions in R&D fraternities are that technological products have intrinsic value and that the best possible product must be developed if it is to influence user choice. In reality, the customer determines value, and choice is determined by what is "good enough to do the job" under given social, economic, physical, and other constraints (Steele, 1989; van Vliet and Gerber, 1992).

Herein lie two important lessons for ecology and conservation.

1. Ecologists pursuing the truth about nature are aiming for the "best possible" understanding of the real world (Fig. 5.2) and so may not be developing appropriate products for conservation.

2. Different ecological products need to be aimed at the different parties and purposes in decision making. The politician needs a very different ecological information package from the conservationist, who, in turn, needs different packages for strategic planning and for management of the landscape mosaic.

Two other common misconceptions about technology transfer have important lessons for ecology and conservation. They are (1) the choice of technology is a consequence of rational planned action on the part of the user and (2) the power of new technology determines its success. In reality, choice is strongly influenced by convention and past practice, and the infrastructure required to support the technology is often the determinant of its success (van Vliet and Gerber, 1992).

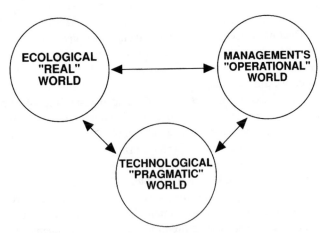

Figure 5.2 Ecological, conservation management, and the technological interfaces have different operational modes and driving forces, but the net result should be a synergistic exchange that enhances the ecological basis of conservation.

Unloading the conventional baggage of the old paradigm from the conservation train will not be a simple or speedy task. We (ecologists, conservationists, politicians, society, etc.) have paid scant attention to the true infrastructure needed to support successful conservation. The sparse literature on how to practice conservation in its broadest sense is evidence of this (Sutherland and Hill, 1995).

Building Consensus

The concept of custodianship of a heritage or resource is as universal to conservation as its multidimensional decision-making environment. An orderly, transparent, and inclusive process of consensus building is the only way to approach custodianship. Without it individual parties can too easily hijack the longer-term vision and scientific base that ecology can provide.

Consensus building is particularly needed in the setting of conservation objectives and evaluation of their achievement. It is important for ecologists to find ways of presenting more unified statements of the problems they see and solutions they propose (Glickman, this volume; Bean, this volume). They must also recognize the need for broader societal input to conservation.

The essence of consensus building should be for the parties to agree on the multiple dimensions of the problems they face and then to develop mutually acceptable solutions. The present general trend of trying to reach compromise on conflicting goals, derived from different perceptions of the problem, is inappropriate and leads to lose/lose outcomes.

The current focus on species conservation has potential to generate conflicting goals, and managing for heterogeneity under the new paradigm is likely to be even more complex. The general concept of ecosystem management (Meyer, this volume) offers a potentially useful basis for scientific integration, but its implementation needs explicit consensus building if it is to avoid the misconceptions about technology transfer outlined here.

Feedback

The literature abounds with the theory of ecology but fails dismally on the theory of its application (Sutherland and Hill, 1995). If ecologists and conservationists were more effectively rewarded for the translation of science into practice and for feedback of management problems to science such a theory would rapidly develop.

The current scientific peer review system is poorly suited to the process of technology transfer and problem feedback because it focuses on the producer not the customer. A review system that judged the utility of ecological research on a par with the quality of the science would provide reward for operating in the interface between science and practice.

The emerging field of conservation biology (e.g., Meffe and Carroll, 1994) is making important contributions in this regard, but it is still primarily a pursuit of scientists and does not act as a technological interface.

Form and Function

A technological interface is essential to enhancing the ecological basis of conservation but what form should it take? How will we recognize it so that it can be both effective and rewarded.

Many publications (Starfield and Bleloch, 1986; Sutherland, 1995; Tisdell, 1991; Costanza et al., 1995) and chapters in this book (Barrett and Barrett, this volume; Karieva et al., this volume; Possingham, this volume; Ralls and Taylor, this volume) provide examples of the emergence of this interface. The challenge is to develop an effective strategy of technology transfer for conservation that will include all participants in the decision-making environment.

Experience in the field of technology transfer (Steele, 1989) has shown that it is seldom appropriate for research and the end user to interact directly. It is far more effective to insert a structured process of product development and client service between the two (Rogers, 1992). This enables the "producer" and "consumer" to maintain their identities and appropriate modus operandi while trained technologists facilitate product development and application. Each of the three parties can then be rewarded for achievement of their appropriate tasks and goals (Fig. 5.2).

The science of ecology and the practice of conservation are distinct processes, and the fields are relatively young. They need to develop their own identities if they are to survive. We therefore need a pragmatic technological interface to link ecological research and the operational world of conservation (Fig. 5.2). On a broad scale, this interface could operate in the same way that pharmaceutical companies form the interface between biochemistry and medical practice, and civil engineering the interface between physics and construction. On a smaller, project scale the interface could take the form of a decision support system that provides structure and process to the interaction between science and management. This is the approach adopted in a program for research and management of rivers on the Kruger National Park.

In the more general field of environmental management, consulting companies operate more and more effectively in the interface but there is not the equivalent service industry for conservation.

Interfacing River Conservation and Ecological Research in the Kruger National Park

The Kruger National Park (KNP) is a rectangular reserve with the north/south long axis (350 km) running perpendicular to the drainage direction (west-east) of all the regional perennial rivers. The catchments of these rivers are therefore beyond the park boundaries and subject to varied development and water resource use and modification (Breen, Quin, and Deacon, 1994).

The geography of the park, a changing political climate, and a new ecological paradigm all interact to present new dimensions to conservation of these rivers. The stakeholders are now more diverse than was historically recognized, resources for conservation and research are few, and most importantly, time is of the essence because the new government must meet its promises to redistribute land and resources, especially water.

The challenge has been to develop a consensus-building management process in which interactions between stakeholders, managers, and research are facilitated by a decision support system (DSS) (Fig. 5.3; Breen et al., 1995). The process consists of three subsets of activities:

1. An operational framework for setting and evaluating attainable and acceptable goals. Attainability must be judged on the basis of ecologically realistic targets and the operational capabilities of management. Acceptable objectives will meet the various stakeholders' broad aims.

2. A predictive modeling framework for prediction of the nature, direction, and rate of system change, and for predicting the consequences of management actions.

3. A system response framework for monitoring system response to management actions and natural disturbance/change.

The DSS (MacKay, 1994) provides, operates, and maintains a range of tools (techniques, protocols, models, and procedures) for servicing the interactive management process. This is the technological interface between management and research. Research results are fed to the DSS and not directly to managers. Similarly, feedback from managers is translated into hypotheses for research that will facilitate the service function of the DSS.

Research is the process by which the data and understanding needed to develop tools for the DSS is generated. As data and understanding are improved so too is the ability of the DSS to service the management process. The needs of the DSS are therefore the main products of the research program thereby preventing it from becoming self-serving. But the DSS does not alone dictate the direction of research within the predictive and response frameworks. Scientific creativity and direction in the interdisciplinary research subprogram are ensured by complimenting DSS demands with

1. conceptual models/hypotheses of ecosystem structure and function that map out current understanding and unanswered questions;

2. predictive modeling that focuses data collection on nontrivial components;

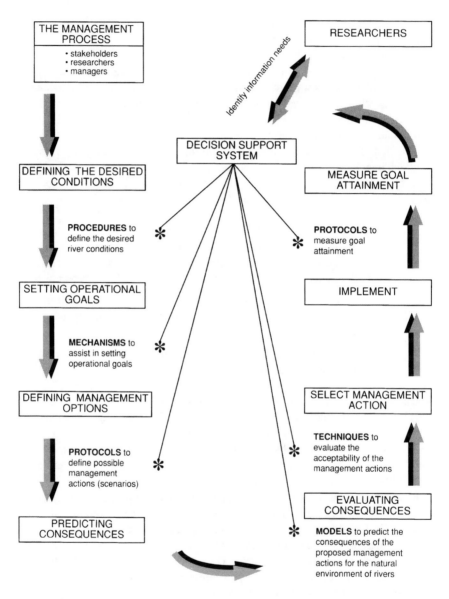

Figure 5.3 Elements of the decision support system for the Kruger National Rivers research program and their relationship to the management process.

3. evironmental and biotic monitoring that measures system response and provides data on ecological events, processes, and phenomena that occur at temporal and spatial scales different from those of the individual research projects pursuing specific hypotheses; and

4. periodic syntheses and revision of conceptual models/hypotheses.

The Kruger Park Rivers Program has been specifically designed to deal with changing ecosystems and with the dynamic nature of trade-offs that need to be made in response to changing socioeconomic and political conditions. The program structure would, however, be of limited value if it were not underpinned by an appropriate scientific paradigm that could deal pragmatically with a very limited database, a small range of management options, and potentially very complex ecosystem responses to management actions.

A hierarchical patch dynamics framework that links the physical, biological, and human landscape forming forces to biotic response has been adopted.

A Pragmatic Approach to Patch Dynamics for Conservation

The basis of the new paradigm of the flux of nature, ecological heterogeneity in space and time, demands that ecology find ways of moving conservation beyond its current focus on species to one that provides context for the interactions of species with all the components of the landscape. Patch dynamics (Pickett and White, 1985; Wu and Loucks, 1996) is an appealing ecological basis for understanding such processes, but can its complexity be reduced to a pragmatic framework for driving conservation strategy and actions?

What follows is a pragmatic reformulation of established ecological principles with the express purpose of providing a useful framework to management. The potential of this framework to also act as an interface with ecological research is demonstrated in the example of the Kruger National Rivers Program.

This framework is based on two premises: The first is that all conservation is goal orientated and requires action to achieve those goals. The second is that to manage patch structure and dynamics, conservators must be able to recognize and manipulate the forces that alter the nature of the landscape mosaic. Conservation is therefore seen as goal-orientated "engineering" of the physical, chemical, and biotic conditions of the landscape. Some of this engineering is passive (allowing natural processes to operate) and some active (culling, feeding, replanting, etc.) but the end result, even in national parks, is a human-defined landscape mosaic.

In an ecological time frame, the historical context of all landscapes is the geological template (Fig. 5.4) that defines the physical and chemical resource base for the regional ecosystems. This resource base is reworked over time by the primary forces of wind, water, gravity, and heat to generate a contemporary geomorphological template. This template represents a physicochemical land-

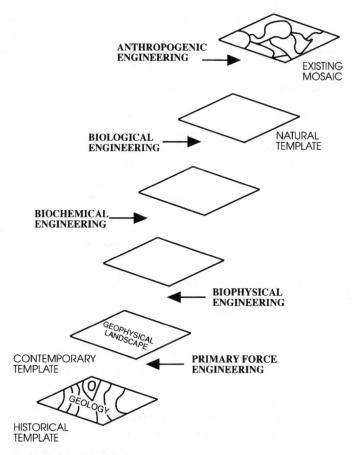

Figure 5.4 A pragmatic perspective of the patch dynamics concept. The empty boxes illustrate that such mosaics could not exist in the real world.

scape mosaic of patches defined by soil type, depth and chemistry, slope, aspect, altitude, and hence potential mesoclimate. The restoration or recreation of this sort of mosaic is becoming an important exercise in modern conservation (e.g., Shachak and Pickett, this volume).

In most instances, the conservation manager is faced with a complex landscape mosaic that has been shaped by many biological interactions, feedback processes, and human manipulations. Ecologists can generate extremely complex models to describe the structure and dynamics of such systems but to extract from these the forces a manager can manipulate is the technology transfer challenge.

Just as the primary forces of nature can be perceived to be engineering the landscape, so too can organisms when their physical and metabolic actions change the form of the landscape (Johnstone, 1995; Naiman and Rogers, submitted) or modulate the supply of resources (Jones et al., 1993). These engineering actions represent fundamental processes in the creation, modification, and maintenance of species habitat and in the realization of the full heterogeneity and integrity of the overall landscape mosaic. They are all processes conservators must manage, either actively (with fire, rifle, bulldozer, etc.) or passively by allowing them to proceed unchecked. The explicit recognition of these biophysical and biochemical engineering processes (Fig. 5.4) can provide managers with a context within which to motivate and justify decisions for manipulation of landscapes.

No landscape that might result from the combined action of primary, biophysical, and biochemical engineering could ever exist, because the organisms would also be reproducing and interacting in various trophic and nontrophic (competition, mutualism, etc.) ways. Ecologists would consider these to typically be the life history, population, and community processes so important in determining the abundance and distribution of organisms. Managers, on the other hand, could see them as another set of processes that "engineer" the full biotic diversity of the landscape mosaic and that they manipulate in one way or another.

In the absence of any human activity, the mosaic would manifest as the consequence of the interactions of primary force and bioengineering. In reality, it forms the template that anthropogenic engineering has moulded into the existing mosaic (Fig. 5.4). This applies equally to conserved and other landscape mosaics, because there is little doubt about the ubiquity of human influence across global ecosystems (McDonnell and Pickett, 1993).

The preceding view of the forces that engineer landscape mosaics is not a conceptual model of the functioning of landscape systems; the ecologist could not decompose a landscape into the various implied levels and so study them independently. Only a contemporary physicochemical template and an existing mosaic are achievable. The ecological integrity of a conserved mosaic will, however, depend entirely on the ability of the responsible managers to emulate the natural engineering in setting and pursuing their goals.

A Patch Dynamics Framework for the KNP Rivers Program

The primary problem facing scientists and managers in this program is to develop the potential to predict and monitor the response of biodiversity in river sections within the park to modifications in hydrology originating at the upstream catchments.

Understanding and predicting the causal links between catchment hydrology and downstream biotic response (Fig. 5.5) is fraught with problems of scale, feedback mechanisms, and contingencies that determine organism response. There are few prospects of managing or monitoring feedback mechanisms and contingency interactions. The most that can be achieved is to manage flow releases from upstream dams and in some cases influence sediment yield from the catchment.

It would therefore be pragmatic to focus attention on the downstream conse-

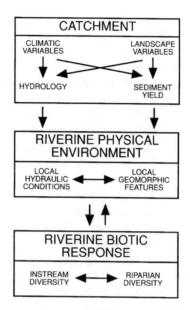

Figure 5.5 The basic conceptual framework describing the research strategy of the Kruger National Park Rivers Research Program.

quences of change in flow and sediment, which will be manifest in the contemporary geophysical template as defined by fluvial geomorphology. The program therefore aims to first develop the potential to predict the local hydraulic and geomorphological response to catchment-generated changes in hydrology and sediment yield (Fig. 5.5; Moon et al., in press). Geomorphic features of the rivers, described in a hierarchical classification from small-scale morphological units (10 m or less) to large-scale reaches (100 m to 10 km) and catchments (10 to 100 km), form the physical basis for a hierarchical description of patch structure (van Niekerk, et al., 1995).

Biological investigations can then be focused, at least initially, on the response of the biota to changes in the contemporary physical template at a scale that is of direct relevance to conservation managers (Breen, Quinn, and Deacon, 1994). In practice, these are not temporally separated phases but rather represent specific research strategies.

The overall management goal, or "desired condition" (Fig. 5.3), is primarily defined by the structure and dynamics of the patch mosaic of different types of river reach. Structure is described in terms of the proportional contribution of different morphological units to the different reach mosaics. General trajectories of change at the reach scale are then mapped for different scenarios of hydrological and sediment yield change (Moon et al., in press). Patch dynamics is therefore initially described at a mosaic, rather than individual patch, scale.

The descriptions of patch structure can be readily expanded with descriptions of the fauna and flora associated with the different patch and mosaic types (van Coller, 1992). Until ecological research can unravel the complexities of organism and community response to patch/geomorphology change, advice to the DSS will be pragmatically based on the assumption that species and community change will be correlated to change in the physical template.

Some biological factors that will modify this template and the biotic response to it must also receive high priority on the research and management agendas. Important among these are the hippopotamus, whose physical activities modify channel morphology, while its trophic activities transport nutrients and detritus across the landscape thereby modifying the biochemical conditions of a river (Naiman and Rogers, submitted). The fact that both these sets of activities will also be directly affected by changes in the contemporary template itself, further increases their research priority rating.

Some rare species, such as the Pels fishing owl, may require special conservation measures. In these cases, the nature of the ecological and management problems will be reconciled within the DSS to determine future research and management requirements. These could entail population or autecological studies at the physical or any other patch template level (Fig. 5.4).

The broad strategy is, therefore, to accept, at least initially, that if management can achieve a known spatial and temporal range of conditions in the contemporary template, the biotic diversity will be catered for in this large protected area. If, however, the objectives require any future focus on specific species or any other problem (e.g., water pollution), this can be done in the context of a changing patch dynamics template.

Iterations of the management process (Fig. 5.2) encourage transfer of updated understanding and predictive potential to managers and feedback of management success and problems to research. The predictive and response research frameworks provide managers and scientists with past and potential future contexts for dealing with present-day heterogeneity.

Acknowledgments

The Foundation for Research Development, the University of the Witwatersrand and the South African Department of Environmental Affairs and Tourism provided financial support and the Institute of Ecosystem Studies a stimulating environment in which to write. Colleagues on the Kruger National Park Rivers Research Program, especially Charles Breen, have generated many hours of fruitful debate. Fiona Rogers, Steward Pickett, Alan van Coller, James MacKenzie, Gina de Ornelas, and Scott Petrie provided valuable comment on earlier drafts.

SECTION II

Foundations for a Comprehensive Conservation Theory

Themes

Steward T. A. Pickett, Richard S. Ostfeld,
Moshe Shachak, and Gene E. Likens

Every science, indeed every human endeavor, has a set of rarely recognized background assumptions. This set of assumptions suggests what things and relationships are important, the problems that are of high priority, the approaches that are trusted, and the way solutions and applications are framed. Such assumptions, in essence, constitute the worldview shared by the people engaged in an endeavor. Historians and philosophers of science call such worldviews "paradigms."

It is difficult to assess and evaluate paradigms, because they are rarely clearly articulated and because they often touch on the deep personal values that attract people to a discipline or a profession in the first place. In addition to those problems, in broad and synthetic disciplines, the breadth and mixture of ideas that are part of the paradigm can be huge. Yet, paradigms, for all their vagueness, are crucial to the structure, content, application, and success of a discipline. The science of conservation is no exception.

What is the paradigm of conservation science like? In what ways does it affect the structure of knowledge in conservation ecology? Does application show the imprint of the scientific paradigm? Are there aspects of the science that are not being effectively used due to some paradigmatic assumption? To answer questions such as these, it is helpful to recognize that paradigms can be considered to be hierarchically organized. For instance, the materialistic paradigm of modern science incorporates more specific paradigms that are associated with specific sciences. Ecology has a master paradigm, which is made manifest only by examining the paradigms of other disciplines. Ecology's master paradigm takes the natural world to be subject to limits to growth. Within that larger ecological paradigm, there are paradigms that apply to different specialties, such as ecosystem ecology, evolutionary ecology, and so on.

There are also historical lineages of paradigms, so that revolution, expansion, or revision of paradigms can be discerned by viewing the history of a science and its application. Paradigm shifts that are less dramatic than something like the Copernican revolution are most likely to occur in the subdisciplines of a field or to be observed over a span of several decades as opposed to a span of centuries. The goal of this section is to examine how the ecological paradigm(s) affecting conservation biology have changed, thus setting the stage to consider how both the science and the practice can be improved by recognizing the change.

A second goal is to examine how the paradigm is manifested in several of the more explicit tools that the science uses. Paradigms may affect the structure of theory, the nature of the models used to translate the general theories for use in specific sites, and the explanations and forecasts made in particular ecological systems, be they landscapes, populations, or ecosystems. One of the key insights of the current ecological paradigm is its exposure of heterogeneity as a fundamental feature of the natural world. Classically, ecologists sought uniformity in their studies, models, and theories. Heterogeneity refers to continuous spatial variation, whereas patchiness refers to discrete spatial variation. Of course, intermediate forms of heterogeneity, such as a graded patchwork, can exist as well. The kind of spatial variation detected depends on the focal system, the scale of observation, and the phenomena of interest.

Spatial variation brings many of the points of contemporary ecological knowledge into focus. Organisms (including humans) and their engineering effects cause or reinforce system openness, dynamism, external regulation, multiple pathways for change, shifting mosaics, and so on. Of course, systems may sometimes be simple, predictable, quasi-deterministic, and stable on particular scales. But ecologists have learned to consider the simple and uniform as a possible special case, rather than the way nature must necessarily be. In other words, looking for patchiness and heterogeneity, and understanding their causes and implications, is an effective way to approach the natural world. Patchiness turns out to be a key tool, not only because of the staggering human-caused fragmentation of landscapes but also because of the commonness of patchiness and the variety of functions it affects. This section makes patchiness, and the more general form of spatial variation, heterogeneity, explicit. Having heterogeneity on the table enables us to examine its role as a tool to unify many of the seemingly disparate scientific specialties and management approaches that are appropriate to the practice of conservation.

The second potential unification is a conceptual or theoretical one. There are linkages between paradigms and theories, between theories and more focused models, and between models and applications. There are feedbacks between expectations emerging from theories and the tests provided by experiment or observation. There are also feedbacks between, on the one hand, the practical manipulations and management activities and, on the other, models, theories, and paradigms. Such feedbacks can be thought of as dialogs between various components of science, between theory and nature, between management and result, and between management and science. This section begins to explore some of these linkages, or dialogs, with an eye to an effective framework for unifying the diverse science behind the practice of conservation.

6

The Paradigm Shift in Ecology and Its Implications for Conservation

Peggy L. Fiedler, Peter S. White, and Robert A. Leidy

Summary

A brief précis of ecology is discussed, emphasizing the central theme of succession in the classical paradigm. This is followed by mention of the nonequilibrium paradigm and its central theme of patch dynamics. Conservation implications of this paradigm shift in ecology include (1) the replacement of a model in which some species are better adapted than others with a model in which all species are simply differently adapted; (2) the population as the fundamental unit, or currency, in conservation, (3) the recognition of the complexity of patch dynamics overlain by habitat fragmentation and the confounding implications of these; and (4) a greater appreciation of multiscalar phenomena. Social implications include some mistrust and skepticism of the "new" conservation, and most importantly, the birth of ecologist as polymath.

Introduction

The environmental historian, Donald Worster (1994), has written that the field of ecology arose out of a new, historicized biology. Specifically, he suggested that the early founders of ecology, such as Ernest Haeckel, Eugenius Warming, and later the Americans Henry Cowles and Frederic Clements, all believed in the importance of a biological and geological past, and in nature's patterns and inherent coherence that persists over time. During the latter decades of ecology's first half-century, however, the relative theoretical importance of historical events versus the ordering of natural phenomenon shifted, such that ecology essentially withered its historical roots.

The centerpiece of the classical paradigm was, of course, the theory of climatic succession. Despite criticism from some of ecology's early theorists, most notably

Henry Gleason, this view of community change remained dominant until the early 1970s. Temporal and spatial heterogeneity, as well as site history, were relegated to less important, even inconsequential considerations of community change. Again, Worster (1994:6) noted that the "history of nature was by and large reduced to a movement of one category [of succession] to the other, after which change normally came to an end. Ecosystems matured but they did not die. They reached a condition of near immortality."

Ecology, not conservation, therefore, was a science born to explore the natural order, that balance of nature, that normalcy within the landscape. And unlike conservation during the mid-twentieth century, ecology's philosophical scaffolding and cultural weft was the belief that, if nature could, through nature's balance, take care of itself in the pristine, protected areas, then ecologists could just study parts of the puzzle and in time assemble the whole. Thus ecologists who studied the basics of nature's balance—competition, predator-prey cycles, energy flux— could continue studying abstract ideas about how nature worked. In doing so, however, they relegated the study of managed areas to the resource managers. As a practical consequence, theoretical ecologists could and did ignore conservation problems, at least in the context of their immediate work without regard to areas perceptibly tainted by human influence. Conservationists—that is, applied ecologists, such as fisheries biologists, foresters, and range managers, among others—focused primarily on the products of either pristine systems or the management of unbalanced, less-than-pristine systems following resource extraction.

It is important to remember that in 1949 Aldo Leopold suggested that there was a split in many specialized fields, such as ecology, forestry, conservation, wildlife, and range management. What he called the "A-B cleavage" holds that "one group (A) regards the land as soil, and its functions as commodity-production; another group (B) regards the land as biota, and its function something broader" (Leopold, 1949:258–259, cf. Leopold, this volume). Perhaps deliberately, this cleavage became institutionalized, codified, following World War II, as the field of ecology became established in the curricula of most major academic institutions. At the larger, particularly land-grant universities, basic (theoretical) ecology was taught in departments of zoology or biology within colleges of arts and sciences, whereas applied ecology was taught in the resource management departments in Colleges of Natural Resources. In these resource management departments, forestry, and wildlife and fisheries management fields were supported philosophically by the A group. As a populace, the American people, through tax-generated funds, supported a curriculum steeped in the utilitarian school of the "resourcists." Today, the emergence of conservation biology, perceived as a distinctive discipline, is a direct result of the failure of resource management fields dominated by the A group to fully embrace the values espoused by Leopold (1933, 1949) and others (Aplet et al., 1992), and a failure to heal the A-B cleavage now endemic to conservation and ecology. It is also a result of a parallel failure of the B group to infuse their land ethic into resource management and the academe. To heal this cleavage is one of the challenges facing conservation biologists today (see also Rogers, this volume).

Of great relevance, too, was the pervasive misperception that applied ecology, conservation, and resource management held little intellectual merit and that those who were involved in these disciplines should be less concerned with the complexities of ecological theory. Although today's resource managers who have maintained an academic perspective have long argued otherwise, in general, students of forestry, soils, and wildlife and fisheries management were considered poor cousins to those of the traditional botanical and zoological sciences.

The Here and Now

Others have written eloquently and succinctly about the "new" or nonequilibrium paradigm in ecology, and it is not our purpose to provide a thorough review. Our use of the term "new paradigm" here and by Pickett et al. (1992) is meant not to imply a wholesale replacement of equilibrium states and conditions as descriptors of ecological phenomena, but the broadening of our embrace of ecological theory accepting equilibrial and nonequilibrial phenomena as scale-dependent, and that as a consequence, equilibrial conditions can exist within nonequilibrial ones, and oddly enough, visa versa. For example, within a land-scape, the number of patches of any one community type may be in dynamic equilibrium with other patch types, but the patches themselves are maintained by nonequilibrium processes. Conversely, within a landscape, the number of patches of any one community type may not be in dynamic equilibrium in part because the processes that create and maintain these patches are now equilibrial in nature. With the acceptance of patch dynamics as an apt descriptor of many natural changes, ecologists came to recognize that more accurate models that better reflected ecological patterns and processes were available.

Beginning over 15 years ago, White (1979) argued incontrovertibly that natural systems are subject to a wide array of disturbances at a wide array of temporal and spatial scales, and that these disturbances are at once destructive and creative. The work by Eric Menges and his colleagues (Menges, 1988, 1990; Menges et al., 1985, 1986; Waller et al., 1987) on the population biology of the rare plant endemic, Furbish's lousewort (*Pedicularis furbishiae*), a patch-dependent species, has demonstrated the great insight of the work of White and other ecologists (e.g., Pickett and White, 1985).

Simberloff (1982b) followed afterward in his characteristically persuasive tone, explaining the deeply embedded, little acknowledged, empirical difficulties with the classical paradigm that had plagued ecology for decades. In short, by the late 1970s, we began to discover that many of our fundamental ecological principles were inadequate for us to understand the natural world. Because suddenly it was clear that pristine natural areas and natural systems experienced great change (i.e., human induced, natural, and both), ecologists were profoundly challenged to produce new theories. Thus basic science, or at least some basic scientists, went back to work on the fundamental problem of understanding nature that now had become, with human pressure, both a fundamental and applied concern.

Christensen (1988), in an elegant and insightful review of succession and its problems in preserving ecosystems, noted that the demise of climatic succession theory as the keystone of the equilibrium school did not follow the typical pattern of scientific revolution sensu Kuhn (1970). Christensen argued that "the demise of this theory . . . has lead to an uncertainty and even cynicism as to whether a comprehensive theory of community change is feasible. As Mark Twain said, about another subject: The researches of many commentators have already shed considerable darkness on this subject, and it is probable that, if they continue, we shall soon know nothing about it" (Christensen, 1988:63). Whether one agrees with Christensen's suggestion that a comprehensive theory of community change is potentially infeasible, we must now address precisely this theoretical challenge if ecology is to move forward together in enhancing the scientific basis of conservation.

In addition to an awareness by ecologists that their empirical results were not thoroughly congruent with their theories, and that disturbances are important processes that create and maintain pattern, the nonequilibrium paradigm shifted to an integration of phenomena at an unprecedented number of temporal and spatial scales, doing so by using an unprecedented number of analytical techniques. Patch, or patchiness, has become both reality and metaphor. Levin and Paine said it first, and perhaps most succinctly, in their classic paper on disturbance patch formation: "This local unpredictability is globally the most predictable aspect of the system, and may be the single most important factor in accounting for the survival of many species" (Levin and Paine, 1974:2744). The conservation implications of patch dynamics were never more clear, then as now.

Of equal importance to the new paradigm in ecology is the notion of contingency, or the acceptance that history very much matters in patterns and processes of community change. Stephen Jay Gould has argued forcefully for the importance of contingency in understanding evolutionary pattern and process. His arguments are equally persuasive, we believe, in the understanding of ecological systems. Following his logic (Gould, 1986, 1989), an ecologist would find that the differences in outcome of successional events within a seemingly similar community or patch is not without meaningful pattern. Itineraries to one of several successional end points, cannot, however, be predicted easily at the outset. Should any early event be altered, slightly or not, then community development cascades down a potentially radically different pathway. In short, that alternative pathways exist to alternative end points within a single system

> represents no more no less than the essence of history. Its name is contingency—
> and contingency is a thing unto itself, not the titration of determinism by random-

ness. Science has been slow to admit the different explanatory world of history into its domain—and our interpretations have been impoverished by this omission. Science has also tended to denigrate history, when forced to a confrontation, by regarding any invocation of contingency as less elegant or less meaningful than explanations based directly on timeless "laws of nature." (Gould, 1989:51)

If we continue this line of thinking under the new paradigm in ecology today, we must recognize that habitats that are relatively small and isolated may contain only a portion of the available species pool for that habitat type. They might have fewer species because they are smaller—the species list might simply be a random subset of the available species, or it might contain only those species tolerant of small areas. Habitats could represent the relics of formerly more widespread habitats and thus they could have fewer species because of either random or nonrandom extinctions of the original, longer species list. These habitats might also have fewer species because of their isolation—that is, less dispersal. If these habitat islands were isolated enough from gene flow or subject to strong enough selection, given enough time, they might evolve their own unique endemics. The picture we are painting is one in which local uniqueness can develop and one in which identical habitats might have different species based on their history and spatial relations among habitat patches. This runs counter to the older paradigm in ecology, which seems to assume that all possible species have access to all possible sites (Pulliam, this volume), with successional competition then sorting among all possible species to produce the final, stable, integated community.

Conservation Implications

As discussed in another forum (Aplet et al., 1992), the 1980 publication of the edited volume *Conservation Biology* by Michael Soulé and Bruce Wilcox marked a formal reaffirmation of the conservation movement in the scientific community. The timing of *Conservation Biology* is not at all coincidental, for it also marks the commencement of a time in ecology when basic scientists were sifting through their intellectual debris for application and relevance in the conservation arena. The healing of the A-B split began during this time as basic scientists began working with applied scientists, in the field as well as in the classroom.

The many consequences to conservation science of the paradigm shift in ecology can be accounted for in social as well as scientific forums. First, we believe that the most exciting implication of the paradigm shift is that the newer paradigm replaced a model in which some species were somehow better adapted than others with a model in which species were simply differently adapted. "Better" in the old paradigm meant somehow more finely tuned by evolution, more dominant, more important. The nonequilibrium paradigm puts species on an equal footing in a sense. In realizing that a diverse world, including a world prone to disturbance and patch dynamics, made for a diverse, indigenous biota, the shifted focus enabled ecologists to view all native species of a region as equally "important." In short, the old view was that some species were better and others still native to the region were just weedy things of little consequence. This, of course, is no way to develop a conservation ethic; surely our attitude in conservation biology is not to consider some species inherently better than others. Now we need to map out the ecological boundaries of species in many dimensions. The newer nonequilibrium paradigm makes all species interesting and unique on various niche axes.

A second consequence, the problem of habitat fragmentation, is even more complex than ecologists had envisioned. For example, in a large landscape, a fire could make its way along, burning cool on some microsites and hot on others. The result? Such a natural disturbance could not only maintain fire-dependent species, but create enough heterogeneity that species sensitive to fire or species that happened to be in a sensitive life stage that particular day, could survive. With a small conservation area, fire might become yet another monotonizing force, particularly if a manager were to use a single fire intensity across the whole conservation area every year at the same time. And although we argue shortly that understanding ecological context is critical, our restoration and conservation efforts sometimes have to resolve the superimposition of a less-than-ideal management scale on nature (Ewel, this volume). Thus one of the problems we now have is that natural variation, perhaps a beaver dam or a hurricane, might impose a monotony on a small patch. This problem will challenge ecologists in the decades ahead in terms of conservation management. However, ecologists could not have gotten to this point without a theoretical shift.

In examining the problems of scale more closely, ecologists have come to realize that perceived persistence or stability of a species or community in a system depends on the type and/or scale of numerical analysis. At the most general level of numerical resolution, for example, presence/absence, many if not most species assemblages appear stable. However, Rahel (1990:333) has demonstrated that most published studies fail to examine persistence across several numerical scales and thus "make implicit or explicit claims about assemblage stability that may be appropriate at only certain analytical scales." The conservation implications of this scalar problem are that a species or community may not persist in the long term when, in fact, apparently incomplete ecological data and analysis suggest otherwise.

Third (and this is certainly not new), ecologists must change the way they think about the world. Ecology as a science is not methodologically equivalent to the sciences of chemistry, physics, as well as mathematics, in which formal Descartian logic generally can more easily be translated into rigorous hypothesis testing. In essence, strict controls in field experiments are rarely if ever possible; constants are elusive and chimeric. But this does not mean that ecology is a soft science. To the contrary, it is extraordinarily rich and complex, with many sources of variation and many possible alternatives. These demand a great deal of innovation and ingenuity on the part of anyone who investigates how nature works (T. Foin, personal communication, 1995).

The translation of this thesis in conservation practice is that we must turn more to simple and prolonged empiricism. Progress in ecology requires iterative exchange between empiricism and experimental analysis. Some ecologists have suggested that true empiricism acts as a "lie detector test" for ecological theory. Thus A. D. Bradshaw's now famous quotation, that "restoration is the acid test of ecology" (Bradshaw, 1987a) might be rephrased to suggest that restoration and conservation practice is the acid test of the application of the paradigm shift in ecology (cf. Christensen, this volume)!

This leads to our fourth point, which is that because we have come to recognize the profound complexity of the natural world and the possibility that studying one small piece of the puzzle does not guarantee that the results will be generalizable to a similar piece next door, we must rethink how we study systems. In short, we must rethink our experimental designs. Specifically, solutions to conservation problems will require many-person teams such that an ecologist is but one among the physiologist, systematist, and atmospheric scientist. In addition, we need to carefully select which systems receive intense scrutiny, rather than react to incidental funding opportunities and environmental crises. By carefully picking on which systems conservation teams work, we might take advantage of the limited large-scale opportunities that are available to conservation scientists (Nott and Pimm, this volume). A corollary to this approach is that if we are more careful in our experimental approaches, ecologists may become more proactive in their conservation efforts.

Fifth, and moving to organizational (i.e., phenomenological) levels, we wish to emphasize that, just as the individual is the fundamental unit in evolution, so is the population the fundamental unit in conservation. This is not to suggest that all conservationists should become population biologists, or that populations or demes are the only legitimate level of organization for the conservation community, only that most of the issues in conservation biology today are focused on population-level phenomena. Accepting the population as the fundamental unit demands that one's understanding of how the population works requires in-depth comprehension both below and above the population level. For example, to understand simple demographic process such as birth, death, and reproduction, one must gather life history data, which derive from study of the individual. A conservationist also must understand the genetics of a population, if patterns of individual fitness and reproductive success are to be accounted fully. Thus simple demographics alone demands investigation at the level of population, individual, and genotype. However, understanding how a population works also necessitates an understanding of the ecological context, and this requires research into natural patterns and processes that describe the community in which one finds a population. Moved to a complete synthesis, context requires an appreciation of the ecosystem or landscape dynamics in which the populations are embedded. For example, the appropriate hierarchical level of study, whether population or landscape, will dictate whether one will detect course-grained or fine-grained ecological patterns (Poff and Allen, 1995).

On the practical level, resource managers manipulate not individuals per se, but populations (patch or otherwise) of animals and plants. Those concerned with the invasion of exotic species are not focused on the successful introduction of an entire species or a single individual, but instead with the possibility that an individual can establish successfully a viable and thus potentially destructive population. Restoration ecologists do not restore an individual or a species, but literally create de novo new populations of various species that, taken as a whole, may comprise a new community. For example, restoration of hydrologic processes that maintain community integrity of river systems often is driven by management objectives to restore declining or extirpated populations of native fishes (Fiedler et al., 1993).

We believe that no other level of biodiversity demands such breadth, and this has been borne out in the newer theories of conservation. The most well known, and certainly the most challenging, include metapopulations, minimum viable populations, and its partner, population viability analysis.

Conclusions

The shift in ecological paradigms presents enormous challenges if conservationists are to utilize effectively the theoretical tools of patch dynamics. Confusion and mistrust among existing resource managers/conservationists over the new conservation has led in some arenas to an immediate rejection of the nonequilibrium approach. Existing environmental statutes were not written within the cultural context of "flux." And, too, so much of conservation implementation necessarily is based on little data, often too little. This has led most recently to some lapses in the credibility of conservationists today, and it is imperative that ecologists replace this scepticism with confidence.

Over 20 years ago, at about the time when ecology's underpinnings were beginning to erode, Theodore Roszak (1972:404) challenged ecologists to revolutionize the sciences as a whole. He felt that ecology is

> the one science that seems capable of assimilating moral principle and visionary experience, and so of becoming a science of the whole person. But there is no guarantee ecology will reach out to embrace these other dimensions of the mind. It could finish—at least in its professionally respectable version—as no more than a sophisticated systems approach to the conservation of natural resources. The question remains open: which will ecology be, the last of the old sciences or the first of the new?

The paradigm shift has given ecology the impetus to be the first of the new, and in the metaphor of the Roman god, Janus, being its partner of conservation.

Today's ecologist must be born again as this century's renaissance scholar, as we believe the conservationist as embodied by Aldo Leopold was a half century ago. Conservation needs have changed radically, however, and in certain ways, there is more darkness than ever. To meet the challenges imposed by conservation now and in the coming millenium, the ecologist cum conservationist must train as a natural scientist, historian, geneticist, economist, public relations specialist, and so on. Given the explosion in relevant technology, at least some ecologists must become the polymaths of conservation's renaissance.

The biggest challenge to ecologists will not be acquiring the necessary but admittedly tangential intellectual tools of the conservationist's tool box, but learning how to communicate the need for conservation as scientifically based resource management, and the critical role of patch dynamics in its practical application. Why must we learn to communicate our science much better than we have in the past? Not very long ago, the great conservationist Raymond Dasmann (1984:10) opined that "The words of conservationists, no matter how well spoken, have shaken no empires thus far." Ecologists, in fact, must shake empires. In essense, if conservation is to be effective, with or without its intellec-tual soul mate, ecology, then we are going to have to be better about articulating our conservation concerns and goals than we have in the past (Chapter 1, this volume; Zedler, O'Neill and Attiwill, and O'Brien, all this volume).

There is precedent and direction, however. The entomologist E. O. Wilson is perhaps better known to the lay public than to professional resource managers through his elegant nature essays that clearly portray the integral link between ecology and conservation. Ethnobotanist Wade Davis (1992) has woven compel-ling prose that, like a bell, brings clarity and compassion for the conservation of indigenous peoples of the non-Western world. Stephen Jay Gould is our generation's clearest writer of natural history essays, and has transformed the seductive stories of evolutionary ecology into family dinner conversation. Where are the ecologists who will speak eloquently to the lay public about natural disturbance and patch dynamics as central features of conservation in theory and practice?

This tack, this change in direction, must also include the way we teach ecology, for in fact, teaching those early paradigms in exhaustive detail is part of the problem. Sometimes we do it just to acknowledge the undeniable importance of history, but often it comes across to students that ecologists just sit around thinking up new ideas and criticizing the past. The overlay of patch dynamics and the spatial template leads to a richer world than would ever have been possible with the older paradigm. Nature is dynamic and locally unique; we need to impart on our students why this is one of many reasons for being fascinated by nature.

In closing, we might argue that if the shift in ecological paradigms is to ring true, if it is to shake empires and lead the battalions of conservationists now training in our universities, then we must have, in addition to the clarity of thought of a trained scientist, the clairvoyance of a poet. In words of California poet Gary Snyder (1980:159–160):

> Clouds of waterfowl, herds of bison, great whales in the ocean, will be—almost are—myths from the dreamtime, as is, already, "the primitive" in any virgin sense of the term. Biological diversity, and the integrity of organic evolution on this planet, is where I take my stand: not a large pretentious stand, but a straightforward feet-on-the-ground stand, like my grandmother nursing her snapdragons and trying at grafting apples. It's also inevitably the stand of the poet, child of the Muse, singer of saneness, and weaver of rich fabric to delight the mind with possibilities opening both inward and outward.

We urge all ecologists to look both inward and outward, and in doing so reform the science of ecology in the name of conservation.

7

The Emerging Role of Patchiness in Conservation Biology

John A. Wiens

Summary

Ecology provides the scientific foundations of conservation biology. Two ecological views of patchiness are particularly relevant to placing conservation issues in a spatial context. Metapopulation models consider the dynamics of populations in patchy environments, especially metapopulation persistence in the face of local extinctions. Several conditions underlie such persistence, and conservation biologists should determine whether these conditions are likely to hold before applying metapopulation theory to particular situations. Landscape ecology emphasizes the structure of spatial mosaics, drawing attention to the effects of variations in patch quality, boundary influences, patch surroundings, and connectivity among patches. The interaction of the spatial structure of the mosaic with the responses of organisms to that structure determines the scale and spatial patterns of ecological patterns and processes.

Whether these complex features of patchiness must always be considered depends on the questions asked, the level of resolution desired, and the spatial patterns and dynamics themselves. Changes in spatial patterns, such as those accompanying habitat fragmentation, may have threshold effects on populations. Knowing when the spatial arrangement of a mosaic is more important than the amount of habitat present requires an understanding of such thresholds, which may be strongly influenced by the movement of organisms. To overcome the traditional focus in conservation biology on simple patch-matrix or island-mainland conceptualizations of patchiness, greater attention must be given to complex landscape mosaics and their effects. Progress in this area requires better landscape theory and better empirical information on how organisms move and scale environmental patchiness.

Introduction

> No general theory about the distribution and abundance of animals should have a chance of
> being accepted as realistic unless it takes full cognizance of the patchy dispersion of animals
> in natural populations—Andrewartha and Birch (1984:184)

Modern conservation biology operates at the interface between science and environmental activism. The issues it addresses are of broad and immediate public concern, yet the perspectives it brings to bear on these issues come from basic science. Indeed, what separates conservation biology from other approaches to environmental issues is its adherence to the intellectual structure, objectivity, and rigor of science. The foundations of conservation biology as a science lie in ecology, so to understand where conservation biology is and how it might develop, we should look to current thinking in ecology and how it has evolved.

Paradigms in Ecology

Ecology has had a number of conceptual themes that have dominated thought and guided research—"paradigms," at least in an informal sense (c.f. Kuhn, 1970, Gutting, 1980, Pickett et al., 1994). Each of these worldviews has, in turn, been superseded by another conceptual framework that, if not a strict alternative to the prevailing view, has revised it extensively. Thus, the tightly integrated, "supraorganismic" view of communities developed by Clements (1916) was largely replaced during the 1950s and 1960s by a less rigid conceptualization of community organization that followed in the tradition of Gleason's (1917) and Whittaker's (1965) more individualistic perspective. The notion that competition was the dominant force structuring many communities gave way during the 1970s to a more pluralistic view of community processes (McIntosh, 1987; Wiens, 1989a). The equilibrium paradigm, which held that ecological systems were normally in a close, deterministic balance with environmental controls (the "balance of nature"), has undergone a shift to a perspective that emphasizes the frequent occurrence of nonequilibrium states and the importance of stochastic effects (Wiens, 1977, 1984; DeAngelis and Waterhouse, 1987; Pickett et al., 1992, 1994). The perception of ecological systems as essentially closed units whose dynamics are determined by local processes has been broadened to focus on the influences of broadscale factors on systems that are open to external influences (Roughgarden et al., 1987; Ricklefs and Schluter 1993). All of these shifts in how ecologists view nature have clear implications for the theory and practice of conservation biology.

Spatial Patchiness

My focus here is on spatial variation—patchiness. Although ecologists have always realized that nature is patchy on at least some scales, many of them at least implicitly adopted a "homogeneity paradigm," in which they focused their attention within relatively uniform chunks of nature (e.g., m² quadrats, small woodlots, watersheds) (Hansen et al., 1988; Wiens, 1989b, 1995a) or assumed that the ecological interactions of interest were homogenized over space ("well-stirred"; Colwell, 1984). This focus was driven by the desire to keep things simple, both in field studies and in theory. Indeed, theoretical progress in many areas of ecology has been possible *because* at least some ecologists were willing to simplify reality. By simplifying, these ecologists did not deny the reality of patchiness, but assumed that it was irrelevant to determining the patterns of interest (Levin, 1992). As Keddy (1991) has observed, "if you begin with a world view based on equilibrium models and homogeneous environments, heterogeneity can appear to be an obstacle to scientific progress."

As ecologists began to incorporate explicit consideration of heterogeneity into their field studies and their models, however, it became apparent that patchiness *does* matter in many situations. Species diversity within an area is often closely related to habitat heterogeneity (Huston 1994; Rosenzweig, 1995); the dynamics of populations or of predator-prey or parasitoid-host interactions may be more stable or persistent in patchy environments (den Boer, 1981; Kareiva, 1990; Taylor, 1991; Hassell et al., 1993); the spread of disturbances, pests, or pathogens is altered by patchiness (May and Anderson, 1988; Turner et al., 1989); or population densities and recruitment are affected by the patch structuring of the environment (Pulliam, 1988, Wiens et al., 1993).

The paradigm shift from homogeneity to patchiness in ecology has been gradual, and it developed on several fronts (Wiens, 1995a). Much of the attention has been focused on the development of *patch theory*, in which ecological dynamics within and among patches embedded in a featureless (or hostile) background matrix are compared with those generated from models assuming spatial homogeneity. Optimal patch foraging theory and island biogeography theory are variations on this theme, as are metapopulation theory and models of source-sink population dynamics (Pulliam, 1988; Shorrocks and Swingland, 1990; Gilpin and Hanski, 1991; Watkinson and Sutherland, 1995). A second emphasis has been on *heterogeneity* and the effects of spatial variance (rather than patch interactions) on ecological patterns and processes (Kolasa and Pickett, 1991). A third, somewhat related area of activity, has been concerned with *patch dynamics*, the ways in which the spatial patterns and relationships of patches in a matrix change through time (Pickett and White, 1985). Shifting-mosaic or wave-regeneration models of ecological succession (Bormann and Likens, 1979b; Sato and Iwasa, 1993) are examples. Finally, the recent emergence of *landscape ecology* threatens to force a synthesis of the other approaches to patchiness, by emphasizing the explicit spatial structure of entire, heterogeneous mosaics, the interactions among the patches in a mosaic, and the dynamics of mosaic structure over time (Urban et al., 1987; Turner 1989; Wiens et al., 1993; Forman, 1995).

These different approaches to dealing with patchiness in ecology have devel-

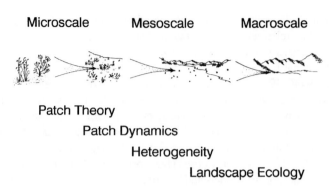

Microscale   Mesoscale   Macroscale

Patch Theory
Patch Dynamics
Heterogeneity
Landscape Ecology

Figure 7.1. The spatial domains of different approaches to patchiness.

oped somewhat independently, and they have usually been applied at different spatial scales. We may arbitrarily partition the scale spectrum into microscale, mesoscale, and macroscale, corresponding roughly to the scales of individuals, local populations, and larger assemblages (Fig. 7.1). In this array, patch theory and patch dynamics have traditionally been focused at the micro- and mesoscales, heterogeneity largely at the mesoscale, and landscape ecology at the meso- to macroscale. It is becoming increasingly evident, however, that what is an "appropriate" scale depends on the kind of organism or system studied and the questions asked (Wiens, 1989b; Kotliar and Wiens, 1990; Haila, 1991). Consequently, the traditional spatial partitioning of approaches to patchiness shown in Figure 7.1 is really incorrect: all of the approaches can be applied at any scale of resolution that is relevant to the organisms and questions. The approaches differ not in scale, but in the aspect of patchiness that is emphasized.

The Foci of Conservation Biology

To understand how these differing perspectives on patchiness may affect conservation biology, it may be useful first to examine briefly what conservation biology is about. Obviously, the discipline emphasizes "conservation," which my dictionary defines as "a careful preservation and protection of something." Fine. But what, then, is the "something" to be protected and preserved? As conservation biology has emerged (or reemerged) as a discipline during the past decade, three foci have become apparent.

The traditional focus of conservation has been on single species, usually ones that are rare or whose abundance is rapidly declining. "Species" are reasonably well-defined biological units, and the availability of a large body of population theory in ecology provides a seemingly strong underpinning for conservation efforts. As Caughley (1994) noted, most of the theoretical attention in single-species conservation has dealt with the effects of small population size rather than the causes and consequences of population declines. Small populations are vulnerable to extinction from a variety of sources: loss of genetic diversity through inbreeding, disruption of social organization, stochastic variation in demography, habitat loss or fragmentation, short-term stochastic variations in environmental conditions, or broad-scale catastrophes such as droughts or fires (Holsinger and Vitt, this volume). The traditional view has been that random catastrophes pose a greater threat to population persistence than does environmental stochasticity, whereas demographic stochasticity becomes important only when populations are very small (Fig. 7.2) (Goodman, 1987; Stacey and Taper, 1992; Nunney and Campbell, 1993; Mangel and Tier, 1994). Lande (1993), however, has shown that the relative importance of both catastrophes and environmental stochasticity depends on the relation between the mean rate of population increase (r) and the variation in this rate due to environmental stochasticity (V_e); if $r > V_e$, the effects of either catastrophes or environmental stochasticity will be great only on relatively small populations (Fig. 7.2). Caughley (1994) has suggested that the distinction between catastrophes and year-to-year environmental variation is artificial (it is largely a matter of scale), but the idea that demographic and genetic factors are generally less important than environmental variation in determining population persistence unless the populations are very small seems (so far) theoretically secure.

The two other major foci of conservation biology are biodiversity and ecosystem functioning. Although "biodiversity" has traditionally been equated with species diversity, it has come to mean much more than that: West (1993) defines it as "a multifaceted phenomenon involving the variety of organisms present, the

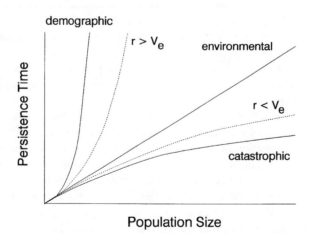

Figure 7.2. Theoretical relationships between population persistence time and population size under the influences of demographic stochasticity, environmental stochasticity, and catastrophic events (*solid lines*). The *dashed lines* represent the effects of environmental stochasticity and catastrophes together for different relationships between the mean rate of population increase (r) and the variance in the rate of increase due to environmental fluctuations (V_e). The diagram is a synthesis of the ideas of Shaffer (1987) and Lande (1993).

genetic differences among them, and the communities, ecosystems, and landscape patterns in which they occur." Insofar as biodiversity is related to species diversity, conservation biologists can draw insights from the large body of community theory dealing with diversity (e.g., Huston, 1994; Rosenzweig, 1995). Some of the earliest efforts to explain patterns of diversity (e.g., MacArthur and MacArthur, 1961; MacArthur et al., 1966) incorporated measures of spatial variation within and between habitats, and Tilman (1994) has more recently related diversity to the spatial subdivision of habitats.

Interest in preserving "ecosystem function" as a conservation objective has developed along with a shift to ecosystem management (Franklin, 1993b; Risser, 1995a). This shift stems partly from a recognition that any area (such as a reserve) contains many interacting elements, and a focus on one or a few species or on biodiversity alone may lead to a neglect of important interactions or a failure to maintain the functional integrity of the ecosystem (Angermeier and Karr, 1994). It also reflects the thinking that species may play similar functional roles in ecosystem processes—there is functional redundancy embedded in biodiversity, which means that some species can be lost from a system without destroying ecosystem functions (Walker, 1992, 1995; Schulze and Mooney, 1994). Redundancy is defined in the context of particular ecosystem functions, however, so species that share redundancy with respect to one function or environmental setting may not in another. Nonetheless, the emphasis of this approach is more on preserving processes than organisms per se.

Each of these foci of conservation efforts has traditionally had a different frame of reference for thinking about spatial variation. Despite the broader importance of environmental stochasticity to population persistence (Fig. 7.2), attention in single-species approaches has usually been concentrated on demography and genetics. The emphasis has been on patchiness of the population rather than of the environment. Correspondingly, patch theory, rather than patch dynamics, has been the predominant approach to patchiness. Efforts focusing on conserving biodiversity, on the other hand, have tended to view patchiness in the framework of heterogeneity. So far as I can tell, there has been relatively little attention given to spatial variation of any sort in deliberations about the conservation of ecosystem function (i.e., homogeneity is implicitly assumed), although references to landscape structure are becoming part of some ecosystem management efforts (Franklin and Forman, 1987; Franklin, 1993b; Ewel, this volume) and "shifting mosaic" models of forest ecosystem dynamics are cast in a spatial frame of reference (Bormann and Likens, 1979b; Sato and Iwasa, 1993).

Patchiness in Conservation Biology

Conservation biology has incorporated patchiness into its thinking in various ways, to varying degrees (e.g., Simberloff, this volume; Christensen, this volume;

Ewel, this volume). Here, I emphasize two approaches, one of which is currently receiving considerable attention, the other of which should, I believe, come to dominate future developments in conservation biology.

Metapopulation Theory

Metapopulation theory views a population as a set of spatially separated subpopulations that are linked by dispersal (Levins, 1970; Gilpin and Hanski, 1991). As a consequence of the factors depicted in Figure 7.2, local subpopulations may suffer extinctions, but under the right conditions, colonization from other subpopulations will reestablish populations in those patches before all of the local subpopulations become extinct. There is thus a balance between local extinctions and recolonizations, with the consequence that the persistence of the regional metapopulation is greater than that of a similar population that is not spatially subdivided.

Conservation biologists have found the metapopulation formulation attractive for several reasons. Because many species have patchy distributions, reality often appears to fit the pattern of spatial subdivision contained in metapopulation models. Habitat fragmentation, which is a major concern in conservation (Wiens, 1995b; Hansson, this volume), only accentuates this subdivision. Metapopulation theory also reinforces the hope that the disappearance of populations from local patches or fragments need not inevitably lead to extinction of populations over broader scales, and conservation biologists find this comforting.

Theoretical and empirical investigations of metapopulations have expanded dramatically over the past decade (see Gilpin and Hanski, 1991; Hanski and Gilpin, 1996; Hanski, this volume), solidifying the expansion of patch theory into the mesoscale region of Figure 7.1. It is tempting to apply this conceptualization of patchiness to any population that exhibits a *pattern* of spatial subdivision (Murphy et al., 1990; Opdam et al., 1993; Harrison, 1994). Before conservation biologists embrace metapopulation models too enthusiastically, however, they should consider whether the *processes* necessary for metapopulation persistence are likely to hold. Metapopulation modeling has indicated four necessary conditions for metapopulation persistence (Hanski and Kuussaari, 1995; Hanski et al., 1995a):

1. Local breeding populations occupy relatively discrete habitat patches. As a result, demography and population interactions are spatially structured. Although many (perhaps most) populations are patchily distributed, it is not a foregone conclusion that this spatial subdivision will by itself produce spatial differentiation in the dynamics of subpopulations (see conditions 3 and 4).

2. No local population is so large that its expected lifetime is long relative to that of the metapopulation as a whole. In other words, there is no large "mainland" or source population that dominates the dynamics of most of the remaining subpopulations. Harrison (1991b, 1994) and Schoener (1991) have suggested

that many natural populations may fit this mainland-island model of spatial subdivision and are therefore not true metapopulations, at least in the strict sense.

3. Population dynamics are asynchronous among local patches. As a consequence, it is unlikely that the stochastic events that may lead to the extinction of one local population will also affect other populations at the same time. This is another way of saying that local population dynamics are controlled by local factors. On the other hand, if patterns at the local scale are influenced by factors acting over larger regional or biogeographic scales (e.g., Ricklefs and Schluter, 1993; Mönkkönen and Welsh, 1994) or by episodic, broad-scale catastrophes (e.g., epidemics, El Niño Southern Oscillations, droughts), then dynamics among the subpopulations are likely to become synchronized. Mangel and Tier (1994) have drawn attention to the underappreciated role of catastrophes in conservation issues.

4. Habitat patches (i.e., subpopulations) are not so isolated that recolonization of empty, suitable patches is prevented by distance alone. Movement of individuals among patches must fall within the "Goldilocks zone": not too much (which would break down the spatial subdivision among local populations and homogenize their dynamics), and not too little (which would lead to an imbalance between local extinctions and subsequent recolonizations and erode metapopulation persistence), but just right. Because we know little about dispersal dynamics in most populations, it is difficult to evaluate how often this condition is likely to be met (but see Hanski et al. [1995a]; Kuussaari et al. [in press] for a well-documented example for the butterfly *Melitaea cinxia* in the Åland archipelago of Finland).

Landscape Ecology

Metapopulation theory is a variation of traditional patch theory: subpopulations are portrayed as patches distributed over a featureless matrix. Landscape ecology, on the other hand, emphasizes the spatial structure of entire landscape mosaics, and therefore admits to the possibility that the "matrix" is itself dynamic and ecologically important (Wiens, 1995a, 1996). Although landscape ecology is frequently practiced at broad spatial scales (Fig. 7.1), its distinguishing feature as a discipline is the focus on explicit spatial patterns and interactions, and this focus is germane at any scale of investigation (Allen and Hoekstra, 1992; Wiens, 1992, 1995a).

Although landscape ecology is concerned with all aspects of the spatial structure of mosaics, four features are particularly important to conservation biology (Wiens, 1996, in press; see also Forman, 1995). First, the elements of a landscape mosaic vary in quality. Nature is not partitioned into "suitable" or "unsuitable" patches (as envisioned in many patch-matrix models or reserve designs), but varies over a spectrum of degrees of relative suitability. For a particular kind of organism, patches may convey benefits, such as food, mating opportunities, or shelter, and costs, such as predation risk, physiological stress, or intense competition. A landscape can thus be visualized in terms of these cost-benefit functions (Wiens et al., 1993): there may be "peaks" where benefits vastly exceed costs, and "valleys" where costs are much greater than benefits. This conceptualization bears a resemblance to Wright's (1932) notion of adaptive landscapes (Fig. 7.3). Wright's "landscape" characterized the relative fitnesses of different phenotypic states, but the translation of this topography into a real-world landscape in which costs and benefits (and thus fitness) vary spatially does not strain the imagination too much. Cost-surface approaches hold considerable promise in modeling the relationship of movements to patch occupancy or other flows in ecological systems. Of course, the patch structure of a landscape and the forms and magnitudes of costs and benefits vary through time, so these cost-benefit functions are dynamic rather than fixed. Patch dynamics (Fig. 7.1) are an integral part of landscape ecology (see e.g., Alvarez-Buylla and Garcia-Barrios, 1993; Pearson et al., 1995).

A second feature of landscapes that is particularly relevant to conservation issues relates to boundary effects (Wiens et al., 1985; Hansen and di Castri, 1992). Since the time of Aldo Leopold wildlife biologists have recognized the importance of "edge effects" for creating wildlife habitat and fostering community diversity (Leopold, this volume), but landscape ecologists have come to recognize that boundaries among elements in a mosaic may play important roles as well in controlling or filtering the movements of organisms, materials, nutrients, or disturbances through a heterogeneous mosaic. The renewed interest in ecotones in ecology (Holland et al., 1991; Gosz, 1993; Risser, 1995b; Ward and Wiens, in press) stems largely from concerns about how landscape structure governs ecological flows and the resulting patterns of spatial distributions of organisms, nutrients, and the like.

The nature of a boundary is dictated by the patches on either side of the boundary. Boundaries between different patch types in a mosaic have differing permeabilities to movements of organisms, and the like. What goes on within one patch in a landscape is therefore contingent not only on features within the patch or the existence of an ecotone, but on features of the surrounding mosaic as well. A given patch may have quite different dynamics or cost-benefit functions, depending on the features of adjacent or nearby elements of the landscape. Studies using artificial nests, for example, have shown that predation risks for forest-nesting birds may differ depending on the nature of the surroundings (Wilcove, 1985; Andrén, 1992, 1995). Patch context matters.

Patch boundaries and context also influence the fourth relevant feature of landscape ecology: connectivity. Connectivity refers to the degree of linkage among elements over an entire landscape (Taylor et al., 1993). Simulation models (e.g., Lefkovitch and Fahrig, 1985) have shown that the pattern of connectivity of a landscape may have major effects on patch interactions and landscape dynamics. A cost-benefit conceptualization of the landscape (Fig. 7.3) leads to the expectation that organisms should eventually accumulate in the patches with the most favorable cost-benefit ratio. Whether or not this expectation will be

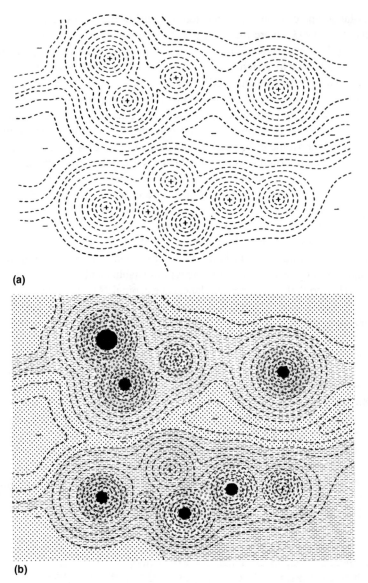

(a)

(b)

Figure 7.3. Landscapes as cost-benefit contours. Diagram (a) is Dobzhansky's version of Wright's (1932) adaptive landscape, in which valleys (−) represent regions of low fitness on the phenotype surface and peaks (+) are phenotypes conveying high fitness. Diagram (b) depicts a patchy spatial landscape in which the contours indicate areas of relatively low (−) or relatively high (+) ratios of benefits (e.g., food, mates, shelter) to costs (e.g., physiological stress, predation risk).

realized, however, depends on whether the favorable patches are accessible, which is a function of patch connectivity. Conservation biologists have tended to think of connectivity in terms of corridors, narrow strips of habitat joining otherwise isolated patches of similar habitat. This view is deficient on two counts. First, although some species clearly benefit from corridors, the evidence that corridors generally produce the expected conservation benefits is meager and inconclusive (Hobbs, 1992). More importantly, in many landscapes organisms move through and reside in a variety of patch types, and a focus on patches and corridors of only certain types (e.g., forest) will produce an incomplete picture of the real landscape linkages.

It is important to realize that all of these features of landscapes—patch quality, boundaries, patch context, and connectivity—are defined by the structure of a landscape mosaic *and* by the responses of organisms to that mosaic structure. What is an impermeable boundary to one organism is not to another, the spatial patterns of costs and benefits differ among organisms depending on which factors of the environment represent costs and which benefits, the degree of connectivity of a mosaic differs as a function of the vagility or organisms and their behavioral responses to patch boundaries; indeed, whether "patches" in a mosaic are recognized as such by an organism depends on its level of resolution of spatial structure (its "grain") and the area over which it ranges during a given time (its "extent"; Kotliar and Wiens, 1990). Thus, what is a highly fragmented landscape to one kind of organism may be relatively homogeneous (or at least continuous) to another. As a consequence of these differences in how organisms scale their environments or perceive heterogeneity, "patchiness" cannot be defined independently of the organism.

When Does Patchiness Matter?

Clearly, real-world environments are patchy mosaics, and patchiness and landscape structure *can* have important effects on individuals, populations, communities, and ecosystems. Does this mean that we must always consider the details of spatially explicit patchiness, that any attempts at simplification are doomed to failure and are worthless or dangerously misleading when applied to conservation or management?

There is no simple yes or no answer to this question. Ultimately, the level of spatial detail required may depend on the organisms and environments studied and the objectives of the study, and the answer must be situation-specific. Nonetheless, some general guidelines can be suggested. Habitat alteration from human activities, for example, changes both the amount of various habitats in an area (habitat loss or gain) and the spatial patterns of those habitats (fragmentation). In landscapes containing a high proportion of suitable habitat, the initial effects of these changes may result primarily from habitat loss and "perforation," which increases the amount of edge and the potential for boundary effects (Andrén, 1994; J. Miller, personal communication, 1995). Under these conditions, it may not be necessary to consider the details of the spatial arrangement of the habitat. At some level of habitat loss, however, fragmentation occurs, and the effects of patch isolation begin to dominate population dynamics. Once this threshold is passed, the spatial structure of the landscape and the connectivity among suitable habitats may have important effects on population subdivision and metapopulation persistence (Lefkovitch and Fahrig, 1985; Gyllenberg and Hanski, 1992; Wiens, 1996). Only when the remaining populations become very small and isolated may demographic effects begin to outweigh those of environmental stochasticity (including spatial variance and patch dynamics) (Fig. 7.2). Percolation models (Gardner et al., 1989, 1992) predict the existence of such thresholds of landscape connectivity, but they also indicate that the position of such thresholds on a gradient of habitat coverage is sensitive to both the structure of the landscape mosaic and how organisms move among patches (Pearson et al., in press; With et al., in press). Knowing when the spatial arrangement of habitats is likely to be more important than the amount of habitat present in a landscape requires an understanding of movement and dispersal (Wiens, 1995b, in press; Noon, McKelvey and Murphy, this volume).

This issue of when spatial patterning is important can be addressed through simulation modeling. By developing models that contain spatial structure and then conducting simulations over a range of parameter values, one can determine when the effects of spatial pattern are or are not present. For example, the effects of spatially explicit patterns seem not to be important when the favored habitat is abundant, when movement distances are large relative to interpatch distances, when movement patterns do not differ greatly among patch types, when organisms can detect patches only over short distances (i.e., their movements are very restricted), or when the habitat patches are ephemeral (Fahrig and Paloheimo, 1988; Fahrig, 1988, personal communication). When both temporal and spatial dynamics are included in models, the effects of temporal variability in the spatial patterning of patches are far greater than those of patch isolation (i.e., dispersal distance) in affecting population persistence and recovery from disturbance (Fahrig, 1990, 1992). This result suggests that the size and spacing of fragments (or reserves) in a landscape may be less important than their persistence over time.

Although the results of such modeling exercises are contingent on the structure and domain of the models used, the approach has the potential to indicate when and to what degree one might need to consider explicit spatial structure. The models and analyses must be tuned to the system studied, the scale of investigation, and the question asked. As a consequence, it is not yet possible to use models to define general conditions in which attention to spatial structure is always needed or never needed. The safest generalization that one can offer at this time is that consideration of spatially explicit patchiness should be included in study designs or management policies whenever there is significant spatial variation in important environmental features, on scales that are relevant to the organisms of interest. This includes most situations of interest to conservation biologists.

Building a Patchiness Perspective in Conservation Biology

To a considerable extent, issues in conservation biology are issues in patchiness. Concerns over habitat loss are closely coupled with fragmentation issues. Fears about the extinction of local or regional populations are usually related to the combined effects of small population size and isolation of populations into remnant habitat patches. Guidelines for reserve design are founded on the tenets of island biogeography theory or the recognition of gaps in coverage of the spatial distributions of species or communities.

Traditionally, these issues have been approached using simple conceptualizations of patchiness. Considerations of the dynamics of small populations have tended to emphasize demographic processes and genetics more than the effects of environmental stochasticity, despite the likelihood that environmental variations may play a more prominent role in the extinction of local populations (Lande, 1993). When environmental effects have been considered, it has most often been in the context of temporal rather than spatial variation. Habitat fragmentation and reserve design have been addressed using island biogeography theory, which has led to an undue emphasis on patch size and isolation as the critical features to be considered (Wiens, 1995b; Simberloff, this volume).

These approaches have reflected the development of patch-related theory in ecology. Because most spatial theory views patchiness in simple patch-matrix terms, it is not surprising that patch theory (Fig. 7.1) has dominated applications in conservation biology. Patch dynamics has received less attention, perhaps because the theory is not so well developed, perhaps because the time frame needed to document the spatial dynamics of many systems is longer than that of most studies (or most management decisions). Although landscape ecology offers a more comprehensive and realistic view of patchiness, it has not yet developed a body of predictive theory (Turner, 1989; Wiens, 1995a). Thus, although recent advances in computer imaging, Geographic Information Systems (GIS), and spatial statistics in landscape ecology have greatly enhanced our ability to quantify the spatial patterns of environments, our ability to interpret the ecological effects of these patterns (which depends on theory) has not kept pace.

What is needed? First, it seems obvious that simple patch-matrix visualizations of patchiness will be inadequate in many situations. Patchwork mosaics are the rule, not the exception, and it is likely that the effects of variations in patch quality, boundary effects, patch context, and connectivity in landscapes are often profound. Reserve design and population or ecosystem management should be elaborated in the context of entire landscape mosaics rather than isolated patches (Harris, 1984; Pressey et al., 1993; Hobbs et al., 1993; Wiens, in press; Barrett and Barrett, this volume). It is no longer adequate, for example, to regard habitat fragments as simple analogs of islands, or to promote corridors as a conservation tool without understanding the broader patterns of landscape connectivity.

Adoption of a landscape perspective takes our considerations of patchiness into new dimensions of complexity, where traditional, analytically based approaches to theory development are insufficient. Instead, greater use must be made of empirically based simulation modeling (Fahrig, 1991; Harrison and Fahrig, 1995; Wiens, 1995a). The continuing development of individual-based computer models (DeAngelis and Gross, 1992), cellular automata (Phipps, 1992), spatially explicit simulations (Liu, 1993; Turner et al., 1995), and the like considerably enhances our capacity to generate neutral models that can be used to explore the effects of complex spatial patterns and their dynamics (e.g., percolation models; Gardner et al., 1987; With, in press).

To understand how environmental patchiness of any sort affects ecological systems, it is necessary to know how and at what scales individual organisms respond to spatial patterns. Movement is a key process in this relationship. Individual movements are strongly affected by the structure of landscapes, and the probability that individuals (or populations) will occur in particular patches is a function of both the structure and connectivity of the landscape and the movement behavior of the organisms (or, for plants, the movements of the dispersal vectors). For example, whether or not metapopulation structure develops in a subdivided population or whether such a population has a high or low probability of persistence depends on the details of interpatch dispersal (Hanski, 1991a; Lamberson et al., 1994; Wiens, 1996). Whether or not a reserve that is established in a particular location will meet conservation objectives may depend on the emigration rates of individuals to the reserve from other patches in the landscape. In the absence of information on movement and mosaic connectivity, there is a possibility of realizing the inverse of the "field of dreams" phenomenon: build a reserve and no one will come.

The "patchiness paradigm" in ecology, if it exists at all, is a very nebulous one without a cohesive body of theory to guide research or management. Nonetheless, we know that the patchiness of environments cannot be ignored. If considerations of patchiness are to become a central part of conservation efforts, the existing elements of patch theory must be extended to a landscape context and must include dynamics as well as patterns. The understanding needed to forge this integrated view of patchiness will require careful documentation of the patterns of spatial mosaics on multiple scales, coupled with information on how individuals move through mosaics and how these movements affect population and community processes. It is an exciting challenge!

Acknowledgments

Rick Ostfeld and Steward Pickett provided the opportunity for me to attend the Cary Conference and to think about patchiness, and I appreciate their help and encouragement. Jim Miller and Steward Pickett offered helpful comments on an initial draft of the manuscript. My research on spatial patterns and their effects has been supported by the National Science Foundation (most recently by Grant DEB-9207010).

8

Linking Ecological Understanding and Application: Patchiness in a Dryland System

Moshe Shachak and S. T. A. Pickett

Summary

A project to restore degraded grazing land in the Negev Desert focuses on biodiversity and productivity of plants. To enhance these two variables, the surface characteristics are manipulated to control water flow and associated nutrient movement in the system. Pits collect resources and permit the survival of trees, producing a savanna physiognomy in the formerly purely shrub desert. Hydrologists, ecologists, and managers interact via a model of the system that is unified around spatial heterogeneity and patchiness. The model suggests manipulations that expand the scientific understanding of the system and point to management options for enhancing and maintaining diversity and productivity. The framework links a scientific paradigm with theories, models, research hypotheses, and approaches to management. The theory is needed to make the general assumptions of the paradigm apparent, the models are needed to operationalize the theory, the hypotheses to generate research, and the management approaches to suggest manipulations appropriate for specific cases of conservation. This framework can be generalized for use in other cases.

Introduction

Scientific management for conservation and restoration depends on relationships between our theoretical constructs and concrete ecological systems. Since the theoretical constructs generated within a discipline are very complex, the definition of the discipline can give some indications of the shape of its theoretical constructs. Analysis of diverse definitions of ecology reveals that ecologists are unified in their view that ecology is the study of *interactions* in the biosphere.

However, their theoretical constructs are divided because ecologists ask three different kinds of question about the interactions in ecological systems:

- What are the interactions controlling the patterns of distribution and abundance of *organisms* in ecological systems?
- What are the interactions controlling the patterns of distribution and abundance of *resources* in ecological systems?
- What are the interactions controlling the patterns of distribution and abundance of environmental *structures* in ecological systems?

A separate suite of conceptual constructs has developed for each of these kinds of ecological question. For organisms, population and community ecology were developed. For resources and environmental structures, ecosystem and landscape ecology were developed, respectively. The conceptual constructs, or theories including their empirical content (Pickett et al., 1994), ideally provide ecologists and those who wish to apply ecological knowledge with expectations of how systems will behave under different conditions, what processes are important in structuring and regulating ecological systems, and how to translate the abstractions of theory to particular situations.

In concrete ecological systems, however, organisms, resources, and structures are inseparable. Thus, to develop more effective tools for understanding and managing tangible ecological systems, we must integrate our multiple conceptual constructs into a unified theoretical construct and then relate it to the concrete system.

In this chapter our objective is to suggest an approach for a "dialog" between tangible and conceptual ecological systems that generates ecological understanding to be applied in conservation, restoration, and ecosystem management of the tangible ecological systems. The dialog is based on linking organisms, resources, and heterogeneity into unified models. Such models indicate the pathways of transformation or influence in tangible ecological systems. Because such pathways can be depicted as flows, we call the models "muliflow ecological systems." We illustrate the general approach using models to explain and predict relationships between productivity and diversity. We demonstrate application of the approach in a project that aims to increase productivity and diversity of desertified landscapes in the Negev desert.

Savannization in the Negev

The open areas of the northern Negev Desert, Israel, have been under an uncontrolled grazing regime for the last several thousand years (Evenari et al., 1983). This has presumably led to a decrease in vegetation cover and increase in soil erosion (Seligman and Perevolotsky, 1994). With the goal of restoring the Negev, and settling certain areas, the Jewish National Fund (JNF), which functions as the national development agency in wild and pastoral lands, decided to develop the desertified area as an ecological park. The aim was to modify the area into a human-made savanna; that is, to add trees and to increase the productivity and species diversity of herbs. Because the management produces a savannalike physiognomy in the desert, the restoration effort is called "savannization." We use this project as a case study to motivate our approach for linking ecological understanding and applications. The project unites scientists, managers, and planners and proceeds based on the interactions among these three groups of contributors. We present the workings of the scientific component. In particular, we demonstrate the use of conceptual and manipulation tools for the modification of the relationships among organisms, resources, and environmental structures for the management of the ecological system.

Ecological Understanding of a Desertified Landscape

The scientific component of the savannization project aims to develop an ecological understanding of the processes controlling primary productivity and species diversity in natural deserts, desertified, and restored landscapes (Fig 8.1). To

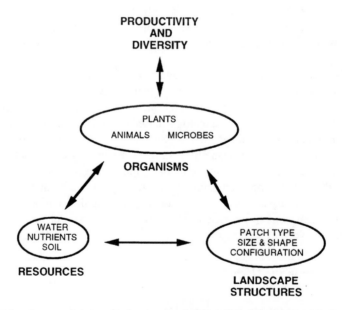

Figure 8.1. A general conceptual construct indicating the desired integration among components of diversity and productivity, resource availability, and heterogeneous structures in the environment. This conceptual construct is generalizable across system types and scales.

fashion a general conceptual construct integrating organisms, resources, and landscape structures, we asked the following question: Does alteration of patchiness change the flux of resources, and hence modify productivity and diversity of the Negev? We approached this question in three stages. First, we identified the relationships among organisms, resources, and patchiness in northern Negev ecological systems. Second, we constructed a conceptual ecological system, which integrates the organisms, resources, and patchiness in the study site. Then we developed conceptual and manipulation tools for a dialog between the tangible and conceptual ecological systems. These conceptual and manipulation tools can be used to communicate the understanding of the tangible ecological system by the managers and planners, and to monitor the ecological success of the management activities.

Relationships among Organisms, Resources, and Environmental Structures

We have studied the relationship between biological patchiness and ecosystem properties in Sayeret Shaked Park, near Beer Sheva in the Northern Negev Desert (31°17' N, 34°37' E). Rainfall, which only occurs in winter, has a long-term annual average of 200 mm. At Sayeret Shaked, as in the Negev as a whole, annual biomass production and plant species diversity is mainly contributed by the annuals (Evenari et al., 1983). On a small spatial scale with a given amount of rainfall, production and annual plant community composition depend on soil surface texture and microtopographic structure, and on soil moisture and nutrient availability. Soil surface texture is either densely packed and cemented, or loose. Microtopography is expressed as flat surfaces, mounds, or pits. Moisture and nutrient availability depend on the interaction of inputs of water and dust with the other environmental features. Therefore, the distribution of species richness, plant density, and biomass in the landscape is the result of patchiness, at various spatial scales, in relation to soil properties and resources availability.

Spatial heterogeneity on the scale of meters in the Negev, is characterized by a "matrix" (White and Pickett, 1985) of soil crusted with microphytic communities, versus distinct patches of perennials with a soil mound and an associated herbaceous understory. The microphytic crust of the matrix consists of cyanobacteria, bacteria, algae, mosses, and lichens (Andrew and Lange, 1986a, 1986b; McIlvanie, 1942; West, 1990). Soil covered with well-developed microphytic crust has a tightly structured surface (Fletcher and Martin, 1948), primarily due to binding of soil particles by polysaccharides excreted by bacteria and cyanobacteria (Metting and Rayburn, 1983). In contrast, the soil of shrub patches lacks a well-developed microphytic crust, and its surface is covered with loose soil particles (West, 1989). Well-developed microphytic crusts are hydrophobic, which reduces rainfall infiltration into the soil and increases surface runoff (West, 1990). Macrophytic patches, due to their loose-textured soil mounds, are able to absorb the runoff generated by microphytic patches up slope. The understory of annuals benefits from this additional water. In addition to water, the loose soil mound provides more safe sites for annual plant seeds than the tight structure of the microphytic soil crust. Germination and growth are also enhanced by the relatively nutrient-rich environment of the soil mound, which is enriched by minerals released from litter and animals feces. The soil mound originates from dust deposition over the landscape which is redistributed by surface runoff water and accumulates under the shrubs. The net effect of the relationships among annuals, shrubs, and microphytes, based on patchy water and soil distribution, is a patchy distribution of annual (macrophytic) plants. Relatively high productivity and diversity exist in the macrophytic patches, whereas these features are low in microphytic patches.

Constructing a Conceptual Ecological System

With the understanding of the relationships among annual plants, water, soil, and patch type just outlined, we developed a conceptual construct for integrating the multiple flows that occur within this ecological system (Fig. 8.2). The model describes the processes that follow the development of a macrophytic patch within a microphytic matrix. Such changes are captured in a transformation pathway, or flow, which converts microphytic patches to macrophytic patches. The higher plants of the macrophytic patch cause increased dust deposition and accumulation, which results in the formation of the soil mound in the patch. We capture this influence as a "controller" of the transformation of dust to soil (Fig. 8.2). The macrophytic patches with their soil mounds act as new sites for seed accumulation in the macrophytic patch. This relationship is captured as a controller on the transfer of species from the regional species pool to a local species pool (Fig. 8.2).

The relatively high soil moisture of the newly created macrophytic patch is due to the high infiltration of rain that falls directly on the patch added to the input of runoff from up slope of the microphytic soil crust. This relationship is designated as the controller of rainfall to runoff (Fig. 8.2). The relationships of seed rain, water, and nutrient accumulation result in germination, establishment, growth, and reproduction of the seeds and is the conversion of the potential from the regional species pool to the actual local species pool. These processes increase species diversity of areas containing macrophytic patches in relation to a similar area covered only by the microphytic soil crust.

The species that colonize the macrophytic patches will respond to the soil moisture content and nutrient availability by biomass accumulation. This will determine the total biomass of the patch, and is illustrated in a controller of solar radiation to biomass (Fig. 8.2). This network of processes increases the productivity of the area compared to one covered only by microphytic soil crust (Boeken and Shachak, 1994).

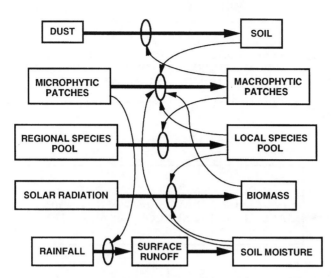

Figure 8.2. A model of the relationships among components of biological diversity, ecosystem productivity, resource availability, and physical structures depicted as linked transformations or flows. The model type is labeled a "multiflow ecological system." The system represents the relationships among four flow types: energy, species, patches, and materials, all essential flows for understanding any ecological system and guiding management. Each flow is shown as an ecological flow chain; that is, the transformation of a given flow between states. Combining the interacting ecological flow chains results in an ecological system. Several critical connections between individual flow chains, in which the donor chain controls the function of the target flow chain, are shown by thin lines running from states of the controlling flow chain to the arrow indicating fluxes in the target flow chain. Such connections are called "controllers" and are depicted by ovals on the arrows representing flows.

Development of Conceptual and Manipulation Tools

To test whether our model is correct requires a dialog between the conceptual construct and the concrete ecological system. In addition, to ensure the utility of the model for the managers and planners in the Savannization Project also requires a dialog based on the model. Here, we illustrate the tools that are used to carry out those two related but different dialogs. The techniques can be thought of as "dialog tools," or specific conceptual constructs that link the basic science to the concrete ecological system or to the concerns of managers and planners (Fig. 8.3).

The first step was to translate the interactions among water, mineral resources, and patchiness into terms of microphytic patches, macrophytic patches, annual plant productivity, and species diversity, using a specific model (Fig. 8.4). The resulting model is a more specific conceptual construct than the loose framework

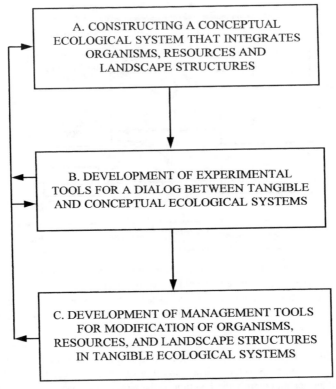

Figure 8.3. General strategy for linking conceptual constructs, tools for translation between general conceptual constructs, and specific management tools. Because the general models or theories representing conceptual constructs can be improved or modified as a result of management, the tools linking management and the general conceptual constructs are labeled "dialog tools."

of ideas embodied in the description of the concrete system outlined in the previous section (Fig. 8.2). Since microphytic patches act as a source of runoff, and macrophytic patches act as a sink, the relationship between macrophytic patches and runoff generation is inverse. This first model represents the relationship between resources generated by patchiness, and diversity or productivity (Fig. 8.4A). A second model captures the relationship between water retention in rainfall and runoff, productivity, and diversity (Fig. 8.4B). Combining these two models summarizes the relationships among patchiness, water availability, productivity, and diversity of annuals in the Negev. This integrated model provides us with a conceptual tool to ask questions concerning the trajectory of productivity and diversity in relation to patchiness and resources in the Negev. For example, the model can answer questions such as these: What is the effect of water enrichment in a macrophytic patch on the productivity and diversity of

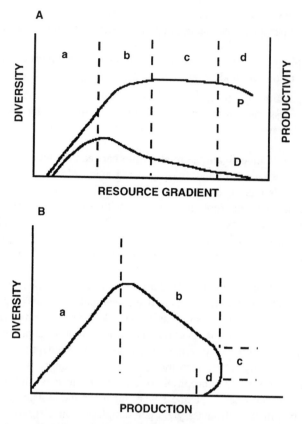

Figure 8.4. A specific model of relationships among diversity, productivity, resources, and structures that can be used for communicating the insights about behavior of the ecological system into manipulations for management. (A) The response of productivity (*P*) and species richness or diversity (*D*) to increased levels of resources. (B) The relationship between productivity and diversity as modified by changes in resource availability that are governed by differing degrees of patchiness. Lower case letters in both panels indicate portions of the trajectory where (a) resource level drives increase in both productivity and diversity, (b) resource level drives a decrease in diversity and an increase in productivity, (c) resource level drives a decrease in diversity while productivity is stable, and (d) resource levels in which both productivity and diversity decrease.

its understory annual plant community? What is the effect of disturbance by digging animals within a microphytic patch on water, soil, and nutrient dynamics, and on productivity and diversity of annuals? The model predicts that increase or decrease in productivity and diversity of annuals is dependent on water availability before enrichment and on the level of enrichment.

To use the model as a conceptual tool for understanding of the interactions among patchiness, resources, and organisms in the Negev, we developed manipulation tools to permit experimentation. The experimental tools include

1. an explicit experimental design that addresses interactions among patchiness, resources, and organisms; and

2. manipulation of one of the main components of the ecological system (patchiness, resources, and organisms) and studying the responses of the other two.

For example, to investigate the relationship between the biological patchiness and water flow, we have studied runoff generation in isolated enclosures containing either a single macrophytic patch, or a macrophytic patch paired with either an adjacent up-slope microphytic patch or with an array of up-slope micro- and macrophytic patches. We measured runoff generation in controlled and manipulated enclosures following rainfall events. Our data confirmed the model of inverse relationship between the cover of the macrophytic patches and runoff generation.

To study the effect of small-scale disturbance of the microphytic soil crust on herbaceous plant communities, we used the manipulation tool of pits dug into the microphytic soil crust (Boeken and Shachak, 1994). During a rainfall, the soil crust up-slope of the pit functions as the contributing area for runoff that collects in the pits. Soil moisture measurements at rooting depth showed that the pits were significantly wetter than the crusted soil. Seventy-seven species were identified in both pits and microphytic soil crust. Of these, 65 (84 percent) were annuals. Density and species richness were higher in the moister pits relative to the undisturbed matrix. Biomass yield per sample was also significantly greater in pits than in the matrix. The increase in total biomass per patch in mounds and pits relative to the matrix was primarily due to increased plant density.

The manipulation tool of forming a pit that modified patch structure showed that the system responded as predicted by the ascending part of the model (Fig. 8.4B); that is, both productivity and diversity increased as a result of patch disturbance.

Ecological Management: Additions and Deletions to the System Model

The applied component of the savannization project is aimed at developing a management methodology, based on ecological understanding, for increasing productivity and diversity of desertified landscapes. The general conceptual construct, which integrates organisms, resource, and landscape structures (Fig. 8.2), clarifies how to modify landscape patchiness and resources flow to increase the productivity and diversity of desertified areas in the Negev. We approached the problem in two stages. First, we identified conceptual and manipulation tools in common to both managers and scientists. Second, we developed specific manipulation tools for modification of organisms, resources, and patchiness in the tangible ecological system to increase productivity and diversity (Fig. 8.3).

The model and its associated tools for manipulation suggest how the system can be managed under different goals and conditions. For example, if, due to natural succession or desertification, the macrophytic patches are unable to absorb the runoff and nutrients that are generated by the microphytic soil crust, then human-made macrophytic patches that can store runoff water and nutrients can be added. This can be done by constructing pits in the landscape, which will increase the number of sinks in the landscape for water and nutrients. Therefore, in the human-made macrophytic patches, productivity, and diversity can be enhanced. In essence, management will add novel features to the system model. Human-made macrophytic patches are added to the patch flow chain (Fig. 8.5,

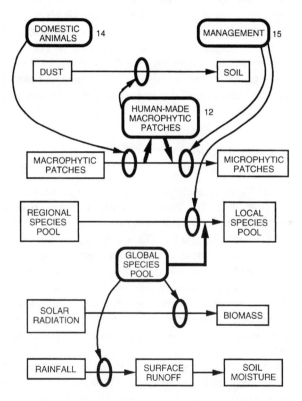

Figure 8.5. A simplified depiction of the flow chains constituting the model of the ecological system (for details, see Fig. 8.2), showing the new components added by humans and only the new controllers added by management. All added state variables are in bold type, and all added controllers are shown by arrows and ovals on the affected flows. Management adds new species from the global species pool, governs the impacts of domestic animals on patch dynamics, and introduces new patches and patch configurations.

box 12). By adding human-made macrophytic patches, we also add an external controller to the human-managed system (Fig. 8.5, box 15). Building macrophytic patches involves the creation of pits that absorb runoff water. How patches are constructed can vary according to the objectives of a specific development project. Thus, we can use the modeling strategy outlined here to conduct different specific scenarios in the responses of the ecological system to the introduction of human-made structure.

In analyzing the effect of grazing on the system it was concluded that uncontrolled grazing will reduce the number and size of the macrophytic patches and will decrease the productivity and diversity of the "natural" area (Seligman and Perevolotsky, 1994). On the other hand, it was assumed that if grazing is removed, then due to natural succession, the number and size of the macrophytic patches will increase and result in a decrease of the source function provided by the microphytic patches. This will reduce the resources available to the human-made patches. Thus, the management decision was not to remove grazing but to modify the grazing regime. Management requires monitoring to adjust patchiness by grazing in such a way that the microphytic patches will supply water and nutrients to both natural and human-made patches. In the development model, the role of grazing is depicted as domestic animal control on the patch flow chain (Fig. 8.5, box 14).

Conclusions: A General Approach

This chapter has exemplified an approach for linking ecological understanding and applications (Fig. 8.3). The first step is the construction of a conceptual construct or model of the ecological system, which integrates organisms, resources, and landscape heterogeneity. The second step is development of tools for a dialog between the tangible and conceptual ecological systems. The last step uses the conceptual constructs and the dialog tools to develop management tools. The management tools modify the relationships among organisms, resources, and landscape structures in tangible ecological systems to produce desired results. Development and refinement of the explicit models and tools for dialog and manipulation can serve as foci for interactions among policy makers, managers, and scientists.

To apply more generally the approach illustrated in the Savannization Project, we abstract the essence from the example. Productivity is the generation of living matter from nonliving materials and solar energy or certain chemotrophic environments, measured over area and time. The process can be modeled as a flow of energy from one form to another. This is the basic process generating the resource base that all members of ecological systems depend on, and is the broadest foundation for the biosphere. Diversity refers to the variety in kinds of organisms and of the structures they generate in interaction with their environments. Diversity is a fundamental property of the biosphere at all levels of organization: There is diversity of genetics and physiologies, diversity of species and various higher taxa, diversity of communities and ecosystems, and diversity of landscapes (Falk, 1990; Franklin, 1993b). These kinds of diversity are products of the unique evolutionary potential of living things, and their interactions with the past and current opportunities in the environment (Wilson, 1992). Diversity can also be affected by dispersal and transformation of dormant propagules or potential colonists into active members of a community. Hence, these processes can be modeled as flows.

These two fundamental features of the living world, productivity and diversity, are the outcome of a network of relationships between the living organisms and their diverse resources in heterogeneous environments. The network of relationships is expressed differentially over space and time. Heterogeneity from place to place can often be discerned as changes in landscape structure in the form of discrete patches. Patchiness results from birth, growth, and death of organisms, biotic interactions, and the underlying heterogeneity of resources with which organisms interact. Patches differ in their suitability to various ecological processes and human uses. Therefore, the quality of patches for specified organisms or functions can be used to characterize an array of patches in the environment or in ecological systems. Specific areas can be transformed from one patch type to another, and thus, can also be modeled as fluxes. Such fluxes can be combined to represent the most general conceptual construct for understanding productivity and diversity as a network of relationships among organisms, resources, and environmental heterogeneity (Fig. 8.1). Therefore, we believe that the strategy of generating multiple flow models of systems, which identify system components and interactions that managers can manipulate, is a promising scenario for planning and management in conservation in all its forms.

SECTION III

Biodiversity and its Ecological Linkages

Themes

Richard S. Ostfeld, Steward T. A. Pickett,
Moshe Shachak, and Gene E. Likens

Societies value diversity, including biological diversity. Endangered species legislation both in the United States and elsewhere in the world has arisen from sets of values that emphasize the importance of protecting the richness of life that has evolved over millions of years. Thus, biological diversity, or biodiversity, is a well-established target for conservation efforts from an ethical or moral standpoint. To ecologists and other conservation scientists, however, biodiversity is a complex and multifaceted concept, and little is known directly about the ecological consequences of diversity per se. For example, how important is biodiversity to the stability and resilience of ecological communities and ecosystems? What aspects of biodiversity (for example, numbers of species, genetic variation, or architectural diversity) are most important to ecological systems? What hierarchical levels of biodiversity (for example, genes, genotypes, species, or community types) are most important to which ecological functions (for example, persistence of populations and communities, flux rates of nutrients, or global environmental change)? Because these questions remain largely unanswered, biological diversity requires much closer evaluation as a scientific target for conservation efforts.

This section begins to evaluate some aspects of biodiversity as a target for conservation. It also points out some other, much less widely recognized conservation goals, such as the protection of ecosystem functions and integrity, and points to links between biodiversity and ecosystem functions. The concept of patchiness is reintroduced with descriptions of how it may affect persistence of populations, biodiversity, ecosystem function, and the interactions among these entities.

Contributions in this section provide concrete examples of the theoretical foundations established in Sections I and II, by describing practical approaches to dealing with habitat patchiness from a management perspective. The concept of habitat fragmentation as a particular, negative type of patchiness makes explicit the notion that heterogeneity can be a destructive as well as a constructive force affecting diversity and ecosystem function. Other concrete applications of patchiness include how to manage it to attain specific goals.

A recurring theme in the conference from which this book arose was the rift between population and ecosystem perspectives within the discipline of ecology. This section evaluates how the concept of patchiness can help unite these two perspectives and emphasize their complementarity. Finally, in the last chapter, this section returns to the moral and ethical considerations in conservation and provides a powerful example of ecological restoration, a discipline otherwise not emphasized in this book.

9

The Evaluation of Biodiversity as a Target for Conservation

M. Philip Nott and Stuart L. Pimm

Summary

One dilemma for conservation biologists is to ponder whether ecological processes are better targets for conservation than are species. In this chapter, we
discuss this dilemma under the assumption that not all species are equal in
terms of the size, shape, orientation of their geographic range, and their density
distribution within the range. Conserving a single species found only in a restricted
area (endemic) may not preserve global biodiversity, but on the other hand,
conserving an ecosystem and the processes therein may still lead to a reduction
in biodiversity within the conservation area, and at the same time leave a large
proportion of "rare" species unprotected. Obviously both approaches should be
used to maintain global biodiversity, but the question remains how we target
those areas that need protection. With a view to setting conservation priorities,
we map an ecological unit of measurement henceforth known as M, which for
any group of species, at any geographic scale (in this case 1° latitude by 1°
longitude), is equal to the sum of the reciprocals of the area of each species'
geographic range.

We map both M and species richness found with reference to their respective
magnitudes. The resultant patterns of endemism and diversity reveal "endemism
hotspots" and "diversity hotspots." Comparing both metrics within and between
two groups of North American passerines, icterids, and New World warblers,
we find three important features in these patterns. First, patterns of species
richness between different groups do not necessarily coincide spatially. We might
expect this result, as warbler and icterid habitats do not generally overlap. It
suggests, however, that the total area required for conservation of both groups
must cover a large areal proportion of North America. Second, we find that
within a group an endemism hotspot may not spatially coincide with a diversity

hotspot, for example, to conserve icterid biodiversity one might target the Midwest states where there are a large number of sympatric species. However, to conserve range restricted endemic icterids we must focus our attention on California. Finally, patterns of endemism between groups may not entirely coincide; California is a hotspot of both warbler and icterid endemism, but the values for icterid endemism in the warbler hotspots of the Appalachians are low.

On a more general but depressing note, a cursory analysis of diversity and endemism reveals North American forest avifauna, South Pacific avifauna, and global flora to be extremely vulnerable to extinction by random habitat loss at a scale of physiographic province.

Species Versus Ecosystem Processes As Targets for Conservation

We have inflicted so much ecological damage on our planet that a real concern is to decide how to protect what remains. Estimates of current and future extinction rates suggest that we will lose double-digit percentages of the Earth's species in the next century (Pimm et al., 1995). Estimates of our impact on ecosystem processes are equally chilling. Humans divert approximately 40 percent of terrestrial primary productivity toward our food supply, that of our domestic animals, and other uses (Vitousek et al., 1986). Outside of barren open ocean ecosystems, approximately 33 percent of marine productivity flows into a food chain of which we are the top predator (Pauly and Christensen, 1995). Worse, we do not merely live off the ecological "interest" that productivity represents, but we destroy the ecological "capital"—the ability of ecosystems to be productive. Many fisheries are overharvested (UN FAO, 1995), and much agricultural land has been degraded, some irreversibly (Daily, 1995). In freshwater ecosystems, only a tiny fraction of major rivers in North America and Eurasia have not been dammed or suffered major changes to their channels (Dynesius and Nilsson, 1994).

So do we place our efforts on protecting biodiversity or on protecting vulnerable ecosystem processes? This is obviously a poor question, because it assumes that we can disentangle species and ecosystem processes. An entire volume has been devoted to documenting their entanglement (Jones and Lawton, 1995). Individual species create and maintain ecosystems, such as beavers and their flooded ponds. The number of species per se may be important in determining the average rates of ecosystem processes (Lawton et al., 1993), and certainly has a major effect on an ecosystem's response to external disturbances (Pimm, 1991; Tilman and Downing, 1994; Lockwood and Pimm, 1994). Descriptions of the effects of fire regimes, hydrology, nutrient flows, and a myriad of other ecosystem processes on species fill our ecological journals. The management of ecosystem processes is necessary for the protection of species, and these processes must often be a target for conservation in their own right.

A different question is: are ecosystem processes a *sufficient* target for conservation? We believe the answer is an unequivocal no. As evidence, we present an anecdote. Figure 9.1 shows an area on the Hawaiian island of Moloka'i in which the forest trees represent a global selection of tropical trees, not native ones. The river running through the forest is clear, unlike many tropical rivers, which run red with sediment from clear-cutting in the surrounding watersheds. The reason for this is that nearly a century ago, foresters in the Hawaiian islands initiated extensive plantings of alien trees to control soil erosion (Moulton and Pimm, 1983).

Such ecosystem *rehabilitation* (Aronson et al., 1993) occurs across the Hawaiian islands, where one can find species-rich tropical rainforests on the windward slopes and dry forests on the leeward slopes, both dominated by introduced trees, housing bird faunas that are similarly nonnative.

We have not one datum on the ecosystem processes in these systems. All we offer is our naturalists' impressions that these alien ecosystems look right and so they probably have the appropriate plant biomass and primary productivity. These alien systems may indeed differ in their ecosystem processes in consistent, subtle, yet important ways from the native ecosystems they replaced. We see that the ecosystem processes are maintained despite the loss of more than 90 percent of the islands' vertebrate and more than 10 percent of the islands' plant species (Pimm et al., 1995).

If the current rate of tropical deforestation persists, we can imagine a future pantropical forest system composed of a few score of tree species, the same introduced insect, bird, and mammal species. Such a world might exhibit acceptable ecosystem processes yet contain a tiny fraction of current tropical biodiversity. Thus, although ecosystem processes may be a necessary target for conservation, they are not sufficient to conserve global biodiversity.

Are species a sufficient target for conservation? We suspect the answer may be yes. We do not consider zoos or botanic gardens—ecosystem–free species management—to be more than intermediate steps in managing species' fates. To manage species effectively, we will certainly need to ensure that the ecosystem processes are suitable.

If Species Are The Target, What Is Their Measure?

Imagine a map with contour lines depicting the priorities for species conservation. What calculation goes into those contour lines? If you replied "the number of species at a given location" you would have part of the answer. The more species an area holds, the more it has to lose. You would not have an adequate answer, however. Maps of species richness fail to provide maps of conservation priorities because all species are not equal. Some are more prone to extinction than others.

Let us anticipate results we present shortly and say that these species are the rare ones. If species-rich areas typically contained the rare species, then maps

Figure 9.1. An area on the Hawaiian island of Moloka'i in which the forest trees represent a global selection of introduced tropical trees, not native ones.

of species richness would suffice. This is not the case, however (Prendergast et al., 1993; Curnutt et al., 1994). Some areas have few rare species, whereas other areas have more species, but these species are more widespread and so less vulnerable.

In this chapter, we survey global "black spots:" endangered areas with high rates of extinction. We then produce maps of species richness weighted by vulnerability to extinction for two groups of North American passerine birds. These maps provide examples of our assessments of species as targets for conservation. We compare and contrast these maps with maps of unweighted species richness. Although these maps represent only a subset of the continentwide maps we are in the process of producing, they show promise for setting conservation priorities.

A Short History of Recent Extinctions

A Survey

In two recent papers (Nott et al., 1994; Pimm et al., 1995), we considered five case histories of high extinction rates. These were assemblages of (1) plants in South Africa, (2) freshwater mussels in North America, (3) freshwater fishes in North America, (4) mammals in Australia, and (5) Pacific birds. Here are brief summaries of those cases.

The fynbos, a floral region in southern Africa, has lost 36 plant species (of approximately 8,500); 618 more are threatened with extinction (Hall et al., 1984).

Of the 297 North American mussel and clam taxa (281 species and 13 geographically distinct subspecies), an estimated 21 have gone extinct since the end of the last century (Williams et al., 1992), and 120 taxa are at risk of extinction.

The freshwaters of the United States, Canada, and Mexico encompass many different habitat types and are home to approximately 950 species of fish. The fish faunas of the spring systems of Southwest states have suffered 23 extinctions, with more than 50 taxa currently at risk (Miller et al., 1989). The Southeast region contains 488 taxa, four of which are believed to be extinct and 80 of which are at risk (Etnier and Starnes, 1993).

Of 60 mammalian extinctions worldwide, 19 are from Caribbean islands. However, Australia's unique mammalian fauna has lost 18 of 282 species (Short and Smith, 1994), and a further 43 species have been lost from over half the area of their former ranges. Some species only survive on protected offshore islands (Burbidge and McKenzie, 1989).

Pacific islands are well known to have suffered large numbers of extinctions of many taxa. The highest body count is for birds. Adding known and inferred extinctions, it seems that with mere neolithic technology, the Polynesians exterminated more than 2,000 bird species, some 15 percent of the world total. We must infer extinctions, because we will not find the bones of every extinct species.

Two approaches have been employed. Applying sampling to the overlap in species known from bones and those survivors seen by naturalists, we infer that approximately 50 percent of the species are unrecorded (Pimm et al., 1994). It is hypothesized that every one of approximately 800 Pacific islands should have had at least one unique species of rail (Steadman, 1995).

Today, only a few remote islands still have rails. Some islands lost their rails to predation by introduced rats in the last century, and large volcanic islands typically lost several rails. There are skins of these species in museums. More accessible islands lost their rails earlier, for every survey of bones from now rail-free islands has found species that did not survive human contact.

Which Species Are Vulnerable?

From this survey of extinction centers, we find high extinction rates in mainlands and islands, in arid lands and rivers, and for both plants and animals. Although we know less about invertebrates, high rates characterize bivalves of continental rivers and island land snails. There is nothing intrinsic to these species' diverse life histories to predict their being unusually extinction-prone. So, what obvious features unite extinction centers?

First, we know the species and places well—as did naturalists a century ago. Second, and importantly, each area holds a high proportion of endemic species; those species found there and nowhere else. Remote islands are typically rich in these endemics: 100 percent of Hawaiian land birds were found only there. Continental areas can also be rich in endemics: approximately 70 percent of fynbos plants, more than 50 percent of North American fish, more than 90 percent of North American freshwater mollusks, and 74 percent of Australian mammals are endemic (Hall et al., 1984, Williams et al., 1992, Miller et al., 1989, Short and Smith). In contrast, only about 1 percent of Britain's birds and plants are endemics (World Conservation Monitoring Centre, 1992). Past extinctions are so concentrated in small, endemic-rich areas that the analysis of global extinction is effectively the study of extinctions in a few extinction centers (Nott et al., 1994). Why should this be?

We can model extinctions due to habitat destruction by imagining a "cookie-cutter" that excises a randomly selected area. Species that are found outside the area may survive and can recolonize the destroyed area, given that the habitat subsequently recovers. Some of the area's endemics will go extinct; the proportion will depend on the extent of the destruction relative to the size and shape of the habitat, and to the spatial distribution of the species in question. The familiar species-area relationship is a good predictor of extinctions when applied to endemics. It is a poor predictor for species with widespread distributions (Pimm and Askins, 1995; Pimm et al., 1995; Brooks et al., in press). Under this model, we do not make additional assumptions, such as island biotas being intrinsically more vulnerable than mainlands.

It follows that by excising randomly selected areas, the number of extinctions correlates weakly with the area's total number of species, but strongly with the number of its endemics. By chance alone, small endemic-rich areas will contribute disproportionately to the total number of extinctions.

This model is consistent with known mechanisms of extinction. Habitat destruction "cuts out" areas as the model implies. Introduced species also destroy species regionally. For example, the brown tree snake eliminated all of Guam's birds (Pimm, 1991). Species need not be entirely within the area destroyed to succumb: the populations outside may be too small to persist. Moreover, across many taxa, range-restricted species have lower local densities than widespread species (Gaston, 1994). The former are firstly more likely to be "cut," and second, their surviving populations will have smaller densities and so higher risks of extinction than widespread species.

This cookie cutter model emphasizes the localization of endemics as a key variable in understanding global patterns of future extinctions. The question is how to map the insights this model encompasses.

Mapping Endemism From Geographical Range Data

Using the Breeding Bird Survey (BBS) for the United States and Canada (Peterjohn, 1994), we recorded whether each species of New World warbler and icterid (blackbirds and orioles) was seen at least once at a BBS survey site during the 23-year interval from 1977 to 1989. We chose these two groups for illustrative purposes; a more detailed analysis of patterns for North and South America will appear elsewhere.

Species were chosen such that more than 75 percent of their entire geographical breeding range was encompassed by the region surveyed. We mapped these BBS locations onto a two-dimensional array of 1° latitude by 1° longitude cells to produce two maps.

The first map is of species richness: the total number of species of warbler or blackbird recorded in each latitude $\times$ longtitude cell. Such maps are very familiar to biogeographers and have a long history.

The second map shows the vulnerability index for a group of species (after Usher, 1986). The vulnerability of an individual species is simply the reciprocal of the species' range; the chance that a species will be excised by a random cookie cutter. In our implementation, we simply count the number of 1° latitude by 1° longitude cells in which a species has been found. Call this index, $1/A_i$, with $i = 1$ to n, the maximum number of species in each taxanomic group. In each geographical cell, j, there is the set of species S_j found there. We then sum the weights of $1/A_i$ for every species found in set S_j, to produce an overall measure of endemism E_j:

Distribution of species richness in 14 icterids

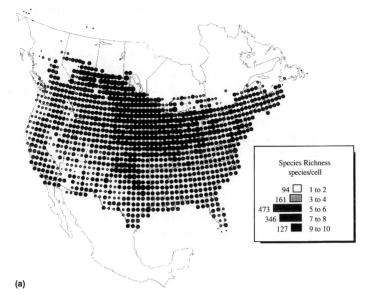

(a)

Distribution of endemism in 14 icterids

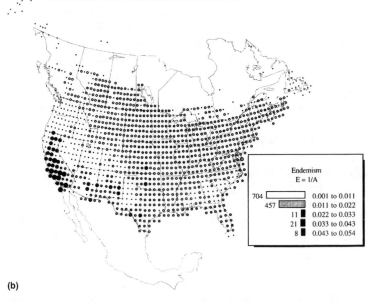

(b)

Figure 9.2. Maps showing hotspots of species richness and endemism in the continental United States, Mexico, and Canada. Two groups of passerines are represented: the icterids or blackbirds and orioles (in parts (a) and (b)) and neotropical warblers of eastern

Distribution of species richness in 24 neotropical warbler

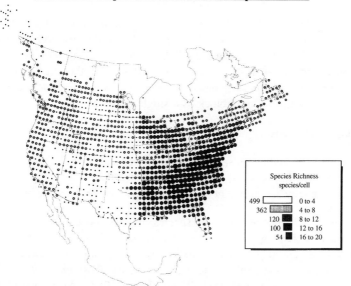

(c)

Distribution of endemism in 24 neotropical warbler

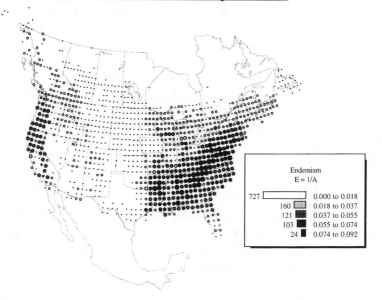

(d)

deciduous forests (in parts (c) and (d)). The size of the filled circles reflects the magnitude of species richness or cumulative "endemism coefficients." The frequency distribution of these measurements is represented by the bar histograms in the legend.

$$E_j = \sum_{i=1}^{n} 1/A_i, \text{ for all } i \in S_j \tag{9.1}$$

The contribution made by a species is constant for each cell. If a species has been recorded at a total of 200 cells, then a value of 0.005 is added to every cell in which the species has been recorded. Species are often recorded in a block more than once, if for instance two survey sites are in the same block or if the species was seen in the same block in successive years. In either case, we treat duplicate sightings within a block as a single occurrence.

Qualitative Patterns

For the warblers and icterids, we display both the species richness and endemism maps in Figure 9.2, parts (a) and (b). We selected these two groups from several taxa because they are sufficient to illustrate the major features we have observed.

1. Areas rich in one group of species are not those rich in other groups. The center of richness for warblers—the Appalachians—is not the same as for icterids—the central plains. This well-known result applies between vertebrate classes. Across a continent, species richness in, say, frogs may not correlate with the species richness in birds (Schall and Pianka, 1978). Worse, the direction of the correlation—positive or negative—may differ between continents.

2. For warblers, the maps of endemism (Fig. 9.2c) and richness (Fig. 9. 2d) are roughly similar, with the highest values in the Appalachians. The differences are subtle, but perhaps important. For endemism, California has relatively higher values than for species richness, and within the Appalachians the most important values for endemism are more localized. For icterids, the patterns are completely different: California is the hotspot of endemism because it includes the severely range-restricted tricolored blackbird, *Agelaius tricolor*. Simply, patterns of species richness in different groups do not necessarily coincide, richness may not coincide with endemism, and patterns of endemism in different groups do not necessarily coincide. The rules for setting conservation priorities are far from simple.

Quantitative Patterns

Let us consider the index of vulnerability defined earlier (Eqn. 9.1). We call the basic unit an M (for Norman Myers) because it quantifies what he has called "hot spots" (Myers, 1988, 1990). An area housing one species that occurs entirely within a range of 10^4 km^2 (approximately the area of a $1° \times 1°$ latitude × longtitude block close to the equator), would have a value of 1M. The area might house many widespread species, their individual contribution to the total M may be insignificant but the sum of their contributions may be very significant. A hundred sympatric species each occupying 10^6 km^2 would produce a value of 1M. Now let us consider how endemism varies spatially.

The total area of North America's eastern deciduous and coastal plain forest is $\sim 2.9 \times 10^6$ km². Only 4 of its ~ 160 forest bird species have gone extinct, despite the loss of about 50 percent of the forest in the last 200 years. All but 28 of these species occur widely across North America and the 10^4 km² blocks have average endemism values of $\sim 0.1M$. The 1.47×10^6 km² of forests in the Philippines and Indonesia (excluding Irian Jaya) hold 544 endemic bird species. If these species were spread evenly across the entire region, the area would average 3.7M—over an order of magnitude higher than eastern North America. Not surprisingly, 119 species are threatened with extinction. The corresponding value for the ~ 130 terrestrial Hawaiian birds would be $\sim 50M$. All but a dozen of these species are extinct or threatened with extinction.

When we weight species richness by vulnerability to extinction, we highlight very different areas from those that are merely species rich. On a global scale, these differences correlate with the very different fates the areas have encountered. What can such analyses predict for the future of global biodiversity? Myers (1988) has noted that 18 areas worldwide are so rich in endemics as to encompass approximately 20 percent of the known species of flowering plants in a total area of 0.74×10^6 km². This yields a value of about 700 M—an index value an order of magnitude higher than even Hawaiian birds. This yields the distinct conclusion that the most vulnerable targets for conservation have yet to be hit by destruction. Simply, the worst is yet to come.

Acknowledgments

We thank Julie Lockwood, Tom Brooks, and Gareth Russell of the University of Tennessee, and Brian Maurer of Brigham Young University, for help in extracting data. We would also like to thank Bruce Peterjohn for providing us with the BBS data.

10

Conserving Ecosystem Function

Judy L. Meyer

Summary

Successful conservation efforts require a consideration of ecosystem function. In this chapter, I present seven principles from ecosystem science to serve as a foundation for ecosystem conservation:

1. Ecosystems are open, which should lead to an emphasis on conserving the fluxes across ecosystem boundaries and linkages with surrounding ecosystems.
2. Ecosystems are temporally variable and continuously changing; the present bears the legacies of past disturbances.
3. Ecosystems are spatially heterogeneous on a range of scales, and essential processes depend on that heterogeneity.
4. Indirect effects are the rule rather than the exception in most ecosystems.
5. Ecosystem function depends on its biological structure.
6. Although several species perform the same function in ecosystems, they respond differently to variations in their biotic and abiotic environment, thereby reducing variation in ecosystem function in a changing environment.
7. Humans are a part of all ecosystems.

A river restoration project provides an example of the use of ecosystem principles to design effective conservation strategies; yet we have little experience with the use of measures of ecosystem structure or function to measure the efficacy of conservation efforts. Research is needed to find appropriate metrics and to determine their range in regional reference systems. A critical first step in using such an approach is to identify and preserve a wide range of ecosystems that could be used for regional references. Unanswered questions remain that limit our ability to use measures of ecosystem structure and function to guide conservation practices. Finding answers to these questions is critical because the public values the goods and services that are a consequence of ecosystem function.

Introduction

Conservation must consider the ecosystem if it is to be successful. Species networks exist in geologic, hydrologic, climatic, and chemical contexts. The total biological diversity on Earth is in part a consequence of heterogeneity in abiotic features of the planet. For example, the heterogeneity inherent in a stream is a consequence of its geomorphology, hydrology, and chemistry: different plant and animal communities occupy bedrock outcrop versus depositional areas (Huryn and Wallace, 1987); hydrologic regime influences the outcome of food web interactions (Power, 1992); and patterns of nutrient upwelling from the sediments alter the distribution and abundance of algae (Pringle et al., 1988). Without the understanding gained from analysis of the physical and chemical setting as well as the biological and societal context, conservation efforts are doomed. Attempting to conserve a wetland bird species without understanding the wetland's hydrologic regime would be courting failure.

To say it as provocatively as possible: conservation will fail if it is the exclusive territory of biologists. Conservation ecology requires expertise from hydrology, geology, geochemistry, meteorology, history, sociology, and other disciplines as well as from the traditional biological sciences.

Many other arguments have been offered for why we must manage at the ecosystem scale (Grumbine, 1990). For example, Franklin (1993b) offers two: there are too many species to deal with them individually and the ecosystem approach offers an effective way to conserve poorly known species and habitats (e.g., belowground).

Despite the declared need for an ecosystem perspective, ecosystem principles are rarely presented in papers and books dealing with conservation; for example, May's (1994b) discussion of ecological principles relevant to the management of protected areas stopped at the level of communities. Hence I begin by outlining a set of basic ecosystem principles that could serve as a beginning foundation for ecosystem conservation. I explore by example how these principles have been and could be used as a basis for effective conservation efforts and then consider possible measures of ecosystem function that could guide future conservation efforts. My focus in this chapter is on ecosystem function, what environmental ethicists term the "instrumental value" of nature (producing goods and services of value to humans) (Callicott, 1989). Nature's "intrinsic values" are equally worthy of preservation.

Basic Principles From Ecosystem Science

What are some principles of ecosystem science on which to base a theory of conservation?

- Ecosystems are open to flows of energy, elements, and biota.

Application of this principle requires consideration of the fluxes across ecosystem boundaries and identifying linkages with surrounding ecosystems. There are examples of successful management based on this principle. One is the Clean Water Act, which, among other things, regulates the inputs of nutrients into our nation's waters. Although it is far from perfect, in part because it does not adequately cover diffuse fluxes across ecosystem boundaries (i.e., nonpoint sources), this act has had a positive impact on the preservation and improvement of aquatic ecosystems in the United States.

There are also situations in which our current conservation policy does not adequately consider fluxes. For example, U.S. Department of Agriculture (USDA) quarantine officials are charged with inspecting material crossing our borders to intercept pests that pose a threat to species of economic importance; as explained to me by a frustrated quarantine officer, there is no mention of ecological or evolutionary importance, only economic. Although we outlaw trade in endangered species, U.S. quarantine officers do not have the authority to block the importation of species that pose a threat to our native flora and fauna.

The Wild and Scenic Rivers Act is another example of a conservation action that did not adequately take into account fluxes across boundaries. The act protects reaches of rivers, but does not protect the river above or below the reach of concern or the watershed as a whole. The reach is vulnerable to poor agricultural practices in the headwaters that add nutrients and sediments or to toxic emissions or dams downstream that restrict the migration of anadromous species. This became apparent to conservation groups who had fought hard for protection of rivers under this act, and then found themselves unable to truly protect the river. Hence many adopted an ecosystem perspective to seek solutions that will protect more than just the reach (e.g., Doppelt et al., 1993).

- Ecosystems are continuously changing; yet the present bears the legacies of the past.

Failure to adequately consider this principle has lead to misguided conservation strategies (e.g., fire exclusion), which have been thoroughly discussed elsewhere (Botkin, 1990; Pickett et al., 1992; Meyer, 1994). Although disturbance is now recognized to be a part of natural ecosystems, human disturbance can not simply be substituted for natural disturbance (Ewel, this volume); logging an old growth forest is not the same disturbance as a lightning-ignited fire.

The legacies of past disturbances (e.g., beaver removal; Naiman et al., 1988) is what Magnuson (1990) has called the "invisible present." Finding that "invisible present" requires us to look in the past, to do some historical reconstruction, and design our conservation strategy accordingly. I provide an example of that process at work in a river basin later.

- Ecosystems are spatially heterogeneous on a range of scales, and ecosystem structure and function depend on that heterogeneity.

To a stream ecologist, spatial heterogeneity is an obvious fact of life (e.g., Pringle et al., 1988). The communities of benthic invertebrates and the relative rates of key ecosystem processes like primary production and respiration vary at scales ranging from the landscape to an individual rock on the stream bed. This variability is a function of the geographical setting of the river basin (e.g., Minshall et al., 1983), where you are in the river network (e.g., Naiman et al., 1987, Meyer and Edwards, 1990), the nature of the stream reach (e.g., constrained or unconstrained), and the nature of the substrate (e.g., bedrock versus a pool). This hierarchy of spatial heterogeneity has been well classified, categorized, and used to guide stream research and conservation (Frissell et al., 1986; Hawkins et al., 1993; Rogers, this volume). Critical ecosystem processes depend on it. For example, the nitrogen cycle depends on spatial heterogeneity in oxygen content: nitrification proceeds in oxygenated environments whereas denitrification occurs where there is little or no oxygen. Elimination of this spatial heterogeneity (e.g., by channelization) alters both structure and function of the riverine ecosystem (Allan, 1995).

- Indirect effects are the rule rather than the exception in most ecosystems.

The conservation message in this principle is that disruption of one part of an ecosystem will have broader repercussions. Ecosystems are not assembled at random. They are the product of a long history of interaction (e.g., Thompson, this volume). It is the province of ecologists to assess the strength of those interactions among both biotic and abiotic components of the ecosystem. This principle has been well documented in experimental manipulations of lakes, where alteration of vertebrate predator abundance impacts not only abundance of their zooplanktivorous prey but also abundance of zooplankton and algae as well as lake temperature regime (Mazumber et al., 1990) and sedimentation rates (Pace et al., 1995). Ecologists are most familiar with the importance of indirect effects in species interactions (e.g., Wootton, 1992); yet there can also be indirect effects in elemental cycles. For example, alterations in sulfur deposition can impact phosphorus burial in lake sediments (Caraco et al., 1991), and the presence of deep-burrowing fauna alters the ratio of pyrite to carbon in marine sediments (Giblin et al., 1995).

- The function of an ecosystem depends on its biological structure; species do not have equal effects on ecosystem function; and an organism's size is not a good indicator of its influence on ecosystem function.

A recent study in a stream experimentally altered by pesticides offers a striking example of the relationship of function to structure (Wallace et al., in press). Values of an index of macroinvertebrate community structure increased and decreased coincidentally with a measure of ecosystem function (seston concentration, which reflects rate of organic matter processing). The second part of this principle is a restatement of keystone species concept, which has been thoroughly explored elsewhere (e.g., Bond, 1992). The third part is a recognition of our dependence on the microbes of the world for essential functions like nutrient regeneration. The application of this principle by conservationists broadens their focus well beyond the charismatic megafauna.

- Although several species perform the same function in ecosystems, they respond differently to variations in their biotic and abiotic environment, thereby reducing variation in ecosystem function in a changing environment.

When lakes were experimentally acidified, primary productivity changed relatively little, despite striking changes in algal species composition (Schindler, 1990). A similar phenomenon has also been observed in a stream being recolonized after experimental alteration of invertebrate assemblages with pesticides: the same rates of leaf decay were observed despite shifts in the species of invertebrates consuming the leaves (Wallace et al., 1986). In this case, recovery of ecosystem function (rate of leaf decay) occurred more rapidly than did taxonomic recovery. In experimental plots in Costa Rica, the first several species added to the plots greatly altered ecosystem function; but after the first few species, there was little change (Ewel et al., 1991; Vitousek and Hooper, 1992). In all of these examples, one is tempted to view the array of species performing the same function as functionally redundant, but they are not. This becomes apparent when ecosystem function is considered in a longer time frame and in the context of environmental change. For example, zooplankton species that appeared after acidification of a lake were those that had been rare earlier (Frost et al., 1995). They may have appeared to be a small and functionally redundant part of the assemblage of grazers before acidification, but were a dominant member of the assemblage when environmental conditions changed.

The conservation message is clear: one goal for conservation is maximization of functional redundancy because that offers the best insurance for maintenance of ecosystem function in a changing environment. The diversity represented by today's rare species relates to future ecosystem function in an environment altered by natural or anthropogenic processes.

• Humans are a part of all ecosystems.

Not only have we altered Earth's ecosystems, we are also dependent on them. This is a principle we ecologists have finally assimilated into our research agendas (Lubchenco et al., 1991; Naiman et al., 1995), which acknowledge the impact of human activity on the biosphere and the need to understand its effect. Ecologists recognize that we can no longer study pristine environments, for there are none (Vitousek, 1994).

This principle is recognized to be of critical importance in conservation ecology (e.g., Meffe and Carroll, 1994). Conservation is essentially management of human activity in the landscape, so to ignore the societal context for conservation efforts, is to invite failure. The success of conservation efforts often rests on their ability to incorporate indigenous peoples, whether they are tribes in the Amazon or ranchers on the borders of Yellowstone National Park.

I offer these as a beginning set of ecosystem principles that can be combined with already elaborated principles from genetics (e.g., Holsinger and Vitt, this volume), population biology (e.g., Nott and Pimm, this volume), and community ecology (e.g., Simberloff, this volume) to broaden the scientific basis for conservation.

Applying These Principles To Conservation

How might an understanding of ecosystem function be used to design better conservation strategies? I offer an example from a river restoration effort on Knowles Creek in Oregon. This is a project conceived and carried out by Charley Dewberry of the Pacific Rivers Council working with staff from the Siuslaw National Forest and Champion International, a timber products company (Dewberry, 1995, 1996). This project is unique in that their efforts have been directed at understanding and recreating long-term ecosystem processes while recognizing that emergency stopgap measures are necessary in the short term.

Knowles Creek is a tributary of the Siuslaw River, draining the western slopes of the Cascade Mountains. It has populations of coho salmon, chinook salmon, and steelhead and cutthroat trout. The project began with a historical reconstruction: What were the key functional processes in the basin prior to European invasion?

The basin can be divided into valley and upland. Sediments from the uplands collected in hollows, which pulsed the accumulated sediment load into the valley in debris torrents at about 6,000-year intervals. The debris torrent stopped in the tributary junction with the mainstem or when it contacted the huge cedars characteristic of the riparian zones of the valley floor. Hence certain sites in the basin were the "geomorphic control points" with extensive flats behind them that provided essential backwater habitat and areas of high aquatic productivity. These areas also helped control stream temperatures by storing large amounts of water in the sediments. But the flats were not permanent features; eventually the debris dams that formed them would be breached, the flat cut down to bedrock, and the material it had held was pulsed downstream to the next flat.

When the first European settlers arrived in the 1870s, the basin was recovering from extensive fires of the previous decade, fires that naturally recur every century or two. Because of increased erodability after fires, it is likely that the uplands were supplying considerable sediment to the valley floor at this time. Flooding of the valley floor occurred frequently: a flood triggered by typical June rainstorm in the 1880s would require a 75-year storm today. Logging in the valley floor altered its sediment storage capacity, and the channel downcut rapidly. Logging and roadbuilding in the uplands was intense from 1950 to 1985, delivering two centuries worth of sediments to the channel in a period of 35 years. With reduction in storage capacity of the valley floor, those sediments were lost from the basin and with them was lost essential habitat for salmonids.

In this stream, where around 100,000 coho smolts would be expected to migrate to the ocean, only 1,660 did in 1982. So in the 1990s we are left with an upland that has recently lost much of its erodable sediment and a valley floor that has little capacity to store sediment, resulting in loss of productive habitat, accentuated low flows and elevated water temperatures.

What conservation/restoration decisions have been made based on this understanding of ecosystem processes?

1. Protect the intact areas to serve as refuges while the basin recovers.

2. Begin the recovery process in the valley floors by replanting cedars, beginning a process that will take a century before its full impact will be felt.

3. Manage uplands to reduce likelihood of major debris torrents in the next century.

4. Simulate debris torrent deposits at sites at the geomorphic control points where they would have naturally occurred. This involved

    a. identifying areas where flats were likely to be formed and would remain at least 50 years,
    b. spacing the flats so that a couple tributaries could contribute sands and gravel to each,
    c. choosing sites that would give greatest immediate storage for the least investment in material, and
    d. choosing sites that posed the least threat to existing roads and bridges.

Ten sites were identified. Crews cabled in a few key pieces of downed timber at each site to mimic the huge immovable trees that would previously have provided the structural stability for the deposit, but the rest of the debris was allowed to move and set up again at the next flat downstream. All major flats have been mapped, and all large pieces of wood tagged so their movements can be followed. The number of coho salmon smolt is being monitored, because that was chosen as the biological response variable. The debris dams have functioned as predicted, trapping sediments during a major storm in January 1995 that approached the storm of record in its magnitude.

This project has elements of many of the principles I discussed: it recognizes the openness of ecosystems and linkages between different ecosystems, the temporal variability of important habitat features (the flats), the importance of an historical perspective, the spatial heterogeneity of systems, the importance of productivity of the smallest members of the food web, and incorporation of the human element.

Using Measures of Ecosystem Function As A Target For Conservation

The previous example demonstrates how we can use an understanding of ecosystem function to guide conservation practices. Yet the tough question remains: Can we use measures of ecosystem structure or function as a target for and a way of assessing the efficacy of conservation efforts? The goal is to find measures of ecosystem structure and function with sensitivity, diagnostic capacity, and ability to offer early warning of problems (Nip and de Haes, 1995). Yet there are examples of extremely useful indicators that do not meet all these criteria; for example, atmospheric CO_2. Its utility as an indicator has been demonstrated; yet it has little diagnostic capacity. We are still arguing about which human activity causes elevated atmospheric CO_2. Finding indicators that meet all these criteria is a challenge that those interested in detecting effects of toxins on ecosystems have been facing for decades. What have they learned?

1. Seek not a single metric. What is needed is a suite of measures that indicate the function of a facet of the ecosystem of concern (Kelly and Harwell, 1990). We are not going to find a single index that will measure the "pulse" of an ecosystem. In our search for measures, it is important to recognize the diversity of ecosystems on the planet and devise measures of local interest so that our management can be particularistic; that is, guided by the peculiarities of the site (Norton, 1992)

2. Consider the structure of the food web as a whole or those portions leading to species of interest, which could be native or endemic species. For example, evidence from a thermally altered river shows the number of links in the food web as the variable showing the greatest change under an altered thermal regime (Ulanowicz, 1992).

3. Look to key geochemical processes. Biological oxygen demand (BOD) has been a useful indicator of threats to stream ecosystems from organic pollution for decades; significant improvement in lakes has been achieved by controlling the supply of phosphorus to the biota. We need additional indicators of system geochemistry because they can tell a manager about the availability of essential

elements to species of concern. In the forests of the eastern United States, the nitrogen cycle seems to be a sensitive indicator of change. Nitrate losses have been seen in response to the human disturbance of logging, as well as to the invasion of defoliators such as the gypsy moth or the fall cankerworm (Swank and Crossley, 1988). Nitrogen mineralization is a key process that supplies biologically available nitrogen to the ecosystem. Hence indicators that consider the forms, stocks, or recycling of essential elements such as nitrogen or phosphorus offer promise.

4. Indicators of productivity and physiological or reproductive function in key species provide an early warning of problems. This has been clearly demonstrated in lake acidification research (Schindler, 1990), where, for example, periphyton productivity was a sensitive and early indicator of acidification. The greatest alterations in ecosystems resulted when species without functional analogs in the system were eliminated (Schindler, 1990); these are the species whose productivity and physiology offer the most promise as indicators.

5. Look to the resource base: has there been a shift in the relative importance of allochthonous versus autochthonous energy resources? In the thermally altered Crystal River, there was a shift from detritivory to herbivory (Ulanowicz, 1992). New techniques of stable isotope analysis offer promise for indicators to detect these kinds of changes: in streams, delta^{13}C signatures of samples of benthic invertebrates could be used to detect shifts in the relative importance of allochthonous and autochthonous carbon sources (Bunn, 1995).

6. Look to key processes. In streams draining forested catchments, leaf litter decay is a key process. Decades ago, Egglishaw demonstrated the connection between nutrient concentration, fungal degradation of a standard C source, and trout growth (Kelly and Harwell, 1990). Decreases in fungal diversity or activity in a system like this indicates a serious threat to the food web.

7. Look to abiotic regulators of key processes. Temperature is of course the most basic, but in lakes and streams, measures of water residence time are critical. For example, transient storage zones in streams (e.g., pools forming behind debris dams, deep gravel beds that exchange water with the surface) alter the length of time water is in contact with stream sediments, thereby affecting the ability of sediment biota to take up nutrients (D'Angelo et al., 1993). An alteration in volume of transient storage zones caused by changes in channel structure will eventually be manifested in reduced capability for nutrient uptake and storage.

8. Look to biotic regulators of key processes. Indicators of the condition of some of the "engineer" species (such as beavers, woodpeckers, earthworms and other burrowers; e.g., Lawton and Jones, 1995) offer insight into the future condition of the ecosystem.

9. Look for integrators. Atmospheric CO_2 is an example of an indicator that integrates human activity over a broad scale. Similarly, aquatic systems can serve as integrators of management of terrestrial landscapes (Naiman et al., 1995).

Indices of the integrity of the aquatic biota offer tools to assess the cumulative impact of watershed management practices (e.g., Rosenberg and Resh, 1993).

10. Expect the unexpected. Disturbances are a feature of the natural world that cause changes in ecosystems, only some of which we are able to predict. Yet they also offer valuable insight into the mechanisms driving observed patterns.

This list is not a primer of measures of ecosystem function that offers useful targets for conservation. We are not yet ready to write that primer. Still needed are further development, testing, and application of measures of ecosystem function that could serve as targets for ecosystem conservation in a wide range of environments. These will add to the effectiveness of our conservation toolbox.

Assuming we add these measures to our toolbox, what numeric values of the measures should management seek to achieve? Here there is a deceptively simple answer: the range of values observed in regional reference systems. A critical first step in taking such an approach is to identify and preserve a wide range of ecosystems that could be used for regional references (Christensen, this volume; Barrett and Barrett, this volume; Thompson, this volume). These ecosystems provide us with the baselines we need for evaluating our compliance with environmental laws and for assessing the efficacy of conservation and restoration efforts. Ecosystems with minimal human impact offer us the moving target we need to design and evaluate restoration and conservation efforts. By referencing our activities to an ever-changing natural system, we are incorporating into our management scheme one of the basic principles presented earlier, the temporal variability of ecosystems.

By considering ecosystem function in addition to evolutionary heritage, conservation will also broaden its list of places worthy of protection. Conservation of systems that provide essential ecosystem services will receive increased attention. We may also want to consider conservation of ecosystems with unique functional attributes, because these are places where we are likely to find species with unusual characteristics that could be beneficial to humankind. Extreme environments, such as hot springs, offer a unique environment selecting for unusual functional adaptations; this is an environment that has already yielded products of economic benefit to humans. It would be fruitful to seek out and conserve environments that are likely to select for organisms with unique functional attributes.

Unanswered questions remain that limit our ability to use measures of ecosystem structure and function to guide conservation practices. Finding answers to these questions is both intellectually challenging and critical to the success of conservation efforts because the ecosystem goods and services valued by the public are a consequence of ecosystem function. Conservation will enjoy wider public support if we are able to relate conservation efforts to services the public cares about, such as providing clean water, clean air, and productive soils (Harte, this volume). These are the products of conserving ecosystem function.

11

The Relationship Between Patchiness and Biodiversity in Terrestrial Systems

Lennart Hansson

Summary

Historical biodiversity in a biome is dependent on the integrity of particular physical environments or interactions between species. As new types of disturbances occur or disturbances increase in intensity or scale, specialized, highly interdependent interactions are disrupted and systems become dominated by more generalized species and interactions. Generalized animals will then also enlarge the area of disturbance, particularly in old biomes. A young biome, such as the boreal forest, is dominated by generalized relationships, sometimes typical for that environment, whereas older biomes, such as temperate grasslands and tropical forests, show more specialized interactions. Specialized interactions occur on overlapping spatial scales; thus landscapes rather than patches should be the focus of management. Specialized patchy systems need to be recognized but at present preservation of large pristine landscapes has priority.

Introduction

The concept of biodiversity is not well defined, although it was stated in the Rio Declaration to include genetic variation, species richness, and ecosystem spectra. My personal view is that our main concern should be to preserve factors and mechanisms that are important in perpetuating the original ("native") fauna and flora of various biomes.

Whenever separate species populations are examined in detail, either strong microhabitat dependence or interactions with many other species emerge. Therefore, I believe that we should be aiming at conservation of functional systems, and especially of those systems that require specific physical and/or biotic environments. Many ecologists regard communities (or "assemblages") just as collections of separate, independent species, whereas others admit more or less strong interactions and dependence between species. Still, detailed examinations indicate such strong interactions with surroundings, including other species, that we should change emphasis from separate species to systems. Biodiversity at species, community, and ecosystem levels would be considered if research and management carefully thought about species networks and necessary physical support.

I focus this discussion on differences and interactions between specialized and generalized species. For the present purpose, I define specialists as species with strong dependence on some physical substrate (sensu latum) and/or strong interactions with one or a few other species. Generalists show low dependence on substrate and many, weak interactions with several other species. Specialization thus concerns habitat or food. When facing profound environmental changes, specialists are argued to be inferior to generalists.

Many species may not have evolved in their present biomes but might have, more or less long ago and by immigration, assimilated into a system that permits their persistence. We should thus consider the physical and biotic conditions, or "life system" (Clark et al., 1967), to which individuals in local populations may have acclimated. This may not mean that such populations will maintain stable densities; indeed cycles or even dynamic chaos may be consequences of the acclimation. Certainly, well-adapted individuals should, at least temporarily, produce more offspring than can be sustained locally. Thus, mobile or dispersing individuals, specialists and generalists, may make up a large part of the general "biodiversity" in a biome. The "minimal structure" (or "minimal system") of Pickett et al. (1989) may be a fruitful basis for analyses: Random community samples may indeed be very random and consist of scattered representatives from various systems or subsystems.

Interactions between a few species and a possible dependence of some of these species on certain physical features at any scale can be called "functional patchiness" (or possibly "community life system" in analogy with Clark et al., [1967]). "Physical patchiness" (or more commonly "patchiness") instead refers to multiple small-scale structures, typically in the hectare-scale or smaller, caused by locally deviating substrates and/or sessile organisms. The relationship between functional and physical patchiness is not always clear. Deductive and experimental analyses of community dynamics have not been very successful (Drake, 1990). I think that, at present and perhaps for the foreseeable future, we will therefore have to base our generalizations and predictions on inductive reasoning from well-selected examples.

To illustrate my points, I give a survey of important community interactions in taiga forest and I present selected examples of interactions between species and mechanisms crucial for persistence. I also compare findings in boreal environments with other terrestrial biomes. Finally, I discuss management implications.

Some General Arguments

The following types of functional patchiness will occur to varying extent in all biomes and constitute bases for ecological heterogeneity:

1. Species dependent on a physical substrate. This substrate can be particular soil type properties, litter accumulation, dead plant parts, or other physical aspects of a habitat. It should have a long history in present or earlier biomes, but may depend on regular disturbances.
2. Dependence on another species. Other species can be limited by physical patchiness but can also occur over vast areas with varying environmental conditions. Interactions between the two species (and additional parasites and predators) may lead to density variations that can affect local persistence. However, persistence should not be endangered in larger pristine areas if the coexistence has a long history. With regard to physical patchiness, sedentary "base" species can be included with the group in type 1, and mobile species with the group in type 3.
3. Interactions that are typical for a particular biome. They may be due to smaller or larger networks of species and generalist species may also be involved. The particular species may be of less interest (also occurring in other biomes) than the specific dynamics and effects of the interactions. Such interactions may not be dependent on some particular structures or patches but may be distributed over a wider landscape.
4. Incidental species relations. Certain herbivores or predators will devour plants or herbivores, respectively, that have drifted from their natural system. This may happen in the matrix between patches or within distinct patchy communities with regard to temporary surplus individuals. It is possible that generalist species have developed adaptations for such situations.

New types of disturbance may cause a complete or partial destruction of various components of these (sub)communities. Specialized species may find less good substrate or food after disturbance, or shelter against weather or predation may have disappeared or diminished. Reinvasion of empty spots will be much easier for generalized mobile species than specialized more or less sedentary species. Generalists (often "r-selected") usually also have higher birth rates and short-term population increases, and such dynamics will be favorable in unsaturated areas (e.g., Crawley, 1987).

Specialists will occur at higher densities than generalists if the latter are restricted to one type of substrate or to interactions with just one other species. This, of course, derives from the general observation that "a jack-of-all-trades is a master of none." When an area is disturbed, generalists will invade, either from within the wider biome or as exotic species, for example, from other biomes or from agricultural areas. Generalist plants will grow on exposed and nutrient-rich soil and generalist herbivores will be able to cope with a new mixture of plants. Special digestive adaptations will not be needed as ephemeral plants often are particularly nutritious or show less developed defense against herbivores (Rhodes and Cates, 1976). With many new species of herbivores and omnivores, generalist predators will enter the scene. There will be an accumulation of generalist species and a new community, however defined.

This invasion by generalists will be limited by the size of the disturbance. A single generalist individual will need a larger area than a corresponding specialist if exploiting the same resource. Furthermore, on the population level there should be a minimum viable population size, small populations being very vulnerable to extinction due to environmental hazards and unbalanced demography (e.g., Soulé, 1987). Patches the size of a few m^2 may be invaded by some single generalist plants, whereas km^2-areas will get complete generalist communities. Furthermore, generalist animals will move far into the surrounding undisturbed community and increase the effective disturbance area considerably.

The movement of generalist species into disturbed patches has been emphasized in an extensive literature (reviews in e.g., Crawley, 1987; Hobbs and Hueneke, 1992). The main topic has been the invasion by exotic generalists. However, the occurrence of native annual plants on mole hills and other small disturbed sites has been demonstrated for many grasslands, and the vertebrates that move into clearcuts from surrounding forests tend to be generalists (Hansson, 1994). In forest plantations in North America, most weedy species are natives, as are the insect pests and pathogens (Pimentel, 1986). Exotic generalists usually establish more easily after disturbances but may also invade undisturbed communities, especially on islands with more simplified flora and fauna (Drake et al., 1989).

The influence of generalist species can be discussed with generalist predators as examples, but equivalent effects will be found on other trophic levels. If new generalist predators appear, or the density of original generalist predators becomes increased by extinction of top predators or by additional food (e.g., the "meso-predator release"), then sparse and naive prey species, often specialists, may be driven to very low levels or extinction as the generalist predator is mainly supported by more common alternative prey at specialist declines. Prey switching will not occur when there is strong prey preference (Murdoch, 1969). Specificity also often indicates physiological limitations in the capture or digestion of prey.

Physical Structures and Disturbance Dynamics in Boreal Forests

The ground and soil properties of boreal forests are very heterogeneous due to the fairly recent glaciation and the highly variable sizes of soil particles in the dominating moraine soils. More homogeneous and thus strongly deviating sediment soils are found along low-altitude rivers.

Forests on several types of moraine soil and generally on sediment soil become dry in summer and are easily ignited by lightning. Thus, boreal forests are dominated by a fire regime with varying intervals (20 to 200, with means around 80 years; Hansson, 1992b). However, forests at high altitude, with northern exposure on hills, and on islands in lakes and mires burn seldom or never (Esseen et al., 1992). Thus, there will be intermingled fire-born environments and fire refugia. Forests that often burn are dominated by pines whereas spruce trees become more abundant in areas protected against fire. Forest fires result in succession with deciduous tree species on nutrient-rich soils and pines on poorer soils. Such succession cause an extensive large-scale patchiness in pristine landscapes (Esseen et al., 1992).

On a finer scale, dead and dying trees constitute important elements in pristine forests. Moist logs are common in fire refugia, whereas fire-scarred and dry pine stems remain, alive or dead, in burned open forests. Fire refugia appear dominated by gap dynamics with single trees dying from storms or insect attacks (Esseen et al., 1992) and are characterized by a regular supply of dead wood.

Forest management has profound effects on the natural dynamics, particularly when it prevents the natural fire regime. The timber is usually completely removed from the forest and little dead wood of large dimensions remains. Clearcuts result in even-aged stands over large areas, which strongly deviates from the variability in size classes produced by forest fires in pristine areas. Within stands, even-aged forests have replaced the old heterogeneity in age classes.

The communities of most boreal forests are young, in Scandinavia hardly 10,000 years for pine and less than 3,000 years for spruce forests. There are many generalist species and very few exotic invaders, a few temporary species in edge areas close to agricultural land being the sole exceptions.

Specialized Relationships in Boreal Biomes and Their Destruction

Specialization on Physical Structures (or Disturbance Effects)

Most boreal plant and animal species do not exhibit any pronounced specialization toward ground structures or soil quality. In Scandinavia, the few dominating tree species (*Picea abies, Pinus sylvestris, Betula spp., and Populus tremula*) actually occur all over the existing soil gradient, except for water-logged places. Still, a few plant species demonstrate a strict habitat dependence, particularly to old growth and burned ground (Table 11.1).

Disappearance of crucial physical structures also leads to disappearance of dependent species. Clearcutting creates a new habitat with drought-tolerant (generalized) herbs and grasses. Plants dependent on fire and old growth disappear.

Dependence of One Species on Another Species

Most ecologists accept close interactions, and even coevolution, in plant-pollinator, plant-seed-disperser and host-parasite systems. Such systems are fairly com-

Table 11.1.   Examples of community interactions in boreal forests.

Species or Effect	Dependent on	Inhibitory Agents or Processes
A. Dependence on physical structures		
Pulsatilla vernalis (spring anemone)	Forest fires	Fire prevention
Geranium bohemicum	Forest fires	Fire prevention
Geranium laniginosum	Forest fires	Fire prevention
Epixylic mosses	Fire refugia	Clearcutting of ancient spruce forests
Dead-wood beetles of deciduous trees	Successional deciduous forests	(Fire prevention) rapid reforestation
Rana lessonae (pool frog)	Wet forests between small lakes	Clearcutting
Microtus agrestis (field vole)	Grassy successions	Rapid reforestation
Dendrocopus leucotus (white-backed woodpecker)	Deciduous successions	Rapid replanting of clearcuts
Resident ("Siberian") passerine species	Old-growth forests	Clearcutting
Birds from temperate deciduous forests	Fire-borne deciduous successional stages	Fire prevention and reforestation by conifer seedlings
B. Dependence on other species		
Babtria tibiale, Acasia appensata (Geometridae)	*Actea spicata* (baneberry)	Changed microclimate after clearcutting
Eupitecia groenblomi	*Solidago virgaurea* (goldenrod) in late successional stages	Early clearcutting
Melampyrum spp. (cow wheat), *Viola* spp. (violets)	Ants	Clearcutting
Thanasimus formicarius	*Ips.* spp. (bark-beetles)	Harvesting of immature trees
Loxia curvirostra (crossbill)	*Picea abies* (Norway spruce)	Might be more common if harvesting was limited to mature trees
Loxia pityopsittacus (parrot crossbill)	*Pinus silvestris* (Scots pine)	Might be more common if harvesting was limited to mature trees
Loxia leucoptera (two-barred crossbill)	*Larix sibirica* (Siberian larch)	Might be more common if harvesting was limited to mature trees
Mustela nivalis (least weasel)	*Microtus agrestis* (field vole)	Very rapid forest regeneration?

Continued

Table 11.1. Continued

Species or Effect	Dependent on	Inhibitory Agents or Processes
C. Landscape-specific interactions		
Moss and vascular plant diversity	Uprooted trees	Fire prevention, dominance by generalist species
Nutritious plant (e.g., *Vaccinium myrtillus* [bilberry]), tissues, insects and, in turn, for birds	Dark, wet forests	Drainage of mire and wet forests, light exposure, and invasion by grasses
Woodpecker nests for boreal mammals and birds	Woodpeckers, especially *Dryocopus martius* (black woodpecker)	Adjoining agriculture, nest occupation by *Corvus monedula* (jackdaw)
"Siberian" bird fauna	Old-growth forests	Old-growth forests destroyed; invasion of common birds from industry forests
Cyclic density variations in wildlife	Thick snow cover; little alternative prey	Generalist predators from agricultural land damping fluctuations in rodents and hares
Low moose densities without overbrowsing	Normal wolf populations; limited early successions	Humans as a generalist from more or less remote landscapes.
Beaver damming effects on biodiversity	Limited predation on the beaver	Overexploitation of the beaver by generalist humans entering the boreal forest

More details can be found, for example, in Esseen et al. (1992) and Hansson (1992).

mon in the present boreal biome (Table 11.1). However, one-to-one interactions might not have evolved in just this biome or even its most recent predecessors. Certain strongly interacting taiga species occur also in other biomes.

Intricate two-species interactions may not necessarily mean that the interacting species as a unit are very specialized. Certain lichens are the first to invade open areas in forests and some tree species interacting with mycorrhizae may be among the pioneers. Species carrying symbionts or physiologically closely interacting species may indeed have evolved to act together as generalists. Such interactions may be very old.

Destruction of a base species or its habitat means that dependent species are also exterminated. They will typically be replaced by very common or generalized species.

Landscape-Specific Interactions

A number of examples, covering both plants and animals, are provided of species interactions that depend on specific landscape features in original taiga areas (Table 11.1). Changes wrought by disturbances and generalist species, which depend on altered or remote landscapes, are indicated. Cascading effects may occur in series of interactions.

Interactions typical for a biome may thus be destroyed at new disturbance regimes or when generalists are introduced, at the edges of the biome when generalists move over borders and cause new interactions, and by movements and eventual invasion from remote landscapes, then particularly by long-ranging predators, including the generalist *Homo sapiens.*

Patchiness, Specialization, and Disturbance In Various Terrestrial Systems

Few authors have considered functional patchiness, however Walker (1989) distinguished between primary diversity and contingent diversity: Contingent diversity depends on species interactions whereas primary diversity is more or less due to sessile organisms so there is some affinity to what I term functional patchiness. I see several reasons why the establishment of important interactions and delimitation of decisive patches has not proceeded further: (1) communities are usually sampled from taxonomical points of view, whereas interactions may mainly be between distant taxa; (2) communities vary both spatially and temporally; (3) different systems interact to the extent that distinct communities may be difficult to discern, (4) community integrity may be characterized as much by absence or low density of certain species as the presence of others; and (5) dynamic effects of the interactions are difficult to study.

At present there is no obvious way of untangling various systems by theoretical analyses. Studies of pristine biomes appear to be the only possibility. For operational establishment of patch-related functional diversity, the following protocol may be useful. First, site-specific relationships are distinguished by focusing on specialized species. Second, supporting patches have to be delimited by physical attributes, for example, the chemical composition or moisture of the soil, physical shelter giving protection from observation by enemies, or substrates for colonial life by plants or animals. Third, adjacent patches, at least when small, should be within dispersal distance for spatial and temporal continuity and there should be species-specific connectivity. It is also important to recognize patches that do not support complete communities.

Influences of generalist exotics appear crucial in certain systems but not in others. Oceanic islands, grasslands disturbed by agriculture or husbandry in the New World and many tropical habitats have been commonly invaded from outside (Drake et al., 1989). They kept originally highly specialized species and communities and there were few generalists. Di Castri (in Drake et al., 1989) suggested that the common mediterranean invaders were selected for their ability to invade due to the long-term human-induced disturbance regimes in early agricultural environments around the Mediterranean Sea. Thus, they could easily invade climatically similar habitats in the New World. Generalists such as rats (*Rattus norvegicus*) and mice (*Mus musculus*) similarly invade areas devoid of generalized predators (e.g., oceanic islands). Conversely, the boreal forest has few invading species. A large proportion of the biota of recently glaciated areas (e.g., the boreal biome) consists of generalized species. The proportion of generalist mammals increases in a pronounced way northward (Pagel et al., 1991). Generalist predators (and herbivores for plants) appear instrumental in preventing other generalists from invading northern biomes.

The following scenarios may therefore be conceived for geologically new and old biomes: In the boreal forest, large natural disturbances have been common and they encourage mainly the development of generalist communities. A few specialists have adapted to fire-born and old-growth environments. In modern forestry, clearcuts are characterized by a new type of partly wet and partly grass-covered ground due to complete loss of trees and bushes, by strong desiccation, and rapid succession due to plantation, and they are almost only exploited by indigenous generalist species. In logged forests, specialists disappear due to the loss of typical old-growth features such as dead wood and snags. In contrast, in tropical forests there are few indigeneous generalists to invade disturbed areas. There, exotic species, if available, will invade the disturbed areas and also may invade the surrounding undisturbed habitat due to limited resistance. These exotic generalists will thus strongly affect specialist species and specialized interactions over large areas and, eventually, the more structured biodiversity of old biota will give way to mixed specialist-generalist communities, or even to generalist-dominated degraded communities.

There is even some theoretical support for such a scenario. As appears from advanced analyses of stability in species composition, coevolution-structured communities tend to be more stable than comparable invasion-structured communities, but more open to invasion from outside (Rummel and Roughgarden, 1985).

Implications For Biodiversity Conservation

Species and ecosystem preservation are approached in very different ways, by different people, and often by different authorities. Species conservation appears at present to have the upper hand when it comes to controversies about action or funding. Still, in the long run there may not be any conservation of specialized species without ecosystem preservation or management. Another example of shortsightedness concerns the species that are examined in applied research and that actually are managed; most efforts are devoted to a fairly limited number of (regionally) very rare species, whereas species still flourishing and important in ecosystem processes but sensitive to environmental changes are hardly considered. Examples of such species are woodpeckers dependent on large trees and providing nest holes for many other species and ant species that are sensitive to clearcutting but strongly affect pest insects and probably also nutrient circulation.

From the present survey it appears that conservation of biodiversity should be focused on the preservation of small- to intermediate-scale interactions specific to a certain biome. There is usually not any need for consideration of native generalist species or biomewide interactions; generalists will take care of themselves under most circumstances. However, for most environments there is a definite need to consider potential invasion of generalistic exotic species. They often are inhabitants of agricultural lands; recent suggestions to reintroduce and conserve traditional weedy species should be treated cautiously, at least for temperate to tropical areas.

Instead, most effort should be devoted to detecting and preserving systems (or communities) of specialist or strongly interacting species occurring on limited spatial scales. However, it is often difficult to protect the specific relationships; we usually need some kind of "real estate" to make protection visible and legally valid. Thus, we need to establish the physical areas (patches) or base species supporting such ecological relations. Overlap in patchiness has to be considered; thus a landscape perspective may be more important than a focus on internal patches (or metapopulations). Isolated patch systems may not be viable outside the landscape context as connectivity between patches also has to be protected (Taylor et al., 1993).

We recognize certain close interactions and their physical bases at present. We can thus first select systems that are particularly representative for study and preservation. However, they may still be destroyed if there are some internal or external changes causing an increase in generalist species. Thus, consideration of patchiness and spatial patterns has to go hand-in-hand with surveys of generalist influences. Even if generalists are already strongly affecting specialized relationships in an area, there are still reasons to recognize functional and physical patchiness. If the physical patchiness is retained then an ecological restoration will be simplified in the future. An original patchiness will be an indicator of the potential value of a modified ecosystem (see Noss, 1990).

It will take a long time both to change emphasis and to establish the most important patchiness. In the meantime, large segments of important nature may be destroyed. With our presently limited knowledge, our main option will be to preserve really large chunks of land in remaining more or less pristine environments, including all the original patchiness and disturbances.

12

Reevaluating the Use of Models to Predict the Consequences of Habitat Loss and Fragmentation

Peter Kareiva, David Skelly, and Mary Ruckelshaus

Summary

Because habitat fragmentation is severe and widespread, it has become the focus of much conservation research. One particularly popular approach involves the development of spatially explicit population models (SEPMs) that are used to evaluate the consequences of different habitat arrangements. These landscape models typically emphasize the importance of habitat clustering to the viability of threatened species or the preservation of biodiversity in general. We caution that the data requirements of these models may often be prohibitive, and argue that alternative approaches should be explored. As examples, we use data sets involving the patterns of colonization and extinction among frog species inhabiting ponds in Michigan, and among ladybird beetles occupying patches of vegetation at Mount St. Helens. We conclude that more attention should be given to the inferences that might be obtained directly from simple monitoring data and to the possibility of alternative explanations that have little to do with fragmentation.

Introduction

Habitat reduction and fragmentation are widely decried as major threats to biodiversity. In several cases, this has led conservation biologists to advocate policies that mitigate the effects of habitat loss via some optimal clustering of remaining habitat (Liu et al., 1995). Although common sense tells us that loss of habitat leads to loss of species, to make subtler predictions regarding the merits of specific land management plans ecologists often turn to computer simulations in which the fate of populations is tracked as a function of landscape design (Liu et al., 1995). Such landscape models typically draw inspiration from the theory of metapopulations, which emphasizes notions of local population turnover due to extinction and recolonization, as well as the possibility of threshold levels of habitat availability, below which populations unavoidably plummet to zero (Hanski, this volume; Kareiva and Wennergren, 1995; Simberloff, this volume). Certainly there are useful insights to be embraced from the theory of metapopulations, source-sink models, and landscape ecology—and for well-studied species, one can find successful applications of spatially explicit models connecting landscape fragmentation to population decline. But for broader concerns regarding the link between biodiversity and habitat fragmentation (or patchiness), we believe two major unanswered challenges are (1) detecting the likely loss of species in fragmented landscapes before it is too late (without the luxury of detailed studies) and (2) examining hypotheses other than habitat fragmentation for species loss, so that spurious correlations between species' distributions and landscape attributes do not lead management policy astray. Before discussing how we advocate meeting these challenges, we first retrace the chain of research that has led us to our position. That chain begins with an investigation of the susceptibility of spatially explicit population models to error propagation. After raising the suspicion that many spatially explicit landscape models are impractical to parameterize, we ask whether more readily collected data in a changing landscape could provide early warnings of population collapse. This exploration relies on manipulative experiments using a small-scale "model system" of plants and associated insects rather than a "real" large-scale conservation problem. Finally, we explore a large-scale data set involving the presence and absence of amphibians in Michigan ponds over a two-decade period. Here we focus on alternative explanations for turnover in species occurrence—contrasting the explanatory power of variables emphasized by metapopulation and landscape theory with more mundane explanations associated with simple directional successional change.

The Data Demands of Spatially Explicit Behavior-Based Models for Species Preservation May Be Prohibitive

If one knew how animals moved around in search of suitable habitat, as well as their demographic rates inside and outside of such habitat, in principle one could predict the ability of different landscapes to support a persistent metapopulation. Thanks to Geographic Information Systems (GIS), detailed maps of landscapes and habitats make the mapping portion of this endeavor straightforward. The hard part is linking "dispersal" or "searching" success to landscape patterns. The link between dispersal and the geometry of habitat patches is the key to "landscape design," because it is this link that results in different landscapes with similar total habitat areas supporting different numbers of animals. Yet this crucial link is resistant to direct experimental manipulation (one cannot easily arrange old-growth forests in different ways and then track the success of owl populations). In lieu of experiments, conservation biologists typically invoke plausible models in which dispersing animals move more readily between close-by patches than between widely separated patches; then with some quantitative parameterization of this dispersal/distance relationship, these dispersal models are embedded in explorations of different landscape scenarios (e.g., Pulliam et al., 1992; Noon and McKelvey, 1992). Unfortunately, any attempt to translate the general trend of declining dispersal success with increasing distance traveled into quantitative predictions about the merits of different landscapes is fraught with errors. Indeed, because our knowledge of dispersal is typically poor, we suspect the errors may be huge. We used computer simulation to explore our suspicion—in particular, we asked if we know the correct dispersal model, how well do we have to estimate model parameters to obtain accurate model predictions?

Our analyses of error propagation used a simple random walk model of dispersing animals, with two parameters that need to be estimated: mortality while moving and total distance that could be moved. Obviously for more realistic models, more parameters would need to be known, and the opportunities for error because of poor parameter estimates would be further amplified. We simulated the fate of animals seeking habitat patches in each of a wide variety of landscapes, in which the proportion, patch size, and shape of suitable habitat were all varied (details are in Ruckelshaus et al., in press). After tabulating the fraction of searchers that successfully locate habitat, we asked what effect errors in input parameters had on the model's ability to predict dispersal success. To do this we assumed our animals still moved via a random walk, but that we had estimated their dispersal distance or mortality with some percentage of error (and then we ran the simulation using these "erroneous" parameters, thereby producing "erroneous" predictions). We restricted our simulations to landscapes that had less than one third of their habitat suitable—larger proportions of suitable habitat would tend to make dispersal less relevent, but such proportions are in excess of the actual values for species typically targeted by spatially explicit landscape models (Doak, 1989; Noon and McKelvey, 1992; Pulliam et al., 1992). Our answer as summarized in Figure 12.1 indicates that prediction errors are most exacerbated for dispering organisms that have low success in finding suitable habitat. Species of conservation concern are likely to fall squarely under conditions promoting huge prediction errors: they typically have low dispersal success because they are searching for habitat in increasingly fragmented landscapes. Unfortunately, gathering data on dispersal mortality and distance is extremely difficult and for few species do we have confident estimates for these dispersal parameters. It would appear then, that the notion of using detailed spatially explicit models of animals moving about in landscapes for quantitative predictions of dispersal success is futile in all but a few cases.

We are not arguing that information about landscapes is irrelevant, but rather that the information needed may be nonspatial, and hence much easier to work with. For example, instead of tracking the arrangement of old growth in the Pacific Northwest and modeling the movement of spotted owls (*Strix occidentalis*

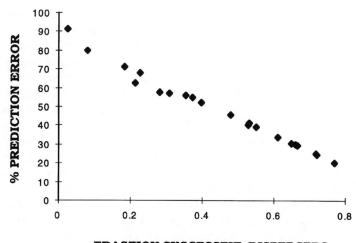

FRACTION SUCCESSFUL DISPERSERS

Figure 12.1. Prediction error increases as the fraction of successfully dispersing organisms decreases. Dispersal success (fraction of organisms that finds suitable habitat) is plotted for organisms dispersing in all landscape types with the lowest dispersal mortality error we examined in simulations (error = 2 percent).

caurina), the fate of owls may well be obvious simply by summing up the average fraction of landscapes that remains as old growth (Lande, 1988a), as well as the mean size of old-growth patches. Subtleties would be involved—but we expect much of the useful subtlety involves aptly weighting the quality of habitats and gathering information on an organism's response to habitat quality, rather than building spatially explicit population models.

Insect-Plant Interactions in the Mount St. Helens Blast Zone as a Model for Consequences of Habitat Fragmentation

In the aftermath of the Mount St. Helens volcanic explosion, large portions of the surrounding landscape are dominated by inhospitable pumice, dotted with small patches of fireweed (*Epilobium augustifolium*). Several species of insects make their living on these fireweed patches—either as specialized or generalized herbivores or as predators and parasites of those herbivores. For the last two years we have been manipulating the spatial distribution of fireweed patches as part of a study of patchiness and predator-prey interactions. The data we collect also provide insight into methods for the early detection of imminent species extinction due to habitat removal. It is worth noting that the duration of these manipulations, which ranged from one to several generations for the insects involved, would correspond to 20 or more years for "poster animals" such as spotted owls or grizzly bears.

One experiment has involved establishing 10 m by 10 m plots surrounded by a 3-m wide buffer zone that is cleared of vegetation. Inside the 100 m^2 plots, fireweed patches were experimentally removed to produce three different treatment levels with respect to number of remaining habitat patches (and because the patches are of similar size, total extent of fireweed habitat): 3, 5, or 10 patches. Our manipulation is thus a model for varying degrees of local destruction of patchy habitats. In 1993 there were only 5 replicates, but all of the patches were sampled in each plot. In 1994 there were 10 replicates for each treatment, and for each plot insects were censused weekly in the same three patches (the additional patches are part of the treatment, but to conserve labor were not sampled). In both years, the average number of species per patch declined significantly with patch removal, but there was a two-to-four-week delay between destruction of the habitat and the onset of a decline in the occurrence of species. One interesting question is whether the species lost as a result of habitat destruction could have been detected as "doomed" before their numbers actually declined? We found the answer to be yes for some species. For example, the ladybird beetle (*Coccinella septempunctata*) disappeared from plots that had been reduced to only 3 patches of fireweed habitat—but this disappearance did not become evident until six weeks after the original perturbation. We found, however, that well before this ladybird beetle's collapse was obvious, it could have been predicted using a simple metric scaling colonization rate by the number of occupied patches that could serve as a source of colonists (Fig. 12.2). Thus, monitoring *Coccinella* with simple censuses was not a good indicator of its ultimate fate in the face of habitat loss; but slightly adjusting the monitoring to include a measure of colonization success did prove to be a good predictor.

Although our observations regarding insects on fireweed do not speak directly to species of conservation concern, they do suggest some methodological options. First, since population responses to habitat destruction often will not be immediate, one cannot simply record abundances as an indicator of how well a species is doing—a species may be en route to doom, but not show this because densities have not yet "equilibrated" to the newly deteriorated conditions. More constructively, it may be possible to foresee a population's collapse by estimating the colonist pool and the rate at which vacant habitat is colonized—which simply requires repeated sampling of presence or absence in the same set of habitat patches. Such an index of population viability is likely to be much easier to obtain than are data on the dispersal behavior of individuals, which then would need to be fed into a model to generate population projections.

Amphibian Turnover in a Changing Landscape—Metapopulation-Type Regulation or Succession?

Amphibians have been heralded as the harbingers of the biodiversity crisis. However, several years beyond the earliest of these reports we still have little

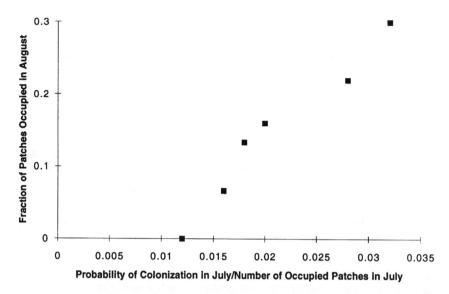

Figure 12.2. Predicting the delayed collapse of ladybird beetles following habitat destruction using colonization rates. Colonization rates are calculated as the weekly probability that a vacant patch of fireweed is colonized by a coccinellid divided by the total number of occupied patches that could be sending forth colonists. The result of habitat destruction is measured by the fraction of fireweed patches occupied by at least one ladybird beetle in August, a month after the colonization data were collected. Prior to experimental destruction of habitats, approximately 30 percent of the patches were occupied by ladybirds.

or no idea how widespread declines have been or why they have occurred (Pechmann and Wilbur, 1994). More generally, there has been little attention to the question of what determines the dynamics of amphibian distributions at large scales of space and time. In this respect, amphibians are probably typical of most organisms (Kareiva and Anderson, 1988). Until such large-scale studies are conducted we will have little idea of the relevance of metapopulation phenomena to real-world populations.

Amphibians should provide a good test case for the ability of metapopulation processes to explain patterns of distributional change because species with aquatic larval stages have discrete populations centered around breeding ponds that are connected by low rates of dispersal (Harrison, 1991a; Sjogren-Gulve, 1994). As an initial attempt to understand distributional dynamics of amphibians we analyzed a large data set collected from 32 ponds at the 540-ha E. S. George Reserve in southeastern Michigan. These ponds were surveyed for the presence and absence of amphibian populations for up to seven years between 1967 and 1974 (Collins and Wilbur, 1979), and again for up to five years between 1988 and 1992 (Skelly et al., in press). The amphibian assemblage of 14 species showed striking distributional shifts between the two surveys. In fact, the average species exhibited nearly 50 percent turnover in distribution among ponds. Despite the degree of overall change in where species were found, there was little indication of species declines (only one species showed a net decline of more than two populations). What factors can explain the marked changes in distributions? We examined three possibilities.

Pond Isolation

Metapopulation models usually assume that the processes of invasion and extinction are influenced by habitat isolation. Increased distance within fragmented landscapes is presumed to decrease the likelihood of a habitat being reached by dispersers that can colonize vacant habitats or bolster the population size of occupied habitats.

Succession

Alternatively, amphibian distributions could change because of changes in habitat quality. Amphibians are known to be specific in their habitat requirements. Some species are known to breed only within permanent ponds, whereas other species may breed in only small, nonpermanent woodland ponds (Conant and Collins 1991). Because succession in ponds and within the surrounding terrestrial habitat can alter important attributes of ponds, succession could lead to large changes in amphibian distributions within even short periods of time.

Null

The change in amphibian distributions could be the result of a random process of invasion and extinction.

We used a type of interacting particle model (Durrett and Levin, 1994b) to evaluate the abilities of the three different hypotheses to explain distributional shifts of amphibians at the E. S. George Reserve. The model was run separately for each of the 14 species. First the surveyed ponds were divided into three classes; rules for division were based on one of the three hypotheses. The isolation model divided ponds according to their distance from the nearest population at the time of the first survey: less than 80 m, 80 to 420 m, and greater than 420 m. These classes resulted in roughly equal-sized groups of ponds across all 14 species. The succession model divided the ponds based on permanence and forest canopy closure: open canopy ponds; intermediate, closed canopy ponds; and temporary, closed canopy ponds. Temporary ponds dried each year they were surveyed between 1988 and 1992; intermediate ponds dried some years but not others. Open canopy ponds included temporary, intermediate, and permanent ponds. Permanent ponds never dried during the five years between 1988 and

1992. The null model assigned ponds to classes arbitrarily. For the null model alone, ponds were reassigned among classes during each model run.

After ponds were assigned to classes the model calculated the rates of invasion and extinction within each class. Then starting with the initial species distribution, invasions and extinctions were assigned stochastically based on the estimated turnover rates for each pond type. For each pond, the stochastic prediction of presence or absence was compared with the actual species distribution during the second survey. The model tallied the number of "mistakes" made: the number of times the model assigned a population where none actually occurred *plus* the number of times the model predicted absence where a population was really present. The model was run 10,000 times for each species to generate a frequency distribution of model outcomes.

We employed the model in two ways. First, we examined the ability of the three hypotheses to explain distributional shifts for each of the 14 species by scoring the number of mistakes made in assigning populations among ponds during each of 10,000 model runs. Second, we combined information across all species to ask how well the hypotheses were able to explain changes in species richness of breeding ponds.

For 8 of 14 species, the succession model provided improved explanation of distributional shifts compared to isolation or null models (Fig. 12.3). The advantage of the succession model ranged from small to considerable: in the best cases,

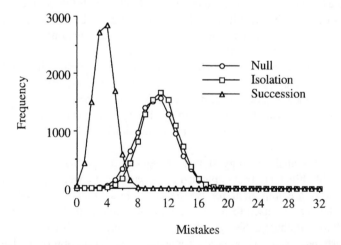

Figure 12.3. Frequency distribution of mistakes made by differing models in predicting the distribution of *Pseudacris crucifer* at the E. S. George Reserve, Michigan. During each of 10,000 runs the model used the initial distribution of populations among ponds plus turnover probabilities associated with a hypothesis for distributional change to predict the final distribution. Three hypotheses of the driving force behind invasion and extinction patterns were evaluated: (1) succession within and around ponds, (2) isolation of ponds from the nearest population, or (3) random variation (null model).

the succession model made roughly two-thirds fewer mistakes than either the isolation or null models. For only one species, *Ambystoma maculatum*, did the isolation model provide the best explanation (but the improvement relative to the null model was minor—only 10 percent fewer mistakes). And finally, for the remaining 5 species the null model was at least as good at explaining distributional changes as either alternative model.

Our interpretation of these patterns is improved when information about the natural history of the organisms is considered. In particular, species for which the succession model provided the best explanation of distributional changes are known to use temporary and intermediate ponds; some are also known to be intolerant of closed canopy conditions (Conant and Collins, 1991; E. E. Werner, personal communication, 1995). Within this largest group, the pond succession model was most successful at explaining distributional shifts of widespread species that are intolerant of canopy closure (e.g., *Pseudacris crucifer*), and less successful at explaining shifts experienced by less common species that were either never found in ponds where canopies closed (e.g., *Notophthalmus virides-cens*) or are relatively tolerant of closed canopy conditions (*A. tigrinum*). In addition, the single species for which the isolation-based model provided the best explanation, *A. maculatum*, is a notably poor disperser. Cortwright (1987) found that among an amphibian assemblage in Indiana, *A. maculatum* was slowest to colonize newly created ponds. Finally, the five species for which the null model was equal to the alternative models tended to be those that use more permanent aquatic habitats that are usually too large to be overtopped by forest canopies and that dry up much less frequently than smaller ponds. These species tended to have more stable distributions, overall.

We also compared the abilities of these hypotheses to explain the final species richness of breeding ponds (Fig. 12.4). The model based on successional changes was almost twice as good as either the null or isolation models at explaining species richness during the second survey. The typical pond contained 5 species during the second survey; on average, the succession model missed actual richness by roughly 1 species.

Overall, our analyses suggest that changes in amphibian distributions at a reserve in Michigan are best explained by successional changes in pond habitats. These results are important for two reasons. First, they show that presence-absence data can be used to resolve among plausible hypotheses for the causes of distributional change. Second, the results are surprising because few animals are superficially more appropriate for a metapopulation approach than pond-breeding amphibians. The current generation of such metapopulation models are practically uniform in their assumption that isolation has strong effects on rates of invasion and extinction. In our analyses, however, isolation among ponds generally provided a poor explanation of changes, doing no better than a random model for most species.

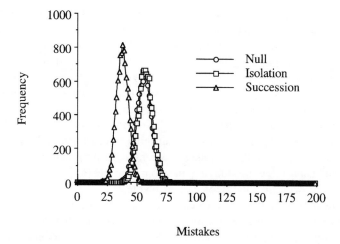

Figure 12.4. Frequency distribution of mistakes made by models predicting amphibian species richness across 32 ponds at the E. S. George Reserve based on hypotheses that distributional change is driven by succession, habitat isolation, or is random (null model). Each form of the model combines responses from all 14 amphibian species to compare the abilities of the three hypotheses to explain patterns of species richness. The models use the initial distribution of species among ponds (measured 1967-1974) plus turnover rates generated from one of the three hypotheses to predict the final distribution of each species and then calculate the deviation between predicted species richness and actual species richness (measured 1988-1992).

What Is the Connection Between the Nonequilibrium Emphasis of Contemporary Conservation Biology and Models Regarding Habitat Destruction?

Our modern appreciation for the importance of turnover, population fluctuations, disturbance, and perhaps even chaotic dynamics implies that little information can be gleened from static surveys. This very dynamic view of nature requires a great deal of information to generate quantitative predictions, and is replete with traps that may produce false conclusions. As we embrace this nonequilibrium view, we need to be very careful to ask how terribly wrong we might be when we apply it to real data and practical questions. In this chapter we offer three examples of *how wrong we can be*. First, when analyzing detailed landscape simulations that include algorithms describing animal movement, the possibility of accurately modeling dispersal success given the quality of existing data must be viewed as remote. Second, if the pace of habitat destruction is rapid relative to the speed with which a population equilibrates to its landscape, then we need to explore alternatives to census data alone as an indicator of a population's well-being. We suggest that a particularly useful metric may be the rate at which vacant patches are colonized, scaled by the number of sources for colonists. This colonization metric reinforces the importance of "unoccupied habitat," which is discussed in the chapters by Hanski and by Simberloff; indeed, according to our view, if one looked only at patches occupied by the species of concern, it would be impossible to anticipate impending doom due to habitat loss. Finally, we saw that for amphibians which would seem to be the perfect candidate for "metapopulation theory," a simple model based on directed successional change best explained the bulk of the distributional shifts recorded over a 14-year period.

When thinking about the mistakes we can so easily make, we should also think about approaches to studying the consequences of habitat loss that are commensurate with the quality of data available.

We believe that the best data we can expect to obtain consistently are records of presence and absence in successive time periods in particular locations. It will be especially important to develop methods of analysis and modeling that focus on presence/absence data when trying to anticipate effects of environmental change. These methods may not need to be spatially explicit, but could instead emphasize relationships between species presence and indicators of habitat quality. Of course, this advocacy of habitat description and analysis spurns many of the more modern ideas in conservation biology (such as metapopulation thresholds, source-sink dynamics, and so forth). But practical concerns—that is, what data are available and an appreciation for multiple explanations of species loss—need to play a larger role in conservation research. Parsimonious explanations for patterns of species decline should be our goal; present approaches to research are too narrowly aimed at theories currently in vogue.

Acknowledgments

We would like to thank Cindy Hartway and Eli Meir for help with the programming, and Ingrid Parker and Ellie Steinberg for editorial comments on the manuscript.

13

Managing for Heterogeneity and Complexity on Dynamic Landscapes

Norman L. Christensen, Jr.

Summary

If spatial and temporal heterogeneity and complexity are critical elements in the function of ecosystems, then it is important that the processes that maintain such heterogeneity and complexity be maintained, even managed. Such processes are often highly variable and unpredictable in their frequency, intensity, and spatial behavior. This variability presents challenges in setting management objectives, developing and executing management protocols and techniques, and evaluating their success.

I have three general messages to convey in this chapter. First, heterogeneity, complexity, and diversity are critical elements to sustained ecosystem function. Natural and, for that matter, human disturbances and the successional processes that derive from them are a critical component of that heterogeneity and complexity. Second, we have created a world in which human influences on and the manipulation and management of natural disturbance and ecological change are ubiquitous and inevitable. Finally, management requires explicit operational goals (Rogers, this volume), the means to know whether management interventions (or lack thereof) are achieving those goals, and the institutional structures that can and will adjust and adapt management techniques and protocols if they are not, or move the goalposts if new data or information suggest that we should.

To set a framework for discussion I first review a few basic elements of ecosystem management. Second, I consider, as a case study, the challenges posed by the management of natural disturbance in wilderness or wildlands. Finally, I consider how, or if, we can know we are managing it right.

Elements of Ecosystem Management

In a recent review on the subject, a committee of the Ecological Society of America (Christensen, et al., in press) identified over a dozen unique definitions for ecosystem management, reflecting differences in the character of resources being managed, the mission of the management organization, or particular philosophical or political agendas. In several of these definitions, ecosystem management is assumed to be nearly synonymous with management for complexity and diversity.

Although maintenance of complexity and diversity are critical components of nearly all definitions of ecosystem management plans, the following definition captures more fully the essence of this approach to natural resource management. Ecosystem management is *management whose goals, policies, and practices are adaptable and based on our best understanding of the ecological interactions and processes necessary to sustain ecosystem function.* Seven themes or elements are central to ecosystem management plans:

- explicit goals set by the capacity of ecosystems to deliver goods and services;
- sound ecological models grounded in ecological principles and understanding of processes and interconnections;
- acknowledgment of the importance of complexity and diversity to ecosystem function;
- acknowledgment of the importance of context and scale;
- recognition that ecosystems are dynamic and that sustainability does not imply maintenance of the status quo;
- realization that humans are ecosystem components, not only as challenges to sustainability but as elements that must be engaged to achieve sustainable goals; and
- management systems that are adaptable and accountable.

Explicit Operational Goals

The need for explicit operational goals seems obvious, after all are not all management plans goal-driven? Although this may be true, such goals are often cast in terms of the goods and services (e.g., natural resources, recreation, water quality, etc.) rather than the ecosystem processes necessary to deliver those goods and services. In the worst-case situations, goals for levels of these goods and services are set by need or desire, rather than a firm understanding of what a system sustainably can provide. Thus, ecosystem management plans must be anchored in objectives for the ecosystems structures and processes necessary to sustain key functions.

Sound Ecosystem Models

By models, I refer to constructs that enable us to map or predict the behavior of one part of a complex system based on knowledge of other parts. Ideally, such models are based on a fundamental understanding of the causal connections among such parts; however, in practice, predictive models are often constructed based on empirically derived correlations whose causal relationships may not be fully understood. Models take the form of simple compartment diagrams that provide a means of organizing information or expressing connections and relationships, or they may be developed as complex computer simulations that enable us to depict processes operating through time and across landscapes.

Complexity and Diversity

Much of exploitive natural resource management strives to simplify ecosystems to maximize what it is humans need or want. Monoculture management with genetically uniform stock in agriculture, aquaculture, and forestry represents the extreme of this strategy. On a general and somewhat superficial level, we might well conclude that this strategy has worked pretty well. Examined more closely, we learn that such management carries significant risks and that simplicity often comes at a significant cost.

In retrospect, it is clear that the debates during the 1960s among ecologists over the connections between diversity and stability may have shed considerable darkness on this subject (cf. Woodwell and Smith, 1969). The debate often pitted those impressed with the "holistic" or "emergent" properties and functions of ecosystems (e.g., Margalef, 1968; Odum, 1969; Savory, 1988) against so-called "reductionists" focused on the fact that the evolutionary mechanisms that produce diversity derive from benefits to individual organisms or even individual genes within a species (e.g., Harper, 1967; Simberloff, 1980).

We now understand that diversity may impart important attributes to ecosystems that may not be tied to the mechanisms of evolution at the level of individuals, and these seemingly conflicting viewpoints have largely been reconciled over the past decade. It is widely agreed that biological diversity contributes to ecosystem sustainability (e.g., Tartowski et al., this volume) in at least five ways.

Different Species Play Very Different Functional Roles.

The functional importance of species within ecosystems has long been recognized and was the basis for the distinction between an organism's habitat (its "address") and its ecological niche (its "job"; Elton, 1927). Functional diversity (e.g., producers, decomposers, consumers, nitrogen fixers, sulfur reducers, etc.) within ecosystems is clearly important, but there is not a "one-for-one" overlay of diversity and function, and notions like "functional redundancy" do not fully describe the role that diversity plays in maintaining function across complex environmental gradients through space and time, as explained later.

Productivity and Efficiency of Material Transport Is Increased by Diversity.

The importance of diversity to the efficient transfer of nutrient and energy is well known. Niche differentiation, such as seen in patterns of root development among prairie grasses (Weaver and Fitzpatrick, 1934), clearly influences efficiency of utilization of resources such as soil water and nutrients. Diversity is directly tied to trophic level feeding efficiency in a variety of ecosystem types (e.g., Nott and Pimm, this volume).

Diversity Provides Resistance to and Resilience from Perturbation.

A variety of studies have emphasized the important connections between diversity and ecosystem resistance to and resilience from a variety of perturbations. Diversity-related resistance that is particularly relevant to the management of agricultural and forest ecosystems with respect to the minimization of spread of species-specific pathogens and "pest" insects. Although monocultures may result in high levels of production of specific products or resources, they present much higher risks from such infestations than more complex systems (e.g., Schowalter and Turchin, 1993). The importance of species diversity to the ability of ecosystems to recover ecosystem function following a disturbance or perturbation has been convincingly demonstrated in long-term studies of productivity responses to drought in grasslands (Tilman and Downing, 1994).

Diversity Provides Adaptability to Long-term Change.

Given everchanging environments, the capacity to adapt is central to the long-term sustainability of ecosystem function. Such changes are obvious in the shifts of species' importance documented in long-term pollen profiles. Relatively unimportant species restricted to particular microsites during one climatic regime may become relatively important and more widespread as the climate shifts (Davis, 1985). The reservoir of genetic diversity within individual species and populations is clearly central to their ability to adapt to environmental change (Antonovics, 1968). In view of this, focus on so-called "improved" genotypes of crop plants and forest trees has raised concern regarding the loss of genetic diversity that might be important to future conditions (Holsinger and Vitt, this volume).

Changes in Species Diversity Can Represent a Very Sensitive Indicator of Environmental Quality.

The loss of diversity, particularly the component composed of relatively rare species, is often an early indicator of ecosystem decline.

Connections and interdependencies among species are complex and not easily predicted, thus, these relationships are by no means simple, linear or universal. In view of this, it is not especially productive to ask how much diversity is enough. Our knowledge of the complex web of interactions that comprise an ecosystem is far too limited to predict the specific outcomes of extinction of even very rare species. Loss of individual species may have only marginal impacts on ecosystem function at a given time, but greatly diminish long-term sustainability (Christensen et al., 1996).

Finally, the numbers and relative abundances of species are not the only important components of ecosystem complexity. Biologically derived structures such as coral reefs and standing and fallen wood are critical to many ecosystem functions (e.g., Franklin, 1993b). Here again, historic management practices, particularly in forests, have had the effect of simplifying or removing entirely these elements. In the past, for example, sivilculturalists removed fallen logs and woody debris from stream channels and riparian zones in forested ecosystems managed for wood fiber. It is now well known that such material (i.e., structural complexity) moderates hydrologic flows and provides critical habitat for many invertebrate and vertebrate aquatic animals (e.g., Gregory et al., 1991).

Context and Scale

Ecosystems depend on processes that operate over a range of spatial and temporal domains. Typically, ecosystem scientists define the boundaries of the system that they study so as to measure or manipulate most easily those processes of greatest interest. Ideally, the jurisdictions of management would also coincide with the behavior of the ecological processes most central to management goals.

The incongruence of jurisdictional boundaries for natural resource management with the temporal and spatial scales of ecological processes presents perhaps the most daunting challenge to ecosystem management (e.g., Pringle, this volume). Consider, for example, how often rivers form the boundaries between ownerships, counties, and countries. A river may have particular strategic advantages as a boundary, but management of water-driven ecological processes is certainly not one of them.

It is tempting to blame the flagrant mismatch of scales of management and ecological process on ignorance or misplaced priorities. Even with the best information and intentions, however, management will continue to be plagued with the "tyranny of boundaries." This is the case simply because the scales appropriate for the management of one process are not necessarily appropriate for the management of others. Thus, watersheds represent an excellent unit for the management of processes connected to the hydrologic cycle, but do not necessarily define appropriate units for management of fire or wildlife and even some processes tied closely to streams.

Ecosystems are Dynamic

Much research, as well as mismanagement, has taught us the futility of equating sustainability with maintaining the status quo. Management determined to "freeze" ecosystems in a particular state has generally proven to be futile and unsustainable. For example, the suppression of wildfires in ecosystems as diverse as chaparral and mixed conifer forests has resulted in widespread accumulation of fuels and larger and more intense fires than would have occurred in the absence of such suppression. Thus, in seeking to define the "desired future condition" for an ecosystem, we must view "condition" as a dynamic concept or, better yet, think in terms of "desired future trajectory" (Christensen et al., 1996).

Early studies of ecosystem change or succession focused on its directionality and predictability, and phrases such as "desired future trajectory" seem to reinforce this deterministic view. Disturbances and the patterns of change that derive from them, however, are often caused by unpredictable phenomena or "forces," with the result that there is always a high level of uncertainty related to patterns of ecosystem change. Given sufficient time and space, uncertain events are likely to happen; surprises are inevitable. Although increased knowledge can diminish the element of surprise, it cannot eliminate it. To be sustainable, management must include a margin of safety for uncertainty. Ecosystem management is not a strategy to eliminate uncertainty. Rather it acknowledges its inevitability and accommodates it (Christensen et al., 1996).

Because the world is constantly changing, one might argue, and indeed some have argued, that human disturbances are in some sense "natural." So far as we can tell, at no time in the past has the Earth experienced change at the rate at which it is currently occurring (Likens, 1992). Furthermore, many changes such as extremes of land fragmentation and certain kinds of pollution have no precedent in the Earth's evolutionary history. It is a tremendous act of faith to suppose that ecosystems can and will respond to these novel insults as they would to natural disturbance processes to which they evolved over millions of years. If our faith is misplaced, the consequences may be disastrous, both in terms of the integrity of ecosystems and their ability to produce sustainably the goods and services on which we depend.

Humans as Ecosystem Components

Human impacts on ecosystems are ubiquitous; certainly no terrestrial ecosystem has escaped the effects of human activities. Whereas, humans may present some of the most significant challenges to sustainability, they are also integral ecosystem components who must be part of any effort to achieve sustainable management goals. Given the growth in human populations, sustainable provision of ecosystem goods and services becomes an even more compelling goal (Cohen, Epilogue, this volume).

As Christensen et al. (in press) pointed out, "many of our most celebrated 'environmental trainwrecks' are not disputes between the 'rights of nature' versus the human demands for resources; rather, they are conflicts among competing human demands." Conflicts may result when practices associated with extraction of one commodity influence the availability of another, as for example, in the effects of logging practices on breeding habitat for migratory salmon. Conflicts also result as a consequence of our limited ability to reconcile human wants and needs across scales of space and time. Thus, the rights of individual property owners on increasingly fragmented landscapes may be in conflict with the collective vision of those same property owners for the entire landscape, and our need to meet the constraints of fiscal years and tax schedules may conflict with our wish to guarantee to our children the same opportunities we enjoy.

Throughout this volume, the need for management at temporal and spatial scales that match those of key ecosystem processes has been emphasized (e.g., Meyer, this volume). Nevertheless, the growth of human populations and the ever increasing complexity and fragmentation of land ownership and resource stewardship responsibility make such management increasingly difficult. Although we have not yet succeeded in placing a price tag on ecosystem processes and functions that are not typically bought and sold in traditional markets (Harte, this volume), the value that society places on such things is certainly influenced by the laws of supply and demand. Wild places and the biological diversity they contain, air and water quality, and the beauty of our landscapes are increasingly valued by society, not only because of our increased awareness of their importance to sustained ecosystem function and environmental quality but also because of their diminished supply. Nevertheless, we are groping to reconcile these large-scale values with aspirations and goals we set at smaller spatial scales (Gordon et al., this volume). The development of mechanisms and institutions devoted to this reconciliation must receive a high priority.

Adaptability and Accountability

Given the uncertainties already described, it is essential that management systems be adaptable. They must be adaptable to variations in environment (including the impacts and needs of humans) from location to location. They must also be adaptable to inevitable changes in those environments through time. Most important, management systems must acknowledge the provisional nature of our models and information base and be adaptable to new information and understanding (Holling, 1978; Walters, 1986). To be adaptable and accountable, management objectives and expectations must be explicitly stated in operational terms, informed by our best models and understanding of ecosystem function, and tested by carefully designed monitoring programs that provide accessible and timely feedback to managers.

As Lee (1993) argued, we must see many of our management protocols and practices as experimental and devise the means to test the hypothesis that these interventions are, in fact, moving the managed system toward intended goals. In the context of ecosystem management, the question of whether management practices are truly sustainable should be central. To conduct ecosystem management, we must clearly explicate objectives in operational terms relevant to sustained ecological function, develop monitoring programs focused on data relevant to those operational objectives, provide for efficient analysis and management of data, and encourage timely feedback of information from research and monitoring programs to managers (cf. Rogers, this volume).

Monitoring focused on operational goals for ecosystem structures and processes is critical to adaptability and accountability. Monitoring programs of one form or another are by no means new; however, three elements, focus, efficiency, and commitment, require special attention if they are to inform management in a meaningful way. Monitoring should be focused on management expectations (operational objectives) and designed to test the success and efficacy of management practices. Efficiency refers to the need for rigorous statistical sampling designs with attention paid to issues of precision and bias in data gathering. Because management situations often offer limited opportunities for replication or are often biased by patterns of ownership and accessibility, sampling designs will often be flawed. Such flaws should not be taken as an excuse to avoid monitoring, but their likely impacts on data quality and uncertainty of conclusions must be explicitly evaluated.

A long-term vision and commitment to the development and maintenance of monitoring programs is critical. Such programs necessarily add cost and can be especially difficult to maintain where personnel turn over frequently. Thus, clear identification of target objectives for monitoring is important; "shotgun" approaches may miss key variables, while incurring the unnecessary cost of gathering irrelevant data.

The community of academic scientists has much to offer the managers with regard to the design and execution of monitoring programs. Research is needed to refine models that will help identify key variables for monitoring, as well as in the design of cost-effective, but nevertheless efficient, sampling protocols. Standards for data gathering are well developed in some areas (e.g., hydrology and climate) and nonexistent in others (e.g, biodiversity or site fertility). The development of such standards should be a high priority for agencies funding research relevant to management.

The adaptive management loop will be closed only if there is timely feedback of results of monitoring information, as well as new insights from basic research programs, to managers. Such feedback will require institutional change in public agencies charged with natural resource stewardship (Peters et al., this volume), and may even require legislative initiatives to eliminate impediments to the exchange of information. Cultural barriers that have traditionally separated communities of managers from those of researchers must be broken down. This will require changes in modes of communication, as well as changes in reward systems for both communities (cf. Zedler, this volume).

Managing for Disturbance and Heterogeneity in Wilderness Preserves[1]

> The National Park Service shall conserve the scenery and the natural and historic objects and the wildlife therein and to provide for the enjoyment of the same by means as will leave them unimpaired for the enjoyment of future generations.
>
> National Park Service Organic Act, 1916

Ecosystem Change and Wilderness Parks—A Case Study

Although not explicit in any printed policies or protocols, management of wilderness preserves a century ago was clearly seen as operationally equivalent to museum curation on a grand scale. As suggested in the Organic Act of 1916, we were preserving "objects"; in the words of the 1963 Leopold Committee Report (Leopold et al., 1963) we were "preserving vignettes of primitive America."[2] If the goals were clear, the means by which we should achieve those goals seemed equally obvious; the mere setting aside of such wilderness areas seemed sufficient.

When the borders of most of our largest wilderness parks and preserves were established and fixed, the North American wilderness was envisioned as a vast array of climax communities, the distribution, structure, and function of which were determined primarily by climate. "Determined" is a key word here. Ecosystem recovery from disturbance was viewed as deterministic, following an inexorable and inevitable path to climatic climax. The critical biodiversity of landscapes was, therefore, reckoned to be contained in their array of climax communities.

Disturbances such as fire were viewed as negative, preventing ecosystems from attaining or maintaining their climax state. In his essay on "Experimental Ecology in the Public Service," Frederick E. Clements (1935), one of the architects of National Park Service policy, argued that fire and other large-scale disturbances were not natural phenomena in the "great climaxes of North America." Furthermore, the emphasis of Clements and many of his contemporaries on the role of dominant species in shaping climax communities provided a scientific justification for management that focused on the behavior and welfare of the most abundant and charismatic species.

Thus, a century ago, the challenge of maintaining "tree museums" seemed a simple matter of delineating preserve boundaries and keeping disturbance espe-

---

[1]Some of the material in this section also appears in Christensen (1995b) and is reproduced here with the permission of the *International Wilderness Journal*.

[2]In fairness to the Leopold Committee, their vision of these "vignettes" was quite dynamic. For example, it was the Leopold Committee that first called the public's attention to the importance of natural fires and the issues of fuel accumulation in the forests of the western Cordillera.

cially fire out. Issues of spatial scale, landscape pattern and heterogeneity, and ecosystem process were only dimly, if at all, understood. The biological diversity of interest was that contained in climax communities, with a particular bias toward dominant organisms. All of this seemed consistent with depictions of wilderness parks as "pleasuring places" and their dedication for "the benefit and enjoyment of the people."[3] If the potential paradoxes of "natural wonder conservation" and human "enjoyment" occurred to anyone, no one seemed especially concerned.

In the case of our public lands, our ignorance of the role of humans in the development of wilderness and of their capacity to alter that wilderness in subtle and not so subtle ways, permitted us to institutionalize profound conflicts between wilderness ecosystem management and obvious legislative mandate that such areas be available for the enjoyment of humans. With undeserved certainty about our understanding of these ecosystems and the task of maintaining them, preserves were established and management protocols were developed with no explicit provision for the acquisition of new knowledge about the functioning of these ecosystems. Certainly there was no mechanism for the systematic incorporation of new understanding into wilderness management protocols (NRC, 1992).

Nowhere have the impacts of the limits of our understanding of wilderness ecosystems come back to haunt us with greater vengeance than with regard to the role of wildland fire. Indeed, fire management serves as a paradigm for the most daunting issues in wilderness management. Fire management during the first half of this century was focused on prevention and suppression, and operationally expressed in the so-called "10 A.M. policy:" any fire started on one day should be out be 10 A.M. of the next (Pyne, 1982).

By the late 1950s, a vast body of evidence contradictory to the Clementsian view of succession to climatic climax had accumulated. Tree ring fire scars dating back hundreds, even thousands, of years reveal that fires were recurring regularly in many ecosystems prior to European colonization (Swetnam et al., 1993). Indeed, studies of lake sediments have extended fire chronologies back tens of thousands of years and demonstrated a clear connection between climate change and naturally occurring wild fires. In some cases, these fires were clearly connected to the activities of Native Americans. However, in other situations they were the consequence of lightning.

Early in this century, ecologists discovered that many plant species were adapted to fire. Thick, nonflammable bark in many conifer species imparts resistance to fire damage, and belowground burls and protected epicormic buds facilitate rapid recovery following fire. However, in other cases, adaptations not only provide *resistance* to the negative effects of fire but actually result in species' *dependence* on fire for successful reproduction. Such adaptations include the

---

[3]Phrases from the 1916 Organic Act and the implementing legislation creating Yellowstone National Park in 1872.

serotinous cones of many pine species, heat-stimulated germination in many chaparral shrub species, and fire-stimulated flowering in many prairie and savanna species (Keeley, 1981).

In addition to its direct importance to many species, ecologists discovered that fire plays an integral role in the functioning of many ecosystems (Ewel, this volume). For example, fire greatly influences the cycling of nutrients, often increasing nutrient availability to immediate post-fire pioneer species. In regions where climate or nutrient availability limits the decay of woody debris, fire is a major agent of organic decomposition. Indeed, in such situations, fire may be viewed as virtually inevitable, with the rate and pattern of fuel accumulation regulating the frequency and behavior of wildland fire. Thus, contrary to the conventional ecological wisdom up to the early 1970s (e.g., Odum, 1969), succession does not necessarily lead inexorably to increased stability.

The historic lack of understanding of the role of fire in wilderness ecosystems, coupled with reactions to such devastating fire events as the 1910 fires in northern Idaho and western Montana and the Tillamook Fires in Oregon during the 1930s, led to management strategies focused on fire prevention and suppression. In many cases, such strategies resulted in dramatic changes in ecosystem structure and function. For example, in the sequoia mixed conifer forests of the central Sierra Nevada, a century of fire suppression resulted in diminished reproduction of the giant sequoia, and increased invasion of shade tolerant species such as white fir and incense cedar. By suppressing or preventing fires, flammable fuels had accumulated on many landscapes to levels considered by some to be outside the range of "natural" (e.g., Leopold et al., 1963; Harvey et al., 1980).

In the late 1960s and early 1970s, recognizing the importance of fire in wilderness ecosystems, the National Park Service and the Forest Service began programs to "restore fire to its natural role."[4] Although an unfortunate misnomer, such programs were often characterized as "let burn" management. In reality, two classes of fire management were initiated. Planned ignition fires were systematically set within predetermined boundaries, and natural ignition fires took advantage of lightning-ignited fires burning within predetermined prescriptions. A key point is that both kinds of fire were "prescribed," that is, allowed to burn only so long as the weather, fire intensity, and fire size remained within preset parameters.

What Have We Learned?

Change is Constant.

In the words of Henry Chandler Cowles (1899), "succession is a variable converging on a variable." Climate change has been constant and has resulted

---

[4]From the U.S. Department of the Interior and U.S. Department of Agriculture Final Report on Fire Management Policy (1989).

in constantly changing patterns of ecosystem disturbance and recovery (e.g., Delcourt and Delcourt, 1991). Thus, we should not study past patterns in the hopes of recreating them, but rather to help us understand how change will determine the patterns of the future.

Change is Complex.

Patterns of change are neither perfectly cyclic nor linear. Rather successional transitions are often complex and patterns of disturbance and recovery are often greatly affected by "chance" events, that is, phenomena such as variations in weather that are controlled by factors external to the system being managed.

The steady accumulation of fuel during succession may result in a predictable increase in the likelihood of fire, but the exact timing and behavior of individual fires are far less predictable owing to variations in climate, weather, and human behavior (Christensen, 1991). Furthermore, the unique patterns of climate that follow any particular fire will likely result in patterns of ecosystem development that are quite different from successional changes occurring at other times.

Human Impacts Are Historically Important and Ubiquitous.

Human interventions in fire regimes and patterns of ecosystem recovery have long historic precedent and are today ubiquitous and inevitable. The romantic vision of wilderness as "nature free of human intervention" has about as much meaning to managers as the concepts of the frictionless plane and the ideal gas have to physicists. Although the details are poorly understood, no one doubts that *Homo sapiens* has significantly altered disturbance regimes and wilderness ecosystems over evolutionary time scales. Across much of North America, Native Americans increased fire frequency by supplementing ignition sources. These changes altered fire behavior and patterns of postfire ecosystem development. It is important to note that the specific patterns of fire used by aboriginal peoples varied considerably through time and from place to place (Pyne, 1982). Furthermore, these variations were very likely influenced by changes in climate and cultural traditions.

Today, direct and indirect human impacts on fire frequency and behavior are obvious. In many areas, humans have provided increased sources of ignition resulting in increased fire frequency. On the other hand, active fire suppression has led to fuel accumulation and resulted in fewer but higher-intensity fires (e.g., Minnich, 1988). Because of the dissection and fragmentation of landscapes, there are few places on the Earth where fire behavior has not been altered.

The inevitability and ubiquity of human intervention are abundantly clear with regard to impacts on our atmosphere. Increasing atmospheric carbon dioxide may result in climatic changes that could influence fire regimes over very large areas.

On a more local scale, urban impacts on air quality influence patterns of tree growth and survival affecting patterns and accumulation of fuels. The results are altered fire behavior and patterns of ecosystem recovery.

Patterns of Disturbance and Recovery Are Uncertain.

The considerable uncertainty associated with determinants of fire behavior and patterns of ecosystem recovery arises from two sources. The first is ignorance; if we only knew more we would be able to make more precise predictions about fire behavior and postfire response. However, the second source of our uncertainty is a direct consequence of the nonlinear relationships among processes that regulate fire behavior and ecosystem change. The complex, chaotic character of fire regimes suggests that there are very real limits to our ability to predict specific behaviors and outcomes. Thus, there will likely always be limits to the precision of our predictions and inevitable surprises regarding fire occurrence and behavior. Although managers should strive to improve our knowledge base, management policies and protocols will not eliminate untoward surprises. Management must acknowledge that surprises will occur and be certain that our preserves are sufficiently buffered not just to endure them but in many cases to prosper from them.

Ecosystem Change—Challenges for Management

Setting Operational Goals.

When they were first initiated, the overall goals of fire management in wilderness areas seemed relatively simple—that is, "to restore fire to a more natural role" (USDI National Park Service, 1978). But, what is natural? The Fire Management Policy Review Team (1988) defined natural as "those dynamic processes and components which would likely exist today and go on functioning, if technological humankind had not altered them." Putting aside the implication that Native Americans lacked technology, this statement seems to suggest that if natural processes are simply allowed to operate, ecosystems will converge to some preferred state. Although the details are far from clear, we now understand that such convergence on "unmanaged" landscapes is unlikely.

There is much that we have to learn about the causes and consequences of variability in fire regimes. For example, the Yellowstone fires taught us that models of fire behavior are not easily transferred among ecosystems (Schullery, 1989). Even within a landscape, interactions between climate and fuels may result in multiple thresholds of fire behavior (Turner et al., 1994). The consequences of variation in such behavior are also poorly known. Fire often results in a pulse of resources that, although ephemeral, may greatly influence patterns of species establishment. The variability in such pulses may have much to do with the biodiversity of landscapes throughout the fire cycle.

Given such variability, it is critical to understand that fire or other disturbance processes cannot, in and of themselves, be the primary goals of management. We do not conserve areas to burn them or otherwise disturb them. Rather, we manage disturbance, for example, we suppress or prescribe fires, in recognition of the key role they play in the sustainability of such key ecosystem elements and processes as mineral cycling, accumulation of fuels and woody debris, hydrology, and species diversity (Christensen et al., 1989). Thus, our goals must be set explicitly with regard to such elements and processes.

Goals for disturbance management must be realistic and "doable." This may seem obvious, but actually represents important challenges with regard to fire. Managers are able to prescribe and manage fires of low intensity and small spatial scale with great precision. However, high-intensity fires in shrublands and forests often defy prescription and control.

For example, natural-ignition prescribed programs allow fires set by natural causes to burn so long as they are within prescribed guidelines. In a sense, such fire management programs substitute knowledge for intervention (Christensen et al., 1989). They assume that threshold levels of fire behavior can be established beyond which fires can and should be suppressed. Serious questions remain as to whether such fire programs are either practical or natural. For example, such plans may deny important, albeit intense, fire events from landscapes and, given the extent of landscape fragmentation and alteration, it is unlikely that fire regimes developed in this manner will simulate the full range of natural processes that would have occurred on pristine landscapes.

Fire management plans for some larger parks and wilderness areas have included large, high-intensity, stand-replacing fires, however, the full range of fire behavior in small wilderness areas is not possible because of the risk of escape. Over 70 percent of Forest Service wilderness areas in the West are smaller than 100,000 acres, and too small to permit naturally ignited prescribed fire programs.

Only in the largest wilderness areas can fire prescriptions be sufficiently broad as to include the full range of natural fire intensities and behaviors. Even then, fuel accumulation resulting from past suppression may first require "restoration" burns. Wildfires that burn out of prescription will occur, and fire management plans must have clear guidelines with respect to specific postdisturbance interventions. One might argue that such interventions following fires in wilderness that burn within the historic range of natural variation should be unnecessary. Nevertheless, pressures for postfire mitigation are often strong where fires in a small wilderness may have effects on adjacent lands. Guidelines for such postfire mitigation must include and be explicit about the wisdom and need for such measures as erosion mitigation, reforestation, and wildlife management interventions. Such postfire interventions should be judged with regard to the benefits of the intervention relative to their environmental and monetary costs, and an evaluation of the likelihood of their success.

The Importance of Sound Ecological Models.

Ecological models are critical to connect management prescriptions for natural disturbance to key ecosystem elements and processes. Nevertheless, we face the reality that relationships between key ecosystem processes and variations in disturbance frequency, intensity, and spatial scale are poorly understood and models are crude at best.

The importance of connections between ecological models and management plans is illustrated with the use of prescribed fire in giant sequoia-mixed conifer stands in the Sierra Nevada. For over two decades, low-intensity surface fires have been set in such stands to restore natural fuel conditions and to encourage reproduction in the giant sequoia. However, in areas where fuels have accumulated owing to fire suppression, low-intensity fires may have the effect of actually increasing the amount of dead woody debris on the forest floor, and models of the reproductive biology of giant sequoia suggest that successful reproduction depends on relatively localized (gap-phase) high-intensity fire events that are considerably outside current management prescriptions (Stephenson et al., 1991).

Complexity and Diversity.

There is much that we have to learn about the causes and consequences of variability in fire regimes. Most prescribed burning protocols have historically been in the context of silvicultural management where the end goals are uniform fuel reduction and discouragement of competitors (i.e., reduced species diversity). Management goals in wilderness areas will likely be quite different and require different burning protocols. In developing burn plans, it is important to distinguish between fires set to restore fuel conditions to some "natural" state as opposed to fires set to simulate a "natural" process. In the former situation, uniform fire behavior may be desirable, whereas in the latter, heterogeneity and variability may be critical.

The challenge of preserving variability is, again, illustrated in giant sequoia ecosystems. Until recently, fire prescriptions specified that fires in burn units of 20 to 30 ha be ignited by "spot" fires set at 30 to 40-m intervals and that such fires be allowed to coalesce into one another. This procedure allowed such units to be burned quickly (over a few hours) and guaranteed that fire would visit the entire unit. A natural fire burning in a similar area starts at a single point and moves at variable speeds and intensities, burning hot in some locations and missing other patches altogether. Such fires might burn a similar size area over a period of days or even weeks. Prescribing fire so as to simulate such behavior carries additional risks, is considerably more expensive, and would in many cases result in less area being burned (Christensen, 1995a).

Management that acknowledges the significance of biological diversity is made all the more daunting by the fact that such diversity is itself a dynamic property of ecosystems affected by variations in spatial and temporal scale.

Context and Scale.

Much of disturbance management explicitly assumes the actions taken at relatively small scales and repeated many times over a landscape are substitutable for processes that typically occurred over large spatial scales. Furthermore, spatial relationships among such patches of management (say prescribed burns) are rarely considered. Much was made of the enormous scale of the 1988 Yellowstone fires; however, we have little information to predict what ecosystem elements and processes would be different if we were to try to simulate such an event with multiple prescribed fires.

Policy makers must understand the potential constraints on management in wilderness preserves. Within the realm of "natural," a wide variety of landscape configurations is possible. However, within the constraints of preserve design, not all these configurations are equally desirable. In an ideal world, preserve borders would coincide with natural divides or boundaries that limit ecological processes such as fire. In reality, we have chosen to preserve relatively little of the once vast expanses of wilderness, and the borders of most preserves are rarely congruent with the natural processes, such as fire, that are necessary for their preservation. In the world we have created, the acceptability of fire events of particular intensities or spatial extents cannot be based solely on whether they are "natural" (defined as having presettlement precedent). Given the constraints of preserve design, many natural events may now be deemed unacceptable or at least undesirable. This is particularly true where we can only preserve small fragments of formerly large landscapes. In these situations, it is important to understand the ecological costs for not allowing large-scale or high-intensity events to occur.

The importance of spatial context has been illustrated in a number of studies. For example, Myers (1985) found that vegetation change following fire in sand pine scrub depended on the availability of seed from species in surrounding patches.

Humans as Ecosystem Components.

The constraints on wildland management posed by liability to other public and private resources are considerable. This is particularly true in wilderness preserves and parks whose goals include recreation or watershed management, and in situations where arbitrary borders separate wilderness preserve from land dedicated to nonwilderness function (e.g., the western boundary of Yellowstone National Park).

The constraints on conservation of wilderness in an increasingly urbanized context are epitomized with regard to issues such as air quality and smoke management. For example, burning in Sequoia National Park contributes particulates and hydrocarbons to the atmosphere of California's Central Valley and Owens Valley. However natural (in the sense that it has happened for millennia) that impact may be, it may be deemed unacceptable by air quality authorities when added to the host of anthroprogenic emissions that now pollute our air.

It is critical to understand that management interventions such as prescribed fire are surrogates for natural processes and in perhaps important ways, may not exactly simulate those processes.

Adaptive Management—How Do We Know If We Are Doing It Right?

"We cannot manage what we do not understand" and "management is an art, not a science" are two aphorisms commonly heard among resource managers. With regard to the first assertion, the complexity and even chaotic behavior of ecosystems are such that we will never be in possession of the knowledge necessary to manage with complete confidence; we will always be managing systems that we do not totally understand. The second statement represents a widespread misconception that science is the domain of very certain laws and theorems and a lack of understanding of the role of uncertainty in science. Given this inevitable uncertainty and ignorance, we wish to assert that science is, in fact, the most appropriate model for the adaptive management of ecosystems.

Managers should not promise and the public should not expect complete or perfect knowledge. The acknowledgment of ignorance is the basis for adaptive management (Holling, 1978; Walters, 1986; and Lee, 1993). Managers, as well as those they serve, must accept that knowledge and modes of understanding are provisional and subject to change with new information. In this context, management goals, protocols, and directives should be viewed as hypothetical means to achieve clearly stated operational goals. Monitoring programs then represent specialized kinds of research programs designed to falsify the hypothesis that current management will achieve the desired goals.

Goals and Expectations

It is often the case that our overarching goal in management is the sustained provision of many goods and services. For example, goals for a large forested ecosystem might include fiber production, water provision or control, hunting and fishing, recreation, and preservation of biological diversity. As a prerequisite for adaptive management, such broadly stated goals must be translated into specific operational objectives and expectations. The phrase "desired future condition" has been widely used in the literature as a euphemism for such operational objectives, although, given the dynamic character of ecosystems "desired future behavior" might be better capture objectives. Thus, to determine whether management activities are leading toward desired goals in our hypothetical forest ecosystem, expectations might be expressed in terms of forest age-class distribution and health, water flows, wildlife population sizes, human visitation, and status of rare and endangered species, respectively. It is essential that such expectations be stated in terms that relate to specific measurements that can be incorporated into monitoring programs.

Models

Knowing exactly what to expect from complex systems is often a nontrivial challenge, and models are essential to meeting this challenge. Models may simply be in the form of simple compartment diagrams that provide a means of organizing information or expressing connections and relationships, or they may be developed as complex computer simulations that enable us to depict processes operating through time and across landscapes.

It is simply not possible to design monitoring programs to measure the dynamics of every species and ecosystem process. Models can be useful in identifying particularly sensitive ecosystem components or in setting brackets around expectations for the behavior of particular processes. They can be especially useful in identifying indices and indicators that provide a measure of the behavior of a broad suite of ecosystem properties. Finally, models often provide useful tools for exploring alternative courses of action (Lee, 1993).

Modeling is often criticized because of its blatant attempt to simplify the complexity inherent in ecosystems (e.g., Ralls and Taylor, this volume), but that is indeed its virtue. Lee (1993) characterized models as "indispensible and always wrong." He goes on to say,

> predictions of [ecosystem] behavior are . . . incomplete and often incorrect. These facts do not decrease the value of models, but they do make it clear that ecosystem models are not at all like engineering models of bridges or oil refineries. Models of natural systems are rarely that precise or reliable. Their usefulness comes from their ability to pursue the assumptions made by humans—assumptions with qualitative implications that human perception cannot always detect.

Monitoring

Monitoring refers here to data gathering and analysis focused on management expectations and designed to test the success and efficacy of management actions. It is naive to think of monitoring in terms of a conventional experimental test of a null hypothesis such as "this treatment has no effect"; few management situations present us with a true control in which no treatment is applied. "Goodness of fit" to expectations provides a much better framework for the design of monitoring systems.

Monitoring programs should pay particular attention to issues of precision and bias in data gathering. True replication of measurements is often impossible and in some cases sample sizes are necessarily small. Bias in data gathering is often unavoidable owing to patterns of ownership, accessibility of areas, or limited sample techniques. These limitations are not an excuse to establish monitoring programs, but they should be reflected in any conclusions regarding the effectiveness of management actions (Holling, 1978; Walters, 1986).

How much departure from expectations should be tolerated? Again, compared to most research situations this is not a simple question. As Lee (1993) points out, scientists are prone to worry about Type I errors of the sort that lead to worry about accepting a proposition as true when it is actually false. Thus, if a management activity results in departures from expectations based on goals at some level of statistical confidence, one might argue that the activity is inappropriate. However, this leaves open the possibility of a Type II error, where a true proposition is rejected. It is often the case that both types of error have significant consequences in management situations (Lee, 1993; O'Brien, this volume).

In most societal situations, our willingness to accept the risks associated with either Type I or Type II errors is conditioned by perceptions of the magnitude of the consequences if we are wrong, as well as value systems associated with those consequences. Thus, we are unlikely to change management protocols, if the perceived benefit is small or the cost is very high. Within the framework of ecosystem management, this decision should focus on which decision presents the greatest risk to long-term sustainability.

The design, development, and maintenance of monitoring programs requires commitment and long-term vision. In the short term, such programs represent an additional cost and are particularly hard to maintain where personnel are not necessarily permanent. In some cases, necessary measurements can be quite costly. The scientific community has much to contribute to the development of programs that will maximize information return relevant to specific management goals while minimizing costs.

Data Management and Timely Feedback

Enormous amounts of ecosystem data are gathered by a wide variety of private and public resource managers, but complaints of inaccessibility and incompatibility are common. Standards for data gathering are well developed in some areas (e.g., hydrology and climate) and nonexistent in others (e.g., biological diversity; e.g., Peters et al., this volume). Furthermore, it is too often the case that budget limitations, institutional structures (e.g., Rogers, this volume), or personnel changes make data files a "write-only memory."

Advances in computer networking and information storage now provide the means to access and move large quantities of data. Institutional structures should guarantee that such information is fed back to managers in a timely fashion and in a form that is directly relevant and accessible to their management activities.

Research

There is much that we do not know or understand about the structure and function of ecosystems. Feedback from monitoring programs will certainly reduce our level of ignorance, but research focused on a more detailed understanding of ecosystem processes is critical. Research programs driven solely by the immediate needs of management risk overlooking new insights and opportunities; however, programs that are innocent of an understanding of research challenges and priorities risk being irrelevant.

Education

There is clearly a need for additional education if successful adaptive management systems are to be developed. Managers have much to learn with regard to the setting of goals and expectations, monitoring, and data handling, and scientists require greater understanding of the priorities of and challenges to ecosystem managers. However, public education is critical. The limited public understanding of how science is done, much less the nuances of specific scientific issues, presents special challenges to adaptive management. Public expectations of both managers and scientists are often unrealistically high, a situation that is sometimes fostered by actions and statements of managers and scientists. It will be unlikely that society will accept "science as a model for ecosystem management" in the absence of a clearer understanding of the importance of uncertainty to both science and management.

14

Integration of Species and Ecosystem Approaches to Conservation[1]

S. L . Tartowski, E. B. Allen, N. E. Barrett,
A. R. Berkowitz, R. K. Colwell, P. M. Groffman,
J. Harte, H. P. Possingham, C. M. Pringle,
D. L. Strayer, and C. R. Tracy

Summary

The division between ecosystem and population ecologists is old, wide, and deep. This schism includes differences in values, history, culture, training, jargon, questions, interests, scale, methods, politics, and prejudices. As more ecologists become involved in conservation, these two contrasting perspectives sometimes produce conflicting conservation priorities. However, conserving ecosystem function while allowing the widespread extinction of species is as unacceptable as preserving species in zoos while allowing the widespread destruction of ecosystems. Species-specific traits are critical to ecosystem processes, and ecosystem processes are critical to population dynamics and natural selection. The successful development of the theoretical basis of conservation will require the integration of population and ecosystem ecology. In practical conservation, theoretical disagreement often gives way to empirical agreement, especially when the usual social, political, and economic constraints are encountered. Furthermore, conservation of spatially heterogeneous, temporally variable, historically contingent species and ecosystems demands context-specific approaches. We propose that ecologists increase efforts to integrate population and ecosystem ecology in the development of the theoretical basis of conservation and use the full range of ecological knowledge and methods to develop the best solutions to specific environmental problems, regardless of subdiscipline. Current differences between population and ecosystem ecologists are minuscule within the broader public discussion of conservation and we should not allow our theoretical or scientific debates to be misused by opponents of conservation, misinterpreted by the press, or misunderstood by the public.

---

[1]This chapter is based on a discussion session held during the Cary Conference. Pringle and Strayer led the discussion and Tartowski took notes.

The Division Between Population and Ecosystem Ecology

Traditionally, population ecologists study the distribution and abundance of organisms. Their investigations often concern biogeography, population dynamics, adaptation by natural selection, and biotic interactions such as predation and competition. The perspective and training of population ecologists are usually biological and mathematical, emphasizing demography, genetics, evolution, and analytical population modeling. Research in population ecology is sometimes characterized as more theoretical and reductionist than ecosystem ecology. Population ecologists tend to approach conservation with the goal of preserving species and have been involved in issues such as endangered species, biodiversity, introduced species, game management, and human population growth.

In contrast, ecosystem ecologists study the structure and function of ecosystems as transformers of matter and energy. Their investigations often concern energy flow, biogeochemistry and biotic/abiotic interactions such as primary production and nutrient cycling. The perspective and training of ecosystem ecologists are usually interdisciplinary, including thermodynamics, chemistry, biology, earth sciences, and simulation modeling. Research in ecosystem ecology is sometimes characterized as more empirical and holistic than population ecology. Ecosystem ecologists tend to approach conservation with the goal of conserving ecosystem structure and function and have been involved in issues such as pollution, energy, global change, natural resource management, and land-use change.

The contrasting perspectives of ecosystem and population ecologists often produce different theoretical approaches to conservation and conflicting conservation priorities. For example, consider how population and ecosystem ecologists might prioritize the following artificial dichotomies. Should we invest our scarce resources in conserving rivers or watersheds that contain few, if any, endangered species, (e.g. Porcupine River in Canada, Hubbard Brook watersheds in New Hampshire) or in saving the last few individuals of an extremely endangered species (e.g. California Condor, Black-footed Ferret)? Which is of higher priority: conserving the last fragments of habitat in areas of high endemism, such as Madagascar (Nott and Pimm, this volume) or conserving large tracts of less threatened landscape in areas of low species diversity and endemism, such as subarctic watersheds and lakes (Schindler, 1995)? Should control of introduced pest species receive higher priority than the control of endemic pests? Which has greater conservation value: an extremely degraded, but partially restorable, freshwater or saltwater marsh compared to a partially degraded, but not presently restorable, cypress dome or mangrove swamp? The resources available for conservation, including time, money, and people, are always insufficient. Thus, although setting conservation priorities is difficult and distasteful, the choice to invest resources in a particular project, policy, or activity has significant consequences for other conservation efforts.

Although we should not ignore the current division between population and ecosystem ecologists, we also shouldn't allow this division to interfere with the inevitable reintegration of ecology. Both the science of ecology and the development of ecological theory underlying conservation will benefit from the rapid integration of population and ecosystem perspectives. Conservation concerns encompass the full range of levels of organization (or criteria, *sensu* Allen and Hoekstra, 1992) from genome to globe. At each level of organization, conservation is justified by ethical, aesthetic, and utilitarian arguments. For example, species contain useful genetic codes, provide useful natural products, or are important to ecosystem function. Similarly, ecosystems contain useful species, provide useful ecosystem services (i.e., ecosystem functions valued by humans), or are important to biosphere function. When we lose a population or species it is a sad and permanent loss. It is also a sad and potentially permanent loss when we lose an ecosystem or ecosystem type (e.g., old-growth forest, coral reef, a prairie soil, or a wild river).

Integration of Species and Ecosystem Approaches to Conservation

We have already made considerable progress in the integration of population and ecosystem ecology (Jones and Lawton, 1995). Many individuals and projects use both approaches and some successfully integrate the two approaches in solving environmental problems (Barrett and Barrett, this volume; Pringle, this volume). The integration of approaches is likely to be more common where the conservation of species depends mainly on the management of habitat (e.g., fire in tallgrass prairie, hydrology in the Everglades), where an endangered species is critical to maintaining desirable ecosystem functions (manatees removing water hyacinth, alligators maintaining "gator holes"), or where species require large areas of habitat (wildebeest, wolves). As habitat loss accelerates, conservation of endangered ecosystems, containing multiple endangered species, will become the norm (e.g. Hawaiian ecosystems, Florida's central sand ridge, California's annual grasslands, Northwest rainforest, Columbia River, Coastal Sage Shrub). Are we approaching the time when most species and ecosystems are endangered or when most endangered species occur in endangered ecosystems and most ecosystems contain endangered species critical to ecosystem function?

Population and ecosystem ecologists have been brought closer together by recent developments in ecology. The implications of open systems, boundary effects, spatial heterogeneity, temporal variation, history, stochasticity, nonequilibrium dynamics, scale, indirect interactions, and human impacts are important to all ecologists. In one example, spatially explicit models of metapopulations, and Geographic Information Systems (GIS) models of ecosystem patch dynamics, combined at the scale of the landscape, inform conservation of both species and ecosystems in the forests of the Pacific Northwest (Marks, et al. 1993). Similarly, understanding the scaling of ecological interactions and processes from organismal physiology to global climate will require the participation of a broad range of ecologists. Historically, autecologists and community ecologists have collaborated successfully with both population and ecosystem ecologists, and as ecosystem ecologists do more process-level research and population ecologists further consider the effects of environment on population dynamics, the opportunities for collaboration across all the subdisciplines of ecology improve.

Obviously, species require habitat and ecosystems have biotic structure. Ecosystem processes such as nutrient cycling, primary production and decomposition are fundamental to the population dynamics of species and species-specific traits sometimes have important influences on ecosystem processes. Does it matter if one species is present instead of another, similar species? We are beginning to understand when, where, and what type of species differences are important to which ecosystem functions and the magnitude and manner in which species are functionally similar (Harte, this volume; Schultze and Mooney, 1994). For example, when cyanobacteria replace non-nitrogen-fixing algae during stream succession we expect changes in nitrogen dynamics. However, the subsequent replacement of one species of cyanobacteria by another might have little effect on nitrogen cycling (Grimm, 1995). There are many clear examples of species that strongly influence ecosystem processes at a particular time and spatial scale (e.g., sphagnum moss, reef-building coral, beaver, sea otter, wildebeest, prairie dogs), and we are accumulating examples of species-level characteristics, such as litter quality (Canham and Pacala, 1995) and nitrogen fixation (Vitousek and Walker, 1989), which cause a particular species to have a disproportionate influence on ecosystem function.

The ecosystem function of species diversity, independent of species composition, is now being investigated. What is the significance of the number of functionally similar species in a functional group (depth of "functional redundancy")? Greater species diversity is associated with greater resistance of primary production to temperature change (Naeem et al., 1994) and possibly with greater resilience of primary production after drought (Tilman and Downing, 1994). When environmental conditions change, ecosystem function may be maintained by compensatory changes in species composition (Schindler, 1987). Perhaps the most critical environmental impact of decreased biodiversity will be decreased resistance and resilience in response to environmental variation.

We are living amidst a worldwide replacement of local endemic species by introduced species. What are the functional differences between assemblages of endemic species and novel assemblages of remnant species combined with introduced species? How important is it to conserve the evolutionary relationships among species and between species and their environments (Thompson, this volume)? How do the processes of natural selection and sorting of species through invasion and extirpation influence ecosystem function? Integration of comparative ecosystem ecology and biogeography, macroevolution and ecosystem development, and other formerly separated research areas will be necessary to address these questions (Brown, 1994).

Even though the current biodiversity crisis has, of necessity, focused conservation efforts on "saving the pieces," the direct and indirect modification of selection pressures by humans must concern us now and over the longer term (for example, industrial melanism, xenobiotics, pesticide and antibiotic resistance, emerging diseases, habitat fragmentation, predator removal, inadvertent domestication, decreased fitness of hatchery fish, gene transfer from feral animals and escaped cultivars, including transgenic organisms, etc.). The maintenance of biodiversity and the process of evolution by natural selection are essential ecosystem services. Surely, it is not sufficient to preserve genomes in liquid nitrogen zoos. Similarly, it is not sufficient to preserve ecosystem services while endemic species are replaced by a simplified assemblage of remnant and introduced species. On occasion, it may be appropriate to bring the last members of a species into captivity or to plant non-native species to halt erosion, but only as emergency responses to extreme crises—not as reflections of conservation goals that neglect either species or ecosystems.

Practical Conservation

Although the division between species and ecosystem approaches to conservation may appear large to research ecologists, these differences are vanishingly small when conservation efforts confront human population growth, economic desperation, greed, or radically different value systems. In the context of the present environmental crisis, the consensus among ecologists is extraordinarily strong. Most of the damage to the environment, including the human-accelerated extinction of species, is not due to lack of ecological or evolutionary understanding. Human population growth, land conversion, and socioeconomic and political forces severely constrain conservation policy and land management. There is little disagreement among ecologists concerning the basic ecological information and limited options relevant to most managers. Many successful conservation efforts are opportunistic and there is usually widespread enthusiasm among ecologists to take advantage of a rare conservation opportunity.

The sociopolitical and ecological characteristics that determine the goals, methods, and success of conservation are usually context specific. Our current understanding of ecology implies that we are attempting to manage spatially heterogeneous, temporally varying, historically contingent species and ecosystems. The specific problems of conservation at specific sites and times demand specific solutions. An *a priori* restriction to one approach is less likely to succeed than choosing the appropriate approach for each circumstance. The difficulty of translating ecological theory into local prediction makes adaptive management critical to developing and implementing the best approaches.

Population and ecosystem ecologists may initially approach conservation and restoration efforts differently. Once involved in the process of solving conservation problems at a particular site, however, the initial differences fade as direct experience informs opinions. Regardless of subdiscipline affiliation, all ecologists seem capable of understanding the lessons of a specific case. In practice, when we begin with a species-based approach, we often become involved in habitat and ecosystem management, and when we begin with an ecosystem-based approach, we often become involved in the conservation of individual species. At the very least, practical cooperation is essential. We all do what we can, when we can, with whatever works.

Proposals

Both the science of ecology and the development of the theoretical basis of conservation benefit when ecologists work directly with managers on a specific conservation issue. If we treat our scientific and managerial colleagues with respect and humility, we will be able to choose from the full range of available knowledge and techniques to develop the best approach for a particular site and circumstance. Our teaching should reflect the diversity of ecological perspectives contributing to conservation and encourage our students to seek problem- and site-specific approaches. There are still managers using prescriptions based on the old "balance of nature" paradigm, and we need to clarify and communicate the management implications of the modern paradigm (Pickett and Ostfeld, 1995).

We should not allow scientific debate within ecology to obscure our shared commitment to conservation. Our public statements should emphasize this strong fundamental agreement and our respect for the different views of our colleagues. We should not allow the term "ecosystem management," which began as an effort to include non-harvested resources in management criteria (Grumbine, 1990), to be used as a cover to avoid the difficult actions needed to conserve endangered species. Nor should we allow the conservation of endangered species to be used as a diversion from the difficult actions needed to prevent widespread environmental degradation and landscape conversion.

15

The Land Ethic of Aldo Leopold

A. Carl Leopold

Summary

An ethic represents an ideal standard intended as a guide to human behavior. Although it constitutes a goal for social benefits, it ordinarily carries some overtones that are self-serving. The Golden Rule dating from the time of Moses, teaches us to do good for others; although this sounds altruistic, it should ultimately do good for us. Showing altruism to one's progeny has self-serving consequences, at least at a genetic level. Loyalty to one's community, or to one's nation, has self-serving overtones. But as one extends restrictive rules of social behavior from family to nation to the human race, the ethical aspect increases and the self-serving aspect diminishes. When restrictive rules of social behavior are extended to the entire Earth, to the biosphere, altruism becomes overwhelmingly the dominant guiding force. The sense of stewardship for the whole community, the sense of pride in providing for sustainable life-systems, and the love for the nurturing Earth, these become the quintessence of altruism; these are the radical essence of the Land Ethic of Aldo Leopold.

Evolution of Radical Ideas

Radical changes in thought will occur only when the social environment is ready. Revolutionary ideas are nearly always met with resistance. But when social thought has advanced sufficiently, ideas that may have previously seemed altogether too radical may then find social acceptance. The gradual changes that precede acceptance of radical ideas can be seen in two somewhat related proposals: Charles Darwin's idea of evolution, and Aldo Leopold's idea of the Land Ethic.

   Each of these radical ideas called for a major extension of ethical concern.

Darwin's evidences for the evolutionary origin of species linked the human race to the spectrum of all animal species. Being related to animal species carries the inference of an ethical as well as a genetic linkage. Thomas Hardy, a contemporary of Darwin, underscored the ethical consequences. He wrote: "Few people seem to perceive that the most far-reaching consequences of the origin of species are ethical: it extends the Golden Rule to the whole animal kingdom" (Nash, 1989:43). The Land Ethic further extends the ethical linkage to include the entire biosphere.

Charles Darwin was a young man when he first generated the idea of evolution and the origin of species in 1840. For approximately 20 years, he studied evidences of evolutionary change before he published this idea. In the meantime, the concept of evolution was very much in the public's eye. Many political radicals were already using the concept of evolution as a way of challenging the power structure of the Anglican Church and its commitment to divine creation. The concept was promoted in 1844 in a book by Robert Chambers titled *Vestiges of the Natural History of Creation*, in which he suggested that evolution was responsible for the entire development of life on Earth. He even posited that evolutionary sequences could account for the origin of the planetary system. These ideas were antithetical to the doctrines of the church. Further challenges to religious dogma appeared in 1850, when Francis Newman published *Phases of Faith*. In this book, he averred that evolution could more readily account for biological systems than could divine creation. At about the same time, a new periodical called the *Westminster Review* was established as a medium for articles on evolution as a continuous process. And of course, in 1855 Alfred Russel Wallace proposed a theory of the origin of species through the process of evolution. All of these precedents prepared the ground for Darwin's theory, which was finally published in 1859. The real impact of Darwin's idea depended in part on the existence of precedents of evolutionary concepts, and in part on his great reputation as a scientist.

Aldo Leopold was a young man when he first stated his ethical idea. It appeared in 1933 in a publication entitled "The Conservation Ethic." Over the subsequent 15 years he reworked and strengthened this statement before its final publication as "The Land Ethic" in his book, *A Sand County Almanac*. As with Darwin's theory, there were many precedents. Among the examples that might be cited, in 1864 George P. Marsh published a book, *Man and Nature*, in which he discussed protection of nature as an ethical issue. In 1880, Henry Ward Beecher called for an expanded ethic which would include all animals. In 1890 Edward P. Evans published *Evolutional Ethics and Animal Psychology*, in which he deplored the anthropocentric nature of Christianity, and called for ethical relations with all life forms. In more recent times, Henry Thoreau, Liberty Hyde Bailey, Albert Schweitzer, and John Muir each taught the protection of all living things, as they contributed to the environmental values of the world's resources. Most of these writers based their teachings on religious grounds. During the last decade of Aldo Leopold's life, many authors were expressing concerns about the depletion of resources and disruption of environment. Examples include Bill Vogt's *Road to Survival*, Paul Sears' *Deserts on the March*, and Fairfield Osborne's *Our Plundered Planet*. The setting was ready for the Land Ethic.

There were important precedents for each of these two concepts, evolution and the Land Ethic. And as society became increasingly aware of the precedents, the potential for receptivity increased. And acceptance was markedly enhanced because each of the proponents could speak from the podium of science.

Emergence of the Land Ethic

Aldo Leopold was painfully aware of the damage that was being wrought by human depredation on earth resources. Beginning in the early 1920s, he wrote often of the damage being inflicted on the land, and the need for conservative stewardship of resources. Being a man who was heavily burdened by worries about the state of the world, he could easily have become a doomsayer. He stated this burden: "One of the penalties of an ecological education is that one lives alone in a world of wounds."

I suspect that there were two factors in his life that turned him toward a more positive outlook in his writings. One was his intense sensibility for the beauty of the natural world. Another was the influence of his friend and former student, Albert Hochbaum.

In the course of an exchange of letters, Hochbaum declared that too many authors were reciting the dismal mismanagement of resources, and he urged Leopold to write something instead that would encourage people to change their attitudes, that would explain the evolution of his own attitudes toward the world, and that offered a positive goal toward which others could strive. I believe that the essay, *Thinking Like a Mountain*, was written specifically in response to Hochbaum's suggestion. That essay was to have major impact, showing how he himself changed from a controller into a dedicated steward. Note the famous line: "I thought that because fewer wolves meant more deer, that no wolves would mean hunters' paradise. But after seeing the green fire [in the wolf's eyes] die, I sensed that neither the wolf nor the mountain agreed with such a view."

Next to "The Land Ethic," I believe that "Thinking Like a Mountain" was his most important essay, for the very reason that it offered a pattern of personal change and growth with which others could resonate.

Restoration As a Component of the Land Ethic

Among the environmentalists preceding and contemporary with Aldo Leopold, it might be said that the dominant stance was for protection and stewardship of the remaining natural resources. In his later years, Leopold became intensely

involved in a very different aspect of environmentalism—the restoration of ecological systems.

The concept of a personal commitment to restoration arose as a logical extension of Leopold's conviction that good stewardship required a sense of affection,
indeed a sense of love of the land. At the end of his essay on the Land Ethic,
he stated: "It is inconceivable to me that an ethical relation to land can exist
without love, respect and admiration for land, and a high regard for its value.
By value, I of course mean something far broader than mere economic value; I
mean value in the philosophical sense."

In addition to fostering appreciation and protection of natural communities,
he struck off in a new direction—a course through which the landowner can find
personal rewards from restoring land that has been degraded through sloppy
exploitation. Love, respect, and admiration for the land are all nurtured by the
very personal actions that are involved in bringing back ecological richness. This
was the beginning of the shack experience.

The story of Aldo Leopold's shack is well known—the purchase of a worn-
out, abandoned farm in the floodplain of the Wisconsin River (Fig. 15.1), and
the establishment of a family project for its restoration. The former empty fields

Figure 15.1. Aldo Leopold at the shack in 1937. The abandoned fields on the banks of
the Wisconsin River in Sand County surround the shack.

have been converted into a splendid blend of pine/hardwood forest, tall-grass prairie, and sawgrass marsh. The Leopold Memorial Reserve stands as a monument to Aldo Leopold's simple application of the Land Ethic through the restoration of wasted land (Fig. 15.2). This type of stewardship enriches the meaning of caretaking of the land. It extends the personal involvement in environmentalism. It yields a spiritual uplift derived from restoring a wreckage of ruined land, and bringing it back to the grandeur of its natural state. It produces rewards both to the landowner and to the community.

The extension of stewardship to include restoration has caught on as a serious component of environmentalism. There are now journals of land restoration, restoration societies, books, and professional meetings on land restoration. It was an idea that found its time.

Using the axe as a symbol of the intensive management of one's property, Leopold wrote: "I have read many definitions of what is a conservationist, and written not a few myself, but I suspect that the best one is written not with a pen, but with an axe. . . . A conservationist is one who is humbly aware that with each stroke he is writing his signature on the face of his land."

I hope that I have given you a close-up perspective of Aldo Leopold, from

Figure 15.2. The shack and its surroundings as they appear today. The woods and fields enfolding the shack are the result of the restoration efforts of Aldo Leopold and his family.

the vantage point of one who was close to him and who could watch and feel the deep commitment that he had for the ethical basis of conservation. His concern was always permeated by a profound resonance with the beauty of the Earth and its workings. He developed a radical idea of personal altruism, through which humans could live in harmony with the environment. And he pioneered the application of such harmony through ecological restoration.

It is remarkable that he brought his entire family into a partnership in his greatest experiment: the reconstruction of quality land at his shack.

And although he could see that precursors of his idea were evolving, he did not live long enough to see how his ideas would spread exponentially in subsequent years. I sincerely believe that he would be astonished at the reverberations of his Land Ethic through the subsequent 50 years. And surely he would marvel at the ranks of enthusiasts who were to catch the excitement that comes from extending ecological conservation to actual land restoration.

Aldo Leopold blended a dazzling literary skill with a profound concern for a gentler manner of coexistence with the natural world. The result: an amazing ethical paradigm from an amazing person.

SECTION IV

Toward a New Conservation Theory

Themes

Richard S. Ostfeld, Steward T. A. Pickett,
Moshe Shachak, and Gene E. Likens

Although conservation biology is a relatively young science, it is based on a solid foundation in disciplines such as population genetics, demographic theory, biogeographic theory, and metapopulation theory. Consistent with this set of underlying theories, the primary focus for much of conservation biology has been populations, and sometimes clearly delimited communities. But the changing paradigms of ecology and conservation, described in Section II, have important implications to the scope of conservation biology. In essence, the scope of conservation biology must be expanded to include larger networks of species, larger spatial scales, ecosystem processes that transcend individual species, disturbance, and diverse sets of ecological interactions.

This section represents an attempt to incorporate some of the traditional disciplines within the realm of conservation biology into a new, broader conservation theory. It also begins to merge some disciplines not traditionally considered a part of conservation biology, such as ecosystem science, into a new theory. For this section, we charged authors to explore how the traditional disciplines are changed by the expansion of the scope of conservation to include larger spatial scales, patchiness, and disturbance. We also asked them to evaluate how their particular discipline would contribute to a new, broader conservation science.

16

The Future of Conservation Biology: What's a Geneticist to Do?

Kent E. Holsinger and Pati Vitt

Summary

Loss of genetic diversity in wild populations is more likely to be a symptom of endangerment than its cause. As a result, changes in the genetic structure of endangered species threaten conservation efforts only in a few cases:

1. Loss of alleles at self-incompatibility loci may pose a direct threat to persistence of self-incompatible plants by limiting seed set.

2. Hybridization may lead to genetic extinction of endangered species if they occur near more common and reproductively compatible relatives.

3. Inbreeding depression, loss of genetic diversity, and adaptation to the captive environment limit the utility of captive breeding in endangered species recovery programs.

As a result, direct manipulation of genetic structure in endangered species may rarely be necessary. In contrast, the techniques that genetics provides (e.g., allozyme, restriction site, and nucleotide sequence analysis) are likely to be broadly useful. They can help to identify evolutionarily distinct populations worthy of conservation concern, and they can provide a window on otherwise unobservable demographic processes. Although conservation at an ecosystem or landscape scale has naturally focused on ecological issues, long-term success depends on the ability of native populations to respond adaptively to environmental change. The evolutionary impacts of habitat fragmentation are uncertain, but it is likely that widespread continuously distributed species are already feeling them. Invasive exotics pose a severe threat to many native populations and communities. The limited success of biological and herbicide control programs, even in asexual species with relatively little genetic variation, only emphasizes the importance of identifying invasives *before* they become widely established.

Introduction

Conservation biologists are faced with several daunting tasks. They must try to prevent the extinction of endangered species, to protect the structural and functional integrity of significant natural communities, and to conserve ecologically or economically valuable genetic resources. Conservation of genetic resources, especially those of crop plants, has been a focus of attention since Vavilov started the germplasm collection at the All-Union Institute of Plant Industry in Leningrad over 75 years ago (Holden et al., 1993). In the last 15 years or so, genetics has also begun to play a conspicuous role in developing strategies to conserve natural populations of plants and animals. In the decade following publication of Soulé and Wilcox (1980), much of the research in conservation biology was directed toward developing techniques to prevent the extinction of endangered species. As part of these efforts, conservation biologists often worried about the threats posed by loss of genetic diversity to both short-term and long-term population viability. Many have regarded investigating the genetic status of target populations as a natural part of population viability analysis (Ralls and Taylor, this volume). The famous, or infamous, 50-500 rule became a standard and widely accepted rule of thumb for judging the short- and long-term genetic viability of populations, and calculations of effective population size were often thought necessary as part of management plans for endangered species (see Allendorf and Leary [1986] and Ledig [1986] for reviews of the relationship between heterozygosity and fitness; see Nunney and Elam [1994] for a recent discussion of methods for calculating effective population size.)

In the past five years or so, however, the emphasis on genetics has waned for two reasons: (1) an increasing emphasis on ecosystem-level approaches to conservation management (e.g., Grumbine, 1994) and (2) accumulating evidence that populations large enough to be ecologically self-sustaining are unlikely to be threatened by loss of genetic diversity (Lande, 1988b). Ecosystem-level conservation emphasizes management approaches that attempt to prevent species from becoming endangered. In contrast, single-species approaches attempt to prevent extinction of species that are already endangered. A natural consequence of this new emphasis on ecosystem-level approaches is an accompanying emphasis on maintaining the structural and functional integrity of native ecosystems— a problem that seems naturally to lie within the realm of community and ecosystem ecology (Ewel, this volume; Meyer, this volume; but see Holsinger, 1995a). The new emphasis on ecosystems would, by itself, tend to reduce the emphasis on genetics, but even within the realm of single-species conservation there is a decreasing emphasis on genetic management. Lande (1988b) argued convincingly that populations large enough to buffer the risks of environmental stochasticity are far larger than necessary to buffer any risks posed by loss of genetic diversity. He convinced us, and at least a few other conservation biologists, that the distinctive contribution genetics can make to single-species conservation efforts may be more limited than we previously believed (see for example, Menges, 1991; Caughley, 1994; Schemske et al., 1994).

In this chapter we explore some of the ways in which spending efforts on genetics may distract us from more important pursuits. We also describe a few of the distinctive and important contributions that genetics can make to the solution of our conservation problems. We argue, for example, that lack of genetic diversity is more likely to be a symptom of endangerment than its cause. As a result, understanding the genetics of natural populations will frequently be more useful as a guide to identifying taxa or populations worthy of conservation concern than as a tool for guiding management strategies. Such understanding may also provide insight into demographic processes that are otherwise unobservable.

The long-term success of ecosystem-level conservation efforts is, of course, likely to depend on the ability of native populations to respond adaptively to environmental change. Their ability to do so will depend, in part, on the extent to which human activities have altered patterns of genetic diversity. To illustrate the potential for human impacts we review the possible genetic impacts of habitat fragmentation and conclude that they are likely to be greatest in species that are widespread, ecosystem dominants. Furthermore, an analysis of some limited data available for old-growth stands of Douglas fir (*Pseudotsuga menziesii*) in southwestern Oregon suggests that these impacts are already being felt (Holsinger, 1993). We also review the limited success of biological and chemical control efforts in plants, even in asexual species with a limited number of genotypes. These sobering results only emphasize how important early detection and control are if we hope to limit the impact of invasive exotics on native communities and ecosystems.

Management of Endangered Species

Any conservation program focused on protection of endangered species will include several components to (1) identify the species or populations in need of attention, (2) determine the threats to those species, and (3) select the management strategies appropriate to mitigate those threats (cf. Holsinger, 1995b). Genetics plays two roles in the design and implementation of such programs. First, it provides laboratory techniques that help to identify evolutionarily distinct populations worthy of conservation concern, and it provides conceptual tools for helping to mitigate the threats associated with life in small populations. Allozyme, restriction fragment, or nucleotide sequence variation, for example, can provide insights into patterns of population differentiation different from those that could have been obtained through analysis of morphological features alone. Second, the principles of population genetics help us to determine the extent to which inbreeding depression and loss of genetic diversity threaten both the short-term and the long-term persistence of unmanipulated populations.

Genetic Threats to Population Persistence

Perhaps the most famous, or infamous, product of the emphasis on genetic aspects of endangered species conservation in the 1980s was the 50-500 rule (Franklin, 1980). This widely quoted rule of thumb was intended as a guide to indicate the minimum population size necessary to ensure short-term viability (an effective size of 50) and long-term viability (an effective population size of 500). Franklin (1980) noted that animal breeders are willing to accept an increase in the inbreeding coefficient of 1 percent per generation in breeding programs, whereas rates of inbreeding greater than 2 percent are typically unsustainable for more than a few generations. Because drift increases the population inbreeding coefficient by $1/2\,N_e$ per generation, where N_e is the population size, this suggests that populations with an effective size less than 50 may be immediately threatened by inbreeding depression.

Over the long term, a population's ability to respond adaptively to environmental change depends on the level of genetic variability it contains. Franklin (1980) argued that the rate at which mutations contribute to additive genetic variance must exceed the rate at which additive genetic variance is lost through genetic drift for populations to retain their potential for adaptive responses. Additive genetic variance is lost from a finite population at the same rate as heterozygosity, that is, a rate of $1/2\,N_e$ per generation (Bulmer, 1985). Lande's (1975) calculations, which were based on observations of abdominal bristle number in laboratory populations of *Drosophila melanogaster*, suggest that mutations add additive genetic variance equivalent to 0.1 percent of the environmental variance per generation. Thus, the loss of additive genetic variance through drift will be more than compensated for by new mutational variance if the effective population size is greater than 500.

These two observations are the genesis of the 50-500 rule and the cause of some concern among conservation geneticists in the mid-1980s. Although originally presented only as rules of thumb, they were the *only* rules of thumb available. As a result, they were widely quoted and applied. Should conservation biologists devote the effort necessary to develop more precise rules that do not involve the enormous extrapolations necessary to apply the 50-500 rule to populations of endangered species? We think not. Lack of genetic diversity is more likely to be a result of the processes that endangered the species in the first place than its cause (cf. Caughley, 1994; Holsinger, 1995a). Populations large enough to be ecologically self-sustaining are also likely to be large enough to remain genetically viable indefinitely (cf. Lande, 1988b; Menges, 1991). As a result, management of most endangered species should focus on establishing ecologically viable populations (cf. Schemske et al., 1994). There are several reasons why loss of genetic diversity is unlikely to threaten ecologically self-sustaining populations.

First, most of the genetic variance that can respond to natural selection (i.e., the additive genetic variance) is accounted for by alleles that are present in relatively high frequency. These high-frequency alleles are the least likely to be lost from a population as a result of genetic drift. For the same reason that levels of heterozygosity are less affected by a population bottleneck than levels of allelic diversity (Nei et al., 1975), small populations are unlikely to lose much of the genetic variation that can respond to natural selection unless they are kept very small for a long time (cf. Templeton, 1991).

Second, adaptation to future environmental changes is likely to involve alleles that are either (1) currently present in high frequency, (2) currently present in low frequency but maintained by natural selection, or (3) not currently present at all. In any case, adaptation is unlikely to involve the alleles most likely to be lost as a result of genetic drift; that is, low-frequency alleles that are currently neutral or selectively disfavored. It is easy to see why if we consider the expected lifetime of a rare allele, which can be calculated from the diffusion approximation for genetic drift (Crow and Kimura, 1970; Ewens, 1979). The number of generations to loss of a neutral allele with a current frequency of 1 percent is less than one fifth of the effective population size on average (Figure 16.1). For a neutral allele with a current frequency of 0.1 percent, the expected time to loss is less than 3 percent of the effective population size. The lifetime of rare alleles that are disfavored by selection is, of course, even shorter. Thus, even if adaptation to changed environmental conditions requires genotypes that are not currently

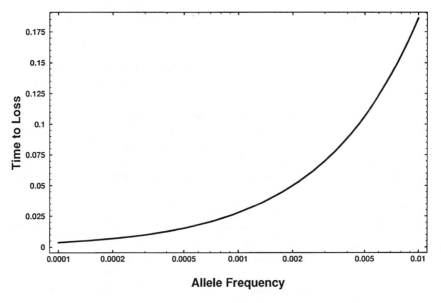

Figure 16.1. Average time to loss of a neutral allele as a function of current allele frequency, given that it is eventually lost. The time to loss is given as the number of generations divided by the effective size of the population. Calculated from standard diffusion approximation results (e.g., Crow and Kimura, 1970; Ewens, 1979).

common in a population, alleles that are currently neutral or are selectively disfavored are unlikely to be the source of those new genotypes. They are likely to have been lost before the new conditions arise. Those alleles most likely to be lost in small populations are the ones that are least likely to contribute to future adaptive responses to changed environmental conditions.

Third, small populations may accumulate deleterious mutations through a process similar to Muller's ratchet (Muller, 1964). As a result of genetic drift, disfavored mutations at any locus may be fixed in the population. Unless mutation from disfavored to favored alleles is common or there are several populations that exchange migrants at least occasionally, each fixation lowers the maximum fitness attainable by any individual in that population. Lande (1994) recently showed, not surprisingly, that small populations go extinct more rapidly as a result of this process than large ones (Fig. 16.2). Although he points out that this effect is greater than many have thought, it is also evident from Figure 16.2 that the frequency and intensity of catastrophes have a much greater impact on persistence times than recurrent mutation to deleterious alleles.

The time to extinction increases much more slowly with increasing population size in the face of catastrophes than if only the effects of deleterious mutation are considered. The mean time to extinction is proportional to $\log (N)$ in the presence of catastrophes, whereas it is proportional to N^{1+1/c^2} when only the effect of deleterious mutation is considered, where N is the population size and c is the coefficient of variation in the selection coefficient. A population with 100 individuals growing at an annual rate of 1 percent and suffering a catastrophe that removes 75 percent of its individuals once every 70 years, on average, has only a 50 percent chance of surviving more than 375 generations. A population of 100,000 individuals subject to the same conditions has only a 50 percent chance of surviving more than 1,000 generations. A 1,000-fold increase in population size is associated with less than a three-fold increase in persistence time in the presence of catastrophes.

These arguments suggest that there is especially little need to worry about genetic threats to the persistence of endangered species that have always been rare. By the very fact of their continued existence they have shown that they have the ability to cope with the genetic and demographic consequences of rarity, whatever those may be (Holsinger and Gottlieb, 1991). Species that were formerly widespread and are now threatened with extinction are more likely to require genetic management than those that have always been rare. Furthermore, it is worth considering the possible genetic consequences of any management action that involves manipulating the reproductive success of individual plants or animals. By ensuring that all individuals contribute equally to the next generation, for example, the effective size of the population can be nearly doubled. There are two other cases in which special attention to genetics may be necessary: (1) In plant species with genetically determined self-incompatibility, loss of self-incompatibility alleles may pose a direct threat to persistence through failure of

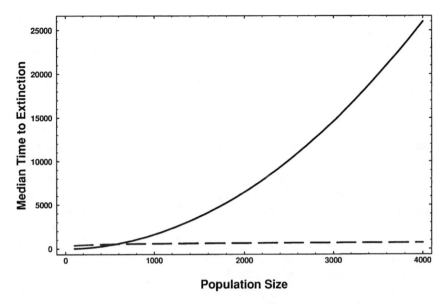

Population Size

Figure 16.2. Median number of generations to extinction as a function of current population size. The solid line is for a model incorporating accumulation of deleterious mutations (Lande, 1994). The dashed line is for a model incorporating periodic catastrophes (Ewens et al., 1987). For both models the mean intrinsic rate of increase is 1 percent. In the model with recurrent deleterious mutations, mutation adds one deleterious mutation per genome per generation, corresponding with the genomic mutation rate estimated for *Drosophila melanogaster* (Mukai, 1979), the mean selection coefficient against deleterious alleles is 0.025, and the coefficient of variation in the selection coefficient is 1. In the model with periodic catastrophes, only 25 percent of the population survives a catastrophe and the time between catastrophes is exponentially distributed with a mean of approximately 67 years. The mean times to extinction given by Lande (1994) for the deleterious mutation model were converted to median times using Gumbel's (1958) theory of extreme values (cf. Ewens et al., 1987).

seed production and (2) in plant or animal species with very small populations, hybridization may lead to genetic assimilation of the endangered species by a more common or more reproductively successful relative with which it is cross-compatible.

Individuals in an Illinois population of the lakeside daisy (*Hymenoxys acaulis* var. *glabra*), for example, failed to set seed for over fifteen years. Experimental crosses showed that the cause was a sporophytic self-incompatibility system, which is found in many members of the aster family (DeMauro, 1993). Investigation of three Ohio populations found only three compatibility groups in one population. The most diverse population had only nine. As a result, many potential outcrosses in the population result in less than 2 percent seed set. Not only is lack of genetic diversity limiting the reproductive capacity of Ohio populations, reestablishment of a self-sustaining population in Illinois is not possible without importing genotypes from Ohio representing new compatibility groups (De-Mauro, 1993).

Hybridization poses a threat quite different from the reduced reproduction associated with loss of self-incompatibility alleles. Repeated crossing with reproductively compatible relatives may result in assimilation of endangered species of plants and animals by more common or more reproductively successful relatives (Cade, 1983), which will usually result in loss of the endangered species. Of the 11 remaining trees of Catalina Island mountain mahogany (*Cercocarpus traskiae*), for example, 5 appear to be hybrids with a common congener (Rieseberg and Gerber, 1995). Similarly, recent range expansion by the weedy annual sunflower (*Helianthus annuus*) poses a threat to many endangered species of sunflowers (Rogers et al., 1982). Although hybridization is often thought to be more common in plants than in animals, both plant and animal species have been endangered through this process (Cade, 1983; Rieseberg, 1991; Ellstrand, 1992; Ellstrand and Elam, 1993).

Off-Site Conservation Programs

Botanical gardens are playing an increasingly important role in plant conservation efforts as integrated approaches to plant conservation (Falk, 1987) become more common (e.g., NEWFS, 1992). Similarly, zoos now often promote themselves as protectors of endangered animal species. To the best of our knowledge, no credible scientists suggest that off-site collections can replace natural populations in their native habitat. Instead they are intended to serve as an insurance policy against catastrophes that might eliminate one or more of the few remaining wild populations of an endangered species. Genetics has two important roles to play in the design and management of such programs: (1) ensuring that a representative sample of genetic diversity is present in the individuals used to establish the off-site collection and (2) minimizing the loss of genetic diversity from the sample and preventing, to the extent possible, adaptation to the captive environment.

The first of these is a far less challenging problem than the second, especially in plants (Brown and Briggs, 1991). Because endangered plant species generally occur in 20 or fewer populations, at least in the temperate zone (Brown and Briggs, 1991; Holsinger and Gottlieb, 1991), samples can easily be taken from a large proportion of the existing populations. By taking them from a relatively small number of individuals per population we are assured of capturing the vast majority of adaptively significant genetic variation present. Similarly, problems posed by adaptation to the captive environment and loss of genetic diversity that occupy the attention of many zoo biologists (e.g., Templeton and Read, 1983) can often be avoided in temperate zone plants by storing seed samples from natural populations in low-temperature seed banks (Eberhard et al., 1991). Because of these characteristics, off-site conservation programs may play a more important role in conservation of endangered plants than in conservation of endangered animals.

Captive breeding programs for endangered animal species are rarely begun until the total number of individuals remaining in the wild is extremely small. Indeed, the goal of animal captive breeding programs is often to re-create a self-sustaining wild population where none currently exists (IUDZG/CBSG, 1993; Ebenard, 1995). Thus, the major task in developing such a program is to design a program that preserves as much of the genetic diversity originally present as possible and prevents, or at least minimizes, adaptations to the captive environment. Unfortunately, this is a difficult task. Controversy continues about methods to avoid inbreeding depression in captive-bred populations. Some authors (e.g., Templeton and Read, 1983; Templeton, 1991; Lacy, 1992) argue that captive-bred populations are often so small that inbreeding depression is unavoidable and that a breeding program must be designed to purge deleterious alleles from the captive population. Others (e.g., Simberloff, 1988; Hedrick, 1994) suggest that the advantages of this purging may be outweighed either by increased risk of extinction during early phases of the breeding program or by loss of adaptive potential associated with homozygosity produced by the suggested breeding programs.

This debate sidesteps an even greater difficulty. No matter how careful managers of captive populations are, inadvertent selection in captive populations may often lead to genetic changes that increase adaptation to the captive environment and decrease adaptation to the native environment. For example, in 1986 Templeton and his colleagues established 39 isofemale lines of *Drosophila hydei* derived from wild-caught females. They were transferred to fresh medium every two to three weeks, the standard transfer schedule used for *D. mercatorum* and other Hawaiian *Drosophila*. Only after several generations did they notice that *D. hydei* takes longer to develop than *D. mercatorum*. This inadvertent selection led to a significant decrease in mean development time from egg to adult (Templeton, 1991). Similarly, Bush (1979) presents evidence that inadvertent selection for reduced flying ability in screwworm production facilities led to decreased mating opportunities for male-steriles released from these facilities. As a result, biological control efforts were unsuccessful until a new male-sterile strain was introduced.

Unfortunately, breeding strategies that have been suggested to minimize adaptive responses to the captive environment often involve a large component of inbreeding (Templeton and Read, 1983; Templeton, 1991) and may result in a substantial change in the genetic architecture of captive-bred populations. Thus, the best way to limit adaptation to the captive environment appears to be to limit the number of generations involved in any captive breeding program. As a result, a captive breeding program is most likely to be a useful component of a conservation program for endangered animals when it can produce large numbers of offspring for reintroduction in only a few generations. These considerations add weight to our earlier suggestion that off-site collections of endangered plant and animal species can only be regarded as insurance policies against catastrophic loss or remaining wild populations, not as a substitute for them.

Identifying Populations and Processes of Concern

It may be necessary to manipulate the genetic structure of endangered plant and animal species when populations are very small, when loss of self-incompatibility alleles reduces reproductive capacity, when hybridization with common relatives threatens assimilation, or when managing off-site collections. These are areas in which genetic *concepts* are important to endangered species conservation. Ironically, the greatest contribution genetics can make to endangered species conservation may have less to do with the concepts it employs than with the *techniques* it has developed. Avise (1994a, 1995) points out that molecular markers often provide substantial insight into patterns of genetic variation within species. In so doing, genetics provides powerful techniques for identifying populations that have evolved independently of one another and so are especially worthy of conservation concern.

Avise and Nelson (1989), for example, present an analysis of mitochondrial DNA (mtDNA) diversity in seaside sparrows of the southeastern United States. Nucleotide sequence divergence between mtDNA haplotypes sampled from different Atlantic coast populations was less than 0.4 percent, as was divergence between sparrows from different populations along the Gulf of Mexico coastline. Divergence between haplotypes from the Atlantic and Gulf coasts, however, was never less than 0.8 percent, and most haplotypes differed by 1 percent or more. These results suggest that the recent evolutionary histories of Atlantic and Gulf coast seaside sparrows have been largely independent of one another. It would make sense to focus conservation efforts on ensuring that multiple populations of both lineages continue to persist. It also suggests, in retrospect, that the extreme attention focused on the dark-plumaged Atlantic coast population in Brevard County, Florida—the dusky seaside sparrow—may have been misplaced. The dusky seaside sparrow appears to have had a history of independent evolution no longer than that of any other seaside sparrow population along the Atlantic coast.

In suggesting the use of genetic techniques to identify populations worthy of conservation concern, however, we do note one caution. Coalescent theory (reviewed in Hudson [1990]) shows that within panmictic populations the genealogical relationships among alleles depend strongly on the selective forces acting on them. Only when populations begin to evolve independently of one another (i.e., only *after* they have stopped exchanging genes) will the genealogical relationships of all genes become concordant (cf. Avise and Ball, 1990). As a result, levels of genetic variation with populations and patterns of genetic differentiation among populations as assayed by molecular markers may be quite different from those that would be detected through quantitative genetic analyses of ecologically significant traits (Holsinger, 1991). Thus, genetic tools should be regarded as providing characters of the sort a systematist would use to delimit species and subspecies. They may not provide useful markers for patterns of variation at loci involved in ecological adaptation.

Genetic techniques may also provide a window into otherwise unobservable demographic processes. In spite of enormous efforts to track individual green sea turtles, for example, debate continues about the extent of natal site fidelity. Restriction site surveys (Bowen et al., 1992) showed that mtDNA haplotypes of individuals from different rookeries are largely nonoverlapping. Because mtDNA is maternally inherited, this pattern indicates that females show a strong propensity to lay eggs where they were born, regardless of where mating occurs (Meylan et al., 1990). As Avise (1995) points out, these results suggest that each rookery should be managed as an autonomous demographic unit. Severe decline or loss of a rookery is unlikely to be compensated by natural recruitment of females from other rookeries.

An even more striking example of the use to which genetic tools can be put is provided by Spix's macaw, which is the world's most endangered bird (Wilson, 1992). Only 32 survive, and since 1987 only 1 of these has remained in the wild. Its sex has been uncertain because Spix's macaw, like many bird species, is not sexually dimorphic. Behavioral observations suggested that the wild individual is a male, but it was essential to confirm these observations before releasing a potential mate. Griffiths and Tiwari (1995) developed a genetic test based on the polymerase chain reaction to determine the sex of the wild individual from DNA traces present in molted feathers. They confirmed that the wild bird *is* a male, and a female Spix's macaw has been released from the captive population as a prospective mate.

Ecosystem-level Conservation

As conservation biologists have shifted their attention to conservation at an ecosystem or landscape scale, they have naturally focused on ecological issues (Grumbine, 1994; Ewel, this volume; Meyer, this volume). They have drawn especially on the disciplines of community and ecosystem ecology in their efforts to refine methods of protecting the integrity of important natural communities and ecosystems (Angermeier and Karr, 1994). As Holsinger (1995a) pointed out, however, there are also some important contributions that population biology can make to the development of ecosystem-level conservation programs. There may even be some limited ways in which understanding the genetics of natural populations can contribute to those efforts.

Genetic Consequences of Habitat Fragmentation

Other than habitat destruction, the fragmentation of existing habitats may be the most dramatic effect human beings have had on the planet. Native communities and ecosystems are increasingly becoming islands in a sea of human-dominated landscapes. The ecological effects of this fragmentation have been widely noted (e.g., Wilcove et al., 1986), but the possible evolutionary consequences have been less widely considered. Given the underlying uncertainties about the genetic structure of widespread species any generalizations must be regarded as tentative, but two points seem clear (Holsinger, 1993): (1) The species most likely to suffer changes in their genetic structure as a result of habitat fragmentation are widespread, more or less continuously distributed species (i.e., ecosystem dominants) and (2) the long-term evolutionary impacts of these changes are difficult to predict, but it is likely they have already begun (see Tilman and Downing [1994], suggesting similar conclusions for ecological impacts).

The genetic structure of rare species is unlikely to be affected by habitat fragmentation unless there has been extensive gene flow between populations that is subsequently disrupted by fragmentation. This could happen to forest-dwelling amphibians, for example, if a forest corridor connecting two populations were converted to cow pastures. Similarly, the genetic structure of widespread species that occur naturally as discrete colonies is unlikely to be affected by habitat fragmentation. Populations may be directly eliminated by fragmentation or the remaining fragments may be too small to support ecologically viable populations, but unless fragmentation disrupts existing patterns of gene flow it is unlikely to have any genetic impact.

Widespread species that are continuously distributed, however, are another story. In such species, isolation by distance may allow distantly separated groups of individuals to evolve independently of one another, even though there are no geographical barriers to dispersal (Wright, 1943, 1946). Quantitative traits will often show the same pattern of variation, even in a population subject to uniform stabilizing selection over the entire range being considered (Goldstein and Holsinger, 1992). As a result of this spatial structure and the fact that many genotypes may produce essentially the same phenotype, spatially structured populations are able to store enormous amounts of genetic variation. Their ability to do so, however, depends on the relationship between dispersal distances and the total area occupied.

A convenient measure of spatial structure in such populations is the point at which the genetic correlation between individuals approaches zero. For quantitative traits, this point is likely to be at least 30 times the root-mean-squared dispersal distance (Holsinger, 1993). For a continuously distributed population to store genetic diversity in independently evolving patches, the population must occupy an area at least three to six times as large as the size of an individual patch (Goldstein and Holsinger, 1992; Holsinger, 1993). Using this observation and data for patterns of pollen dispersal in Douglas fir (*Pseudotsuga menziesii*), Holsinger (1993) concluded that the largest fragment of old-growth forest remaining in the Siuslaw National Forest of southwestern Oregon was less than one tenth the size necessary to preserve the genetic structure characteristic of the species prior to logging. It is not unreasonable to think that a similar pattern will be found when the same sorts of analyses are applied to other widespread species.

There is a little good news in this grim scenario. Change in the genetic structure of remaining old-growth stands of Douglas fir requires generational turnover. Because the generation time in old-growth Douglas fir is probably more than a century, any genetic effects of fragmentation may not be seen for a millennium or more. Moreover, by using local seed sources to replant clear cuts, especially those near remaining old-growth stands, we may be able to lessen even this long-term impact. Of course, if we fail to act soon, it may also take a millennium or more to recover from our error.

Genetics and the Management of Invasive Exotics

In many areas, introduced species are among the greatest threats to the persistence of native ecosystems. Exotic plants pose a severe threat to over 7 million ha of federal lands in the western United States (McCleese, 1994), for example, and the economic impacts associated with these threats are enormous. A recent study estimated that economic losses attributable to leafy spurge (*Euphorbia esula*) invasion in rangelands is responsible for more than $144 million in losses annually in North Dakota, Montana, and Wyoming (cited in Jensen, 1994). Daehler and Strong (1994) argue that invasive exotics may pose the most severe threats to native plant diversity in nature reserves. The zebra mussel (*Dreissena polymorpha*) not only spread through most waterways in the northeastern United States in less than seven years (Ludyanskiy et al., 1993), its population densities are so great and the quantity of water its colonies filter so vast that Campbell and Kenyon (1994) suggest that they may have a dramatic impact on fish species diversity in Lake Erie. By the year 2000, the cost of industrial, utility, and municipal water-use reductions plus the zebra mussel's impact on navigation, boating, and sport fishing could reach $5 billion in the Great Lakes alone (Ludyan-skiy et al., 1993).

Reviewing the literature on genetic structure of invasive plant species reinforces the common-sense impression that efforts focused on preventing invasions are more likely to be successful than those focused on reversing them. Rush skeleton-weed, for example, is an herbaceous perennial native to Eurasia that is now naturalized in the Mediterranean (Moore and Robertson, 1964). It was first noticed in Australia in 1917, and it is still one of the most significant weeds in southern Australia, in spite of extensive control efforts (Heap, 1993). Over 300 distinct forms of this predominantly apomictic species are currently found in Eurasia and the Mediterranean (Cullen, 1991), but only 3 are found in Australia (Hull and Groves, 1973; Burdon et al., 1980). In 1971 a biological control effort began with the introduction of a rust fungus (*Puccinia chondrillina*) and two host-specific arthropods (*Cystiphora schmidti*, a gall midge, and *Aceria chondrillae*, a gall mite). It initially appeared that the effort had been successful. Plant density declined in many areas to levels typical of western Europe, where it is rarely an agricultural pest (Wapshere et al., 1974; Cullen, 1978; Supkoff et al., 1988). Unfortunately, the rust, which was the most effective biological control agent, was effective only against the narrowleaf form of skeletonweed. The distribution and abundance of other forms have since expanded to the extent that the potential effect of the species has not been diminished (Heap, 1993).

The differences in susceptibility to biological control presumably reflect genetic differences among the forms. If the relatively stereotyped differences between three asexually reproducing genotypes can limit the success of biological control efforts so greatly, it is reasonable to conclude that similar efforts on self-incompatible outcrossers like purple loosestrife (*Lythrum salicaria*) are unlikely to succeed (e.g., Malecki et al., 1993). Such differences, moreover, might even limit the success of herbicide control. The broadleaf form of skeletonweed, for example, may be less susceptible to clopyralid than the other forms (Heap, 1993). If the patterns we have just described are general, if genotypes of invasive species commonly differ in susceptibility to chemical and biological control agents as much as genotypes of skeletonweed in Australia, it only emphasizes the importance of identifying invasives *before* they become widely established (see Ruesink et al., 1995). Once established, eradicating an invasive species may prove nearly impossible, even with a combination of biological, chemical, and physical controls.

Conclusions

Caughley (1994) pointed out that two paradigms coexist uneasily in conservation biology. The small-population paradigm is concerned with managing threats to small populations. It is, of necessity, concerned with stochastic threats to population persistence, be they demographic or genetic. Because the magnitude of stochastic threats depends more on population size than on life-history attributes of a particular species, a general theory about these threats can be and has been developed. The declining-population paradigm, on the other hand, is concerned with mitigating or reversing threats to population persistence. It is concerned with deterministic processes that have caused populations to decline. Because the processes responsible for decline are likely to be as different from one another as the life-history attributes of each threatened species, it may not be possible to construct a universal theory of population decline comparable to the stochastic theory of small populations. Nonetheless, it is these deterministic processes that have brought threatened species to the brink of extinction, and it is they that must be reversed if persistence is to be assured (cf. Holsinger, 1995a, 1995b). The most important lessons from small-populations theory, therefore, may be that (1) managing small populations requires enormous effort and (2) to ensure long-term persistence managers should attempt to reverse deterministic threats to persistence and increase the size of threatened populations as rapidly as possible (Caughley, 1994; Schemske et al., 1994; Holsinger, 1995b).

These arguments are complementary to those we presented earlier concerning the proper role of genetics in conservation programs. Retaining genetic diversity is undeniably important in ensuring the long-term evolutionary potential of *all* species, endangered or not. It may be, however, that conservation managers need to manipulate the genetic structure of populations only in a limited set of circumstances: when loss of self-incompatibility alleles poses a direct threat to continued reproduction or when crossing with more common relatives threatens genetic assimilation, for example. This is not to say that the possible genetic consequences of management actions can always be ignored. To take just one example, it makes sense for genetic reasons as well as ecological ones to design reserve systems that support multiple examples of populations and communities and to ensure that they encompass as much of their historical range as possible (cf. Simberloff, this volume). Similarly, careful consideration of possible genetic impacts is warranted whenever demographic management involves manipulating individual survival and reproduction.

It may be, however, that there are only two important contributions that geneticists can make to conservation biology, *as geneticists*, that differ from those that ecologists, systematists, or biogeographers would also make: (1) geneticists can develop techniques to identify populations worthy of conservation and to provide insight into otherwise inaccessible demographic processes and (2) geneticists can reassure conservation managers that if they are able to conserve ecologically self-sustaining populations and communities, the genetic diversity within species is likely to take care of itself. Of course, if managers are unable to conserve such populations and communities, they are likely to need the advice of geneticists, and we are confident that many will be willing to offer it.

17

Habitat Destruction and Metapopulation Dynamics

Ilkka Hanski

Summary

The rapid loss and fragmentation of natural habitats worldwide has spurred an expanding theoretical literature on metapopulation dynamics, which is expected to answer questions about the minimum viable metapopulation size and the minimum amount of suitable habitat necessary for long-term metapopulation persistence. The simplest rule-of-thumb approaches are unlikely to have much practical value, whereas complex simulation models can only be applied to a few exceptionally well known species. Spatially realistic metapopulation models with a limited number of parameters may represent the right compromise between realism and generality. These models can be parameterized with a limited amount of information to make practical predictions about metapopulation lifetime in arbitrary networks of habitat patches. Such predictions represent an analogous but probably more useful contribution to conservation than the well-known rules of reserve design based on the dynamic theory of island biogeography. One severe limitation of metapopulation modeling is the common assumption of stochastic steady state at the metapopulation level. This may be a bad assumption for most metapopulations in rapidly changing landscapes. In the worst scenario, many still existing rare and endangered species are already "living dead," committed to extinction because extinction is the equilibrium toward which their metapopulations are moving in the current fragmented landscape. To conserve them we must reverse the process of habitat loss and fragmentation.

Introduction

The dynamic theory of island biogeography (MacArthur and Wilson, 1967) captivated the deep ranks of population biologists in the 1970s in a manner that perhaps no single idea had done before. Conservationists received the theory with enthusiasm, as it promised to lay a solid foundation for the process of designing nature reserves (IUCN, 1980; Simberloff, 1988). The timing was apparently just right, as the global reserve area increased dramatically from 2 million km2 in 1970 to more than 5.5 million km^2 10 years later; growth has been substantially slower since 1980 (Groombridge, 1992). To what extent, and in which ways, the island theory actually affected the practice of conservation remains an unexplored but worthwhile subject for an interdisciplinary study. Quite suddenly, though, this period ended.

In the 1980s, ecologists became markedly less excited by community approaches, of which the dynamic theory of island biogeography is a prime example. Rigorous studies of single-species populations were regarded as a healthy cure for the careless ramblings of community ecologists. In the new discipline of conservation biology, interests shifted from large-scale explorations of species richness patterns to viability analyses of individual species. In North America, particularly, a handful of species have received disproportionate attention.

With continuing fragmentation of natural habitats, questions about spatial structure refused to go away, but the emphasis moved from communities in large reserves to species in ordinary fragmented landscapes. The dynamic theory of island biogeography was replaced, as a conceptual framework, by the theory of metapopulation dynamics (Gilpin and Hanski, 1991), as is described by Simberloff (this volume) and in greater detail by Hanski and Simberloff (1996).

The shift in the spatial scale may have helped to popularize the metapopulation notion. Not many of us are familiar with oceanic islands, which supplied key examples for the theory of island biogeography, but all of us are familiar with the loss and fragmentation of some natural habitat. Species distributions become fragmented with their habitat, and metapopulation models of more or less isolated local populations connected by dispersal become fatally attractive. It is not a coincidence that whereas birds provided the bulk of examples, at least initially, for the island theory, butterflies have gained a somewhat similar status in current metapopulation studies especially in Europe (Thomas, in press; Hanski and Kuussaari, 1995; Thomas and Hanski, in press). We are now concerned with matters at the scale of ordinary bugs!

In ecology, the formal theory of metapopulation dynamics was established as a simple abstract model (Levins, 1969), but during the past quarter century the theory and the practice associated with it have successfully radiated in many directions (Gilpin and Hanski, 1991; Hastings and Harrison, 1994). The original concept of a metapopulation as "a population of populations," with a critical role for population turnover, has swelled to encompass other sorts of spatial population structures, for instance the mainland-island (Hanski and Gilpin, 1991) and source-sink metapopulations (Pulliam, 1988). Habitat fragmentation and patchiness may affect various individual=, population=, and community-level processes, of which only some are explicitly addressed by the metapopulation concept (Wiens, this volume).

In this chapter, I elaborate on the question of whether the current metapopulation theory is helpful in assessing the chances of survival of species living in increasingly fragmented landscapes. It is worth making clear at the outset that metapopulation theory cannot provide the full story, it can only make a contribution toward a better understanding of population dynamics in fragmented landscapes. The good news is that, in qualitative terms, we now understand many facets of metapopulation persistence in fragmented landscapes. We may even be able to produce useful short-term predictions, for instance, to compare the probability of metapopulation survival in alternative scenarios of changing landscape. The bad news is that the time scale of environmental change is presently so fast that many metapopulations must be tracking their changing environment with a noticeable lag. This is bad because, in practice, it foils our attempts at long-term prediction. For conservation, the message is even worse because it implies that many of the currently rare and endangered species may be slowly but inevitably approaching metapopulation extinction—an equilibrium created by past habitat fragmentation.

I have organized this chapter under three headings. In the section on *minimum viable metapopulation size* I review two rules of thumb about the dependence of metapopulation persistence on the amount of suitable habitat. In the next section on *spatially realistic metapopulation models* I comment on current efforts in the other extreme, in the construction of detailed models for the purpose of making quantitative predictions about particular metapopulations. In the final section on *nonequilibrium metapopulations* I return to the bad news.

Minimum Viable Metapopulation Size

Assuming the classical metapopulation scenario of an assemblage of extinction-prone local populations connected by some dispersal, it is natural to inquire how the persistence of the metapopulation might depend on the number of such local populations. Answering this question is not quite enough, though, because the local populations are by assumption ephemeral and new populations are being established at currently empty but suitable "habitat patches." This means that the question about the minimum viable metapopulation size cannot be properly dealt with independently of the question about the minimum amount of suitable habitat for metapopulation persistence. These two elements are brought together in the following formula, albeit just a rough approximation, for the expected lifetime of a metapopulation, T_M:

$$T_M = T_L \exp\left[\frac{H\hat{p}^2}{2\,(1-\hat{p})}\right]. \tag{17.1}$$

Here $\hat{p}$ is the fraction of H habitat patches that are occupied at a stochastic steady state, in the balance between extinctions and colonizations, and T_L is the expected lifetime of a local population (Gurney and Nisbet, 1978).

Some explanation is needed. The formula tells us how the properties of the species (encapsulated in $\hat{p}$) and the properties of the landscape (H) combine to scale the local persistence (T_L) to metapopulation persistence (T_M). (Note that, as written, Eq. [17.1] does not provide a complete solution, because $\hat{p}$ depends on T_L.) The theory assumes that all populations have identical dynamics, and hence that no single population persists for much longer than the others. This assumption may often be unrealistic because metapopulation persistence may be due to the survival of one or a few extinction-resistant populations. The theory also assumes no regional stochasticity (spatially correlated environmental stochasticity; Hanski, 1991a), which could wipe out even a large metapopulation by causing simultaneous extinctions of all or most local populations.

Hanski and Thomas (1994) report an average $\hat{p}$ value of 0.65 for three species of butterflies in three well-connected patch networks. Assuming this value of $\hat{p}$, Eq. (17.1) suggests that a metapopulation is relatively safe from extinction $(T_M > 100T_L)$ if the network consists of roughly 10 or more suitable habitat patches. The empirical example on the butterfly *Melitaea cinxia* in Figure 17.1 is in broad agreement with this prediction. Thomas (in press) and Thomas and Hanski (in press) discuss other butterfly examples also supporting the theoretical

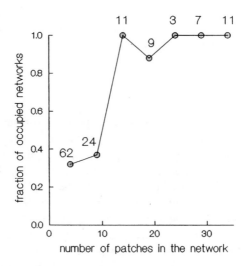

Figure 17.1. Fraction of patch networks that were occupied as a function of the number of patches in the network in the Glanville fritillary butterfly *Melitaea cinxia* on Åland islands in southwestern Finland (based on material described in Hanski et al., 1995a, in press b). The networks have been classified in groups with 1–5, 6–10, and so forth, patches (number of networks is given by the numbers in the figure).

prediction. Adding a safety margin to account for inevitable environmental vagaries, the first rule of thumb says that a network of at least 15–20 well-connected patches is required for long-term persistence of a metapopulation. One or a few small habitat fragments is not enough. Warren (1992) makes the same point forcefully by demonstrating a high rate of population extinctions of nationally rare species of butterflies on small and carefully managed but isolated reserves in Britain over a period of only 20 years.

It goes without saying that Eq. (17.1) is useless for the purpose of deciding how many habitat fragments, which kind and where, are needed for the protection of some particular endangered species in some particular region. It would be foolish and irresponsible to answer such questions with a back-of-an-envelope calculation, ignoring practically all that is known about the species and its environment. Equally, Eq. (17.1) gives no justification for abandoning conservation efforts if in some landscape only a few patches remain. Equation (17.1) and similar results may have some value in providing strategic guidelines for the management of ordinary fragmented landscapes for the benefit of biodiversity in general.

Metapopulation survival is threatened by a small number of patches in a network, but also by great isolation among the patches, regardless of patch number, as isolation lowers the recolonization rate of empty patches and thereby lowers $\hat{p}$ in Eq. (17.1). Patch isolation typically increases with habitat loss and fragmentation. The question then arises whether we can predict the critical minimum patch density necessary for metapopulation persistence, assuming that the properties of the species remain unchanged (a question first asked by Lande, [1987]). Equation (17.1) cannot be applied without having a model to calculate the value of $\hat{p}$ in the increasingly fragmented landscapes. But it turns out that an answer is provided by another rule of thumb, not requiring any knowledge of the details of metapopulation dynamics. Applying this principle to the mother of all metapopulation models, the Levins model, the fraction of empty but suitable habitat patches at steady state is given by

$$\text{fraction of empty patches} = h - \hat{p} = \frac{e}{m}, \tag{17.2}$$

where h is the fraction of patches that remain suitable (fraction $1 - h$ has been destroyed), and e and m are two parameters setting the rates of extinction and colonization, respectively. Here $\hat{p}$ is calculated as the fraction of occupied patches out of all patches, including the destroyed ones. The practical aspect of this result is that the same quantity, e/m in the Levins model (Eq. [17.2]), gives both the fraction of empty patches in all less fragmented ("pristine") states as well as the threshold patch density necessary for metapopulation survival (obtained by letting $\hat{p}$ approach zero; Nee and May, 1993; Nee, 1994; Lawton et al., 1994).

Although Eq. (17.2) has heuristic value, I am not enthusiastic about this rule of thumb. Apart from lacking any empirical support, there are theoretical reasons to be doubtful about Eq. (17.2). This result is based on the assumption that the metapopulation closely tracks the equilibrium (given by Eq. [17.2]) even in a changing landscape. Close tracking of the equilibrium in a rapidly changing environment requires that the turnover rate is high, which, in turn, implies a high immigration rate. But if the immigration rate is high, it becomes hard to defend the Levins model assumption that metapopulation dynamics occur slowly relative to the time scale of local dynamics. Thus, the various assumptions underlying Eq. (17.2) simply do not square well with each other. Note that in Lande's (1987) model, "habitat patches" are individual territories and hence "local dynamics" occur instantaneously, rescuing the model from the preceding criticism. But his model can be applied only to the dynamics of highly mobile species.

A high immigration rate will lower the rate of local extinctions. Taking such a rescue effect into account in a discrete-time version of the Levins model, Hanski et al. (in press a) demonstrated that the fraction of empty patches does not remain constant with increasing h, as happens in Eq. (17.2), but it decreases with increasing h. In this case, a naive application of Eq. (17.2) would lead to underestimation of the minimum amount of suitable habitat for metapopulation persistence, the worst mistake that the manager could make.

Metapopulation models that include the rescue effect predict the possibility of alternative stable equilibria in the fraction of patches occupied (Gyllenberg and Hanski, 1992), essentially because of the positive feedback generated by the rescue effect between metapopulation size and local population size. Figure 17.2 gives a theoretical example as well as a putative empirical example, in which one of the equilibria is metapopulation extinction. The possible occurrence of multiple equilibria in metapopulation dynamics has important implications for conservation. Metapopulations with multiple equilibria may go abruptly extinct even in only slowly degrading environments, and once extinct they may be difficult to reestablish. Multiple equilibria impose an obvious constraint on our ability to predict the occurrence of species in fragmented landscapes.

Spatially Realistic Metapopulation Models

Space is unquestionably an integral element of all concepts and models of meta-population dynamics, but space has been treated in different manners in the models. I have distinguished between spatially implicit, spatially explicit, and spatially realistic metapopulation models (Hanski, 1994a). Spatially implicit models, like the Levins model, make the assumption that all populations and habitat fragments are equally connected to each other, hence the actual spatial locations do not matter and do not need to be modeled. Spatially explicit models typically assume that dispersal is spatially restricted and hence interactions are localized. These models have yielded major new insights about the dynamics of spatially

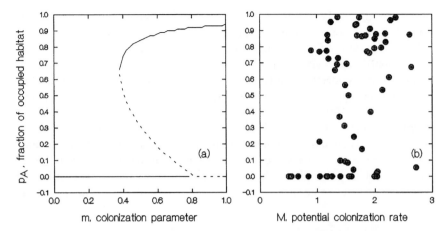

Figure 17.2. Bifurcation diagrams for the fraction of occupied habitat in patch networks (p_A) with increasing colonization rate (m). Panel (a) gives a theoretical example of multiple equilibria from Hanski and Gyllenberg (1993). The continuous lines represent stable equilibria, and the broken lines are unstable equilibria. Panel (b) gives an empirical example from 65 semi-independent patch networks for the Glanville fritillary butterfly *Melitaea cinxia* in southwestern Finland. The networks vary in the number of patches, average size of patches and total area covered by the network. The potential rate of colonization M, corresponding to the colonization parameter m in panel (a), was calculated as $[\sum_j\sum_i e^{-d_{ij}}\sqrt{A_j/n}]^{1/2}$, where d_{ij} is the distance between patches i and j in kilometers and A_j is the area of patch j in hectares (square root transformation was used to scale patch area to population size; Hanski et al., in press b). Here M gives the expected rate of migration from the neighboring patches to the focal patch on the assumption that all patches are occupied (because M has to measure the potential, not actual, rate of colonization). For more details see Hanski et al. (1995b).

structured populations (Durrett and Levin, 1994a; Hassell et al., 1993; Kareiva and Wennergren, 1995).

The metapopulation models that most conservationists think about are spatially realistic models that make quantitative predictions about the behavior of real organisms in real landscapes. The recent coupling of metapopulation models with GIS-based landscape descriptions has created a new wave of excitement. In spatially realistic models, interactions among populations are not only localized but they occur among populations with particular sizes, spatial locations, and other attributes.

One danger with spatially realistic models is the temptation to make them too realistic, to forget that they are models. It may be possible to make good use of a model with tens of parameters depicting the idiosyncrasies of particular species, but one should not assume that such an exercise could be successfully accomplished without a major commitment of resources, not available for most species of interest. I cannot cite a single example in which a complex metapopulation model would have been fully documented, justified, parameterized, tested, and applied (though some such examples may soon become available).

My own preference is for spatially realistic models with relatively few parameters, say only four to six parameters. The advantage of such models is that their behavior can be understood fairly completely and there is some hope of estimating the parameter values for many species. One example is the incidence function model (Hanski, 1994b, 1994c), which can be parameterized with "snapshot" information on the presence or absence of species in a network of habitat patches or with data on population turnover. This model assumes that the metapopulation from which the parameter values are estimated is not too far from equilibrium. As many details are omitted, the model predictions cannot be assumed to be very accurate, but I argue later that this may not be as critical as might first appear.

Nonequilibrium Metapopulations

It is instructive to visit places in nature that one has not seen for two or more decades (I realize that some readers are in a better position to do that than some others). Not always, but often enough, one's principal sensation is change. What used to be a meadow 20 years ago is a thicket today. In areas not much affected by humans, landscape structure may have reached an approximate equilibrium at a larger spatial scale, but in our modern landscapes this is unlikely to happen. Changes in landscape structure are likely to be directional even at large scales.

Changing landscape structure poses no conceptual difficulty for metapopulation theory. We could easily develop spatiotemporally realistic models in the spirit of spatially realistic models. The practical problem, though, is that it all remains hypothetical, because we do not have sufficient knowledge about the environmental changes caused by humans in the past or about the changes that will take place in the future. The following conclusions apply widely:

1. Because we do not know, and cannot predict accurately, the changes that have taken and will take place in most landscapes, long-term prediction of metapopulation dynamics is more of academic than practical interest.

2. Metapopulations track, but with a delay, the changing landscape mosaic. Any model predictions that assume an equilibrium should be treated with caution. Equation (17.2) is an example.

3. For many rare and endangered species, the metapopulation equilibrium *in the present landscape* may be extinction. These species are still around and add to local biodiversity but only because it takes some time for an entire metapopulation to go extinct. Figure 17.3 gives an empirically based example. Unfortunately, we remain utterly ignorant

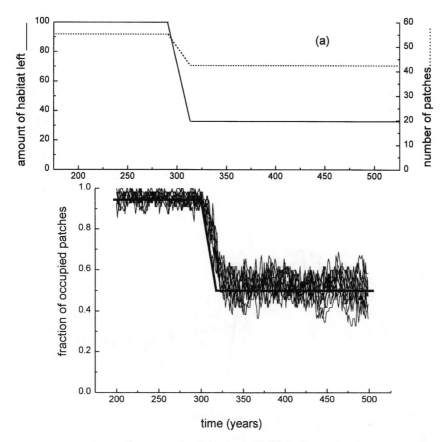

Fig. 17.3. Metapopulation dynamics of the Glanville fritillary butterfly *Melitaea cinxia* in a declining patch network on Åland islands in southwestern Finland. Panel (a, lower) gives the predicted equilibrium metapopulation size (thick line) with 10 replicate predicted trajectories before, during and following an observed reduction in habitat area over a 20-year period (shown in the upper panel; for a map of the observed change in landscape structure see Hanski et al. [in press a]). In this case the metapopulation tracks closely the equilibrium, apparently because the patch network remains well connected and most of the turnover occurs in small habitat fragments. Panel (b) shows similar results for a hypothetical scenario of further loss of 50 percent of the area in each of the remaining patches (upper panel). In this case the equilibrium moves to metapopulation extinction, even though a substantial amount of habitat still remains. The simulated trajectories of metapopulation change show a slow decline with much variation, and it may take tens or even hundreds of years before the metapopulation has reached the equilibrium. Many metapopulations in severely fragmented landscapes may represent this case: they are currently approaching extinction, which is the equilibrium in the present landscape. The trajectories were calculated with the incidence function model (Hanski, 1994b) using field-estimated parameter values. This example is described in a greater detail in Hanski et al. (in press a).

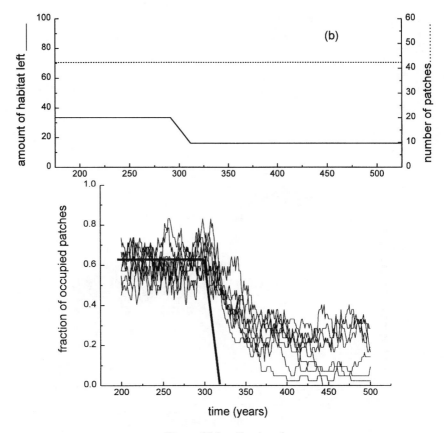

Figure 17.3. Continued

about the frequency of metapopulations that are already such "living dead."

Given that long-term prediction is difficult, what use do we have for models? In conservation, the main use of metapopulation models is probably to predict the consequences of alternative scenarios of landscape change, where it suffices to correctly rank the alternatives. For instance, other things being equal, is it more harmful to the survival of a metapopulation if one large habitat patch is entirely removed, or if the areas of several patches are halved (Hanski, 1994b, 1994c)? The dynamic theory of island biogeography inspired the set of well-known rules about the design of nature reserves (Simberloff, 1988). I suggest that the analogous contribution that one may expect from the metapopulation theory is qualitative predictions about the performance of particular metapopulations in particular landscapes based on relatively simple but spatially realistic models. There are two reasons to expect that the latter sort of predictions are more helpful than the rules of reserve design based on the island theory. First, the rules of reserve design as used in practice are static, although justified by the dynamic theory of island biogeography. The fundamental concept is the species-area relationship, which in the applications means a fixed number of species in a fixed area. In contrast, the metapopulation predictions explicitly address the dynamics of species survival. An example is the prediction of expected metapopulation lifetimes in networks A and B. Second, the reserve design rules contrast fixed alternatives, whereas with the spatially realistic metapopulation models one is practically forced to make predictions for specific fragmented landscapes. It remains to be seen how widely any use can be made of such predictions in practice. So far, the main contribution of the metapopulation theory to conservation has been to help focus attention on population dynamics and persistence in small-scale fragmented landscapes. This contribution should not be belittled either.

Acknowledgments

I thank Atte Moilanen for providing the results for Figure 17.3, and Russell Lande and Daniel Simberloff for comments on the manuscript.

18

How Viable Is Population Viability Analysis?

Katherine Ralls and Barbara L. Taylor

Summary

Population viability analysis (PVA) uses data in a model to estimate the risk of
extinction for a population. Ideally, PVA would assess and integrate the effects
of all factors affecting the persistence of a population, which are usually stated
as the demographic, genetic, environmental, and catastrophic risks and the interactions between them. This goal, however, is still beyond our capabilities.

Three types of PVA have been developed: analytical and both custom and
generic simulation models. Each type has strengths and weaknesses. PVAs have
been criticized because (1) they are a single-species technique, (2) they omit risk
sources that are difficult to estimate, (3) they are sometimes inappropriate for
endangered species management, and (4) they project current conditions long
into the future.

In spite of its limitations, PVA is useful in a variety of situations as long as
we emphasize our limited abilities as scientific prophets and make uncertainties
explicit. When PVA is used to estimate relative risk, it may be acceptable to
omit some risk factors. However, some applications of PVA require the estimation
of absolute extinction risk, and in these cases, the estimate of extinction risk
should attempt to include all sources of risk and uncertainty.

Even in an era when management emphasizes an ecosystem perspective, some
species will need individual attention. We predict that PVA will continue to
evolve as a useful tool for single-species management.

Introduction

The boundaries of population viability analysis are fuzzy. We consider that PVA
uses data in a model to estimate the risk of extinction (or "quasi-extinction";

Ginzburg et al., 1990) for a population. By defining PVA in this way, we exclude viability assessments based on verbal models (Ruggiero et al., 1994) and/or expert opinion (Thomas and Fema, 1993). Another problem in defining the limits of PVA is that PVA-like models, especially when used to make short-term projections, may not be called PVA (e.g., Starfield et al., 1995). General reviews of PVA are provided by Boyce (1992), Burgman et al. (1993), and Nunney and Campbell (1993).

Soulé (1987) envisioned using PVA to help define the minimum conditions for the viability of natural systems. He suggested that multiple PVAs could be performed for this purpose on various types of candidate species. Other goals for PVA are species-specific management and the ranking of species according to risk. Different authors have suggested a variety of types of candidate species for PVA in relation to these goals:

1. Reserve design/system viability:

   a. Species that create critical habitat[1, 2]
   b. Mutualist species[1, 2]
   c. Predators/parasites that regulate populations[1, 2]
   d. Well-studied species[2]
   e. Indicator species[2]
   f. Large carnivores[3]

2. Species-specific management:

   g. Species of special cultural value[1,2]
   h. Rare species or narrow endemics[1, 3]
   i. Threatened or vulnerable species.[1, 2, 3]

3. Ranking species according to risk:

   j. Threatened or vulnerable species.[1, 2, 3]

However, due to the realities of funding, most PVAs have been conducted on rare or endangered species. PVAs on threatened species have been used to help assess reserve design to a limited extent, particularly in the context of habitat conservation planning (D. Murphy, personal communication, 1995). However, species-specific management has come to be the most common use. Using PVAs to rank species according to risk is still impractical (Taylor, 1995).

Will PVA—a single-species technique—remain useful as conservation policy

---

[1]Soulé 1987.
[2]Possingham et al. 1993.
[3]Noss and Cooperrider 1994.

increasingly stresses the preservation of functioning communities, ecosystems, and landscapes (Edwards et al., 1994; Noss and Cooperrider, 1994)? We explore this question by considering PVA's methods and uses, its critics' objections, and its potential future contributions.

PVA Techniques

There is no standard method for conducting a PVA. The classical PVA technique is a case-specific simulation model that attempts to maximize the use of available data, consider the life history of the organism, and identify the most likely threats to its continued existence. PVAs are generally based on existing data, rather than data specifically collected for a PVA.

Three types of PVA have been developed: analytical and both custom and generic simulations. Analytical models (such as Dennis et al., [1991]) are not commonly used so we concentrate on simulation models. Although custom models potentially provide the best fit to the problem, they require an expert modeler, a good deal of time (and hence money), and could suffer because there is no quality control over the complex programs. Such case-specific models, however, probably represent a minority of the PVA analyses currently being performed. The relative proportions of different types of PVA are difficult to estimate because many PVAs are never published. In the United States, government agencies often develop models for a specific species, whereas PVA work in other countries has often used generic models (R. Lacy, personal communication, 1994).

Generic models have different assumptions and different capabilities (Akça-kaya, 1992; Lacy, 1993; Possingham and Davies, 1995). The limited expertise required to run these models can be both a plus and a minus. Nonmodelers can assemble all data, obtain quick results, and use a quality-controlled program. However, nonmodelers may give too much weight to results and not appreciate limitations set by model assumptions (Harcourt, 1995). At its worst, an attempt to use a generic model "can be an exercise in the data-blind (the programmer, through the program) leading the algorithm-blind (the field biologist or other user of the program)" (Lacy, in press). This brings us to consider interpretation of PVA results.

Interpretation of PVA

Ideally, PVAs would assess and integrate the effects of all factors affecting the persistence of a population, which are usually stated as the demographic, genetic, environmental, and catastrophic risks and the interactions between them. This goal, however, is still beyond our capabilities, and this ideal analysis is only approximated to different degrees by different case studies. In the worst cases, the basic natural history of the target organism is unknown. Genetic risks are almost always unknown. Long-term data to estimate the environmental risks are very rarely available, and catastrophes may be of major importance (Mangel and Tier, 1993b) but their nature, frequency, and impact are often unknown (Young, 1994). So, what use is an estimate of risk based on data of questionable quality?

Within a species, the estimate of risk can be considered a relative measure. For species of conservation concern, PVA can be used to help decide among management options by identifying the options that maximize risk reduction. Similarly, PVA models can help assess the benefit of different research proposals. When used in this manner, PVAs are not predictive but rather used as tools in single-species management. If a number of different models lead to the same ranking of management options, we may have some confidence in these rankings, even though the models give different quantitative predictions.

The probability of extinction has been proposed as a measure to rank species according to risk (Mace and Lande, 1991). This requires that the measure be absolute rather than relative. This is problematic because of difficulties in model choice and in incorporating uncertainty in parameter estimates. Mills et al. (1996) have shown that the generic models vary in their predictions when applied to the same case, in particular, regarding the effects of density dependence. Taylor (1995) showed that uncertainty in estimating such parameters as population growth rates, contributed more uncertainty to estimates of extinction probabilities than did inclusion of environmental variability.

Assessments of reserve or system viability may use PVA extinction probabilities as either an absolute or a relative measure. If management is reduced to choosing among limited options, then PVA can be used to assess the relative risk for one or preferably several species (Lindenmayer and Possingham, 1994). Alternatively, if a very large system is currently extant and the question is how far can it be reduced, the extinction probability should be absolute, that is, reflect the real risk of extinction. The latter is the most demanding use of PVA because the desired output is a real prediction about the chance of extinction following a course of action. If we omit potential catastrophes or genetic problems from our model, we should choose conservative criteria for minimum viable population size to compensate for this bias. Alternatively, we could attempt to include some hypothetical level of risk for these unknown factors. The limitations of our present ability to predict extinction risk imply that many system designs based on current models may ultimately prove inadequate and thus require active human intervention to compensate for their deficiencies.

When PVA is used to generate a relative measure, it may be acceptable to omit effects of genetics and catastrophes to the model. Alternatively, sensitivity analysis could explore the effects of different assumptions regarding these risks. Discussions of PVAs that omit risk factors should emphasize that the extinction probability is an underestimate but the bias would be constant as long as those factors do not differ between management options. However, when a decision is made that is irreversible in the short term, such as destruction of slowly regenerating habitat, the estimate of extinction probability should attempt to include all sources of risk.

Criticisms of PVA

Before discussing criticisms of PVA, we note that PVA is a process that produces a product. Criticism is generally directed at the product. The process, however, may be useful even when the product is not outstanding. Existing data, many of which are unpublished and scattered among various agencies and researchers, may be evaluated and synthesized for the first time. Research may be undertaken to fill the data gaps identified. One of the major benefits of modeling is that assumptions must be articulated and reasoning made explicit. Thus, those participating in the PVA process are likely to develop a clearer perception of the problems facing the species and begin to attack them in a more unified manner. Furthermore, documenting the problem, the data, the uncertainty, the model of extinction, and the management options under consideration enables others to criticize the decision-making process. Such challenges often lead to improvements in the process.

PVAs have been criticized because (1) they are a single-species technique, (2) they omit risk sources that are difficult to estimate, (3) they are sometimes inappropriate for endangered species management, and (4) they project current conditions long into the future. Most PVAs do not attempt to explicitly incorporate the dynamics of other species. They do not, however, ignore interspecies interactions because the models incorporate variance in birth and death rates. These rates vary because of changes in the target organism's biotic and abiotic environment. Thus, "environmental variance" implicitly incorporates interspecies interactions. Because most studies are short term and small sample sizes generally result in imprecise estimates, it is very difficult to quantify environmental variance. Therefore, it is often necessary to repeatedly use the few long-term studies as surrogates. For example, Barlow and Boveng (1991) present a technique to estimate age-specific survival based on minimal information. Their technique relies on using expected patterns of mammalian survivorship, based on a few well-documented species, to define a general model of age-specific mortality. Long-term studies also provide estimates of catastrophic effects. The value of these studies should encourage more long-term studies on carefully chosen PVA candidates.

Omitting risk sources, such as genetics (Frankham, 1995) or catastrophes (Mangel and Tier, 1993a) could result in overly optimistic risk assessments. Other reasons PVAs may underestimate risk are that probable interactions between different sources of risk are poorly known and not reflected in most existing models (Gabriel and Bürger, 1992) and that existing models model random events as "white" rather than "colored" noise (Lawton, 1988; Halley, 1996).

For most species, very little is known about genetics or the nature of catastrophes. Furthermore, catastrophes have different dynamics: some may be density independent, such as fires or frosts, whereas others may be density dependent, such as disease. The type of catastrophe is important in deciding between different management options. For example, if fire is a threat, corridors between reserves may allow population escape. If epidemic diseases are frequent, corridors may accentuate the problem. When designing a reserve, different options may be best for different species. Although PVAs are potentially limited by parameters that are difficult to estimate, at least the models can be employed to assess the importance of further research on these risk factors. Whether or not particular factors are relevant to the available management options will differ depending on the case. If ranking of management options is extremely sensitive to some risk factor, we clearly need more information on that factor.

Often, PVAs have been perceived as the only or best model for endangered species management. These mistaken perceptions have led to unnecessary PVAs and various inappropriate uses of PVAs, which in turn have led to criticism. Caughley (1994) feels PVAs are inappropriate for most species facing extinction because they do not directly address the causes of decline. We agree that, if the problem is clear, such as an introduced rat eating eggs, no modeling may be required. If a species is declining from unknown causes, highest priority should be given to models that help prioritize research to determine such causes. For example, a model may be used to determine that reduced fecundity alone could not account for an observed decline. Further models may indicate that reduced juvenile survival is the most likely problem, and research efforts should estimate juvenile survival, identify mortality causes, and assess different strategies for reducing juvenile mortality. If a simple demographic model will suffice, a PVA is unnecessary. In fact, conducting a PVA that requires estimation of many parameters, with few data, may only cloud the conclusions.

This does not mean, however, that PVAs are never useful for declining populations. PVAs can be used to develop objective listing and delisting criteria for an endangered species, which may be required under protective legislation. Once a problem has been identified, different management options can be compared using the risk of extinction generated via PVAs (Possingham et al., 1993).

The appropriate complexity of a model must depend both on the objective of the modeling exercise and on the availability of data. A model has served its purpose if it provides useful insight into a problem. It is unnecessary and unrealistic to believe that a model perfectly mimics a natural system. However, if a generic model forces users to make unrealistic assumptions, then the model is probably inappropriate. The scientist must choose a model appropriate to the problem, which will not always be a PVA.

Projecting current conditions (or conditions over whatever length of time we have data for) into the future is not unique to PVA but is a general characteristic of simulation models used for forecasting. It poses a problem because the future "is a combination of the known and the unknowable" (Rosenhead, 1989). We can predict that surprises will occur in the future with confidence—it is only the nature and timing of these surprises that will be a surprise! Even if we had good estimates of conditions over the past decade, it is unlikely that we could use these data to accurately predict conditions 50 to 100 years into the future. The likelihood of serious errors increases as we project simulations farther into the future and when we attempt to make absolute rather than relative risk estimates. It is important that we emphasize our limited abilities as scientific prophets, attempt to make uncertainties explicit, and incorporate a margin of error in the interpretation of results (Ludwig et al., 1993).

PVAs also suffer from a lack of standards. There are no consensus guidelines that can be used to judge the quality of a PVA (Warshall, 1994). This is a practical problem because even "state-of-the-art" PVAs can be legally challenged, and because many poor-quality PVAs are never subjected to scientific review.

The Future of PVA

PVA was developed as part of the field of conservation biology, which, from its inception, viewed nature as a dynamic process (Soulé and Wilcox, 1980). PVA models can be compatible with an ecological paradigm emphasizing that systems are open, frequently non-deterministic, and rarely at equilibrium (Pickett et al., 1992). Factors originating either inside or outside the modeled system can cause variance in reproductive and survival rates. PVA models are stochastic rather than deterministic. And modelers are ingenious. Thus, we envision PVA developing in accordance with emerging ecological paradigms.

Even in an era when management emphasizes an ecosystem perspective, some individual species will require detailed study and specific attention (Soulé, 1994; Noss and Cooperrider, 1994). Today's cutting-edge models may be destined to join the "broken-stick models, . . . discarded alpha matrices, and other strange and wonderful debris" that Soulé (1987) pictures littering the field of mathematical population biology. However, population models will continue to evolve as useful tools for species management. Most of these models, whether or not they are called PVA, will have been influenced by PVA's emphasis on stochasticity, uncertainty, and risk.

Existing PVA models are most suitable for vertebrates; the future should bring models that are more suitable for plants and invertebrates. Existing models address single species; the future will see multispecies models as well. New approaches include the use of Bayesian techniques (USFWS, 1994b), combining PVA-like models with structured decision-making methods (Ralls and Starfield, 1995), and spatially explicit population models (McKelvey et al., 1992; Dunning et al., 1995; Turner et al., 1995; Lindenmayer and Possingham, 1994). Bayesian models (Smith and Gelfand, 1992; Berger and Berry, 1988) provide a framework for including uncertainty. Uncertainty in model parameters has already been incorporated in a PVA (USFWS, 1994b) and uncertainty in model choice could also be similarly incorporated. Structured decision-making approaches can help groups organize the massive amount of computer output typically generated by simulation models and facilitate consensus. Spatially explicit population models, which combine a population simulation with a landscape map, require even more data than other types of models and thus can be developed for even fewer species (Conroy et al., 1995). Kareiva et al. (this volume) suggest that simple presence/absence data may often be more useful than spatially explicit models. Custom models, whether or not they are spatially explicit, will be used as part of an adaptive management process designed to increase understanding of the system being modeled and provide data for improving the model (Christensen, this volume).

Although PVAs can be used to help evaluate management alternatives for endangered species, models more directly designed to address the current decision may prove useful as well. PVA models often include factors on which decisions are rarely based, such as the rate at which a population is losing genetic diversity, and omit factors that heavily influence management decisions, such as the amount of money available and the cost of various management alternatives.

Because PVAs cannot be conducted on many threatened and endangered species due to lack of data, resources, or suitable models, the development of simple criteria to rank species according to risk would be extremely useful. PVA could be used to evaluate the robustness of proposed criteria. A Bayesian approach that incorporated uncertainty in parameters and model choice may yield measures of risk usable to rank species, but these techniques have not yet been developed.

PVA may also help define the minimum conditions for the viability of natural systems. PVAs of large carnivores are still recommended to help determine area requirements for system viability (Noss and Cooperrider, 1994). Conducting multiple PVAs on various types of species within the same system, as envisioned by Soulé (1987), might provide additional insight.

In view of the many potential future uses of PVA, we predict that PVA is not likely to become extinct in the near future but will survive and likely speciate: many of its descendants may be known by other names.

Acknowledgments

We thank Bob Lacy, Marc Mangel, and Hugh Possingham for comments on the draft manuscript.

19

Reserve Design and the New Conservation Theory

Nels E. Barrett and Juliana P. Barrett

Summary

Reserve design is one of the most fundamental tools conservationists have to protect, maintain, or enhance ecosystem function, heterogeneity or patchiness, and ultimately, biological diversity. Unique objects of our ecological heritage, such as species, communities, and sites, once the traditional focus of reserve design, are now perceived as integral components of ecosystems demonstrated to be open, dynamic, and heterogeneous. From a contemporary conservation view, reserve design encompasses ecological processes governing the ecosystem, including the activity and turnover of organisms and the structure of natural communities. Reserve design also acknowledges the landscape context defining the spatial and temporal relations of ecosystems bounded by natural, functional limits, and anthropogenic constraints.

The challenge facing contemporary conservation practice is to design a reserve that will effectively preserve and manage samples of natural and seminatural landscapes that are representative of a larger, regionalized landscape, and yet, still maintain the integrity of ecosystem function, heterogeneity, and biological diversity. Accomplishing adequate representation of a regionalized landscape requires establishing an integrated, landscape-scale, reserve design. This type of reserve design, referred to as a "bioreserve," is a regionalized network of individual preserves as core sites integrated within a series of concentric zones acting as corridors or successive buffers to the core sites. The level of protection in each zone is based on the direct relationship of the zone to the dominant core and any perceivable threats or constraints due to conflicting land uses. An integrated, reserve network is described to protect the Connecticut River Tidelands ecosystem.

Reserve design is as much a process as it is a product. To achieve the goals of reserve design, key ecological issues must be conveyed between three interrelated, but different disciplines: science, management, and policy. A matrix approach is proposed to structure the interactions within and among disciplines. To illustrate the utility of the interactive matrix approach, a case study is presented for the establishment of an integrated reserve design for preserving the ecological integrity of the Connecticut River Tidelands ecosystem.

Introduction

Recent advances in ecological knowledge and understanding have been described as a shift in ecological perception from the classical "equilibrium worldview" to a contemporary "nonequilibrium worldview" (Fiedler et al., this volume; Pickett et al., 1992). The implications of such a paradigm shift have forced a reexamination of conservation principles related to reserve design (Table 19.1). Three important prerequisites of contemporary reserve design include (1) an enhanced knowledge of ecological systems; (2) a bioreserve approach to preserve and protect the integrity of ecosystem function, heterogeneity, and ultimately biological diversity with representative samples of the regional landscape mosaic; and (3) modes of transdisciplinary interaction and cooperation.

An Enhanced Knowledge of Ecological Systems

Traditionally, reserve design has always been justified on the grounds of preserving valued objects of our ecological heritage, for example, rare and endangered species, critical habitats, unique communities, and areas deemed outstanding for reasons of their natural beauty. According to the classical equilibrium viewpoint, any object or unit of nature of ecological interest, in and of itself, was considered conservable because natural systems were assumed to be closed, static, and fixed (Pickett et al., 1992). Reserve design under the equilibrium paradigm typically required an appreciation of nature, but little or no ecological understanding. Valued objects of our ecological heritage were enclosed within a separate and often arbitrarily defined "natural area," which was simply set aside for posterity as a "diorama" of sorts. Epitomized metaphorically as "the balance of nature," the once prevailing belief was that such natural areas, if left alone, could persist indefinitely. Ecologists soon realized that such simple conservation measures were not always adequate. Threats originating outside the natural area remained unchallenged. Even natural changes (i.e., succession and natural disturbances) invariably impacted the preservation status of conservation elements of interest. In such cases, a strictly hands-off management approach amounted to no less than "benign neglect." Conversely, attempts to manage for biological diversity by techniques of isolation amounted to no more than managing for biological antiquity. Accumulating ecological evidence at variance with the traditional equilibrium paradigm has eventually dispelled the notion of nature-in-balance as a

Table 19.1. *Contrasting conservation principles derived from a shift in ecological thinking from the classical viewpoint or "equilibrium paradigm" to the contemporary viewpoint or "nonequilibrium paradigm."*

	Conservation Principles	
	Classical Viewpoint or "Equilibrium Paradigm"	Contemporary Viewpoint or "Nonequilibrium Paradigm"
Goal	Preserving valued *objects of ecological heritage*; unique individuals, communities, habitats, and vistas	Preserving the *ecosystem processes and context* (ecological integrity and biological diversity) by representativity in natural and seminatural landscapes
Focus	*Fixed* natural areas surrounding *closed* and *static* communities, i.e. "dioramas"	*Heterogeneous landscape* mosaic sustaining *open* and *dynamic* communities, i.e., "functional landscape units"
Theme	The *consequences* of life; a fixed and final outcome	The *circumstances* of life; multiple causation
Emphasis	*Stability* and *persistence of objects*, structural *completeness*	Structural *context,* and dynamic *processes,* historical *contingency*
Humans as	Cultural landscapes and humans *excluded* components	Seminatural landscapes, and humans *incorporated*
Scale	Generally *small*; set by size of object (fine grain; small extent)	Generally *large*; set by range of processes (variable grain; large extent)
Metaphor	"The balance-of-nature," nature is constant or self-sustaining, i.e., *"equilibrium"*	"The flux-of-nature," nature is manifold and dynamic, i.e., *"nonequilibrium"*
Knowledge	Ecological understanding *not essential*	Knowledge of ecological systems is *critical*
Partnerships	*Competitive* or *isolated* "party lines," cooperation not emphasized	*Transdisciplinary communication* and *cooperation* vital
Management	From non-intervention "benign neglect", to *passive* or limited management	*Active* management of processes and implications context (structure and linkages)
Application/ example: overview of reserve plan	Preserve the biological diversity of the Connecticut River tidelands by establishing several reserves	Preserve and protect ecosystem processes and context that maintain ecological integrity and perpetuate the biological diversity of the Connecticut River tidelands ecosystem by establishing and managing a regionalized, watershed-based network of integrated core-dominant reserves with concentric buffers, i.e., a "bioreserve"

tenable foundation for conservation practice (Botkin, 1990; Egerton, 1993; Pickett et al., 1992).

Alternatively, the nonequilibrium viewpoint, typified as "nature-in-flux" (Pickett et al., 1992), offers a prospective basis for contemporary conservation (Fiedler et al., this volume; Shachak and Pickett, this volume) and an incentive to reexamine traditional goals of reserve design. Valued objects of our ecological heritage such as unique species, communities, and sites, once the traditional emphasis of reserve design, are now perceived as integral components of ecosystems, demonstrated to be open, dynamic, and heterogeneous. From the contemporary nonequilibrium viewpoint, reserve design emphasizes ecological processes that govern the transfer of energy and cycling of materials as well as circumstances that organize the structure of natural communities. Reserve design also acknowledges the ecological context that defines the spatial and temporal relations of ecosystems restricted by natural, historical, and functional limits including anthropogenic constraints (Pickett et al., 1992). Presently, the challenge facing contemporary conservation practice is to design a reserve that will effectively preserve and manage samples of natural and seminatural landscapes that are representative of a larger, regionalized landscape and yet still maintain the ecological integrity of ecosystem function and heterogeneity, and ultimately, biological diversity.

Applying a Bioreserve Approach

Contemporary reserve design must be justifiable on the basis of preserving two major aspects of ecological systems, biological diversity and ecological integrity (Angermeier and Karr, 1994). Biological diversity generally refers to the biotic basis of the Earth, that is, "the diversity of life" (Wilson, 1992), which encompasses multiple levels of organization, for example, genes, species, and ecosystems (Angermeier and Karr, 1994; Noss and Cooperrider, 1994). Ecological integrity is defined as the state of ecosystem development that is representative of the historical range of ecological processes and constraints governing the system (Woodley, Kay, and Francis, 1993). The task of preserving biological diversity and ecological integrity by managing ecosytems in natural and seminatural landscapes is to take utmost precedence in conservation (Eidsvik, 1992; Grumbine, 1994; Western et al., 1989; Woodley et al., 1993).

For the purposes of reserve design, both biological diversity and ecological integrity are contextually defined by region, that is, encompassing a specified geographic area and organized by the processes and constraints that integrate communities and habitats representative of the landscape mosaic. Accomplishing adequate representation of a regionalized landscape requires establishing an integrated, network of reserves hereafter refered to as a "bioreserve." Analogous to biosphere reserves (UNESCO, 1974) and reserve networks (Noss and Cooperrider, 1994), the bioreserve design is a regionalized network of individual preserves as core sites integrated within a series of concentric zones acting as corridors or successive buffers to the core sites. The level of protection in each zone is based on the direct relationship of the zone to the dominant core and any perceivable threats or constraints due to conflicting land uses.

Modes of Transdisciplinary Interaction and Cooperation

Reserve design is as much a process as it is a product. The task of reserve design is not strictly a scientific enterprise. A number of different disciplines, including but not limited to science, management, and policy, have considerable influence on the ultimate success or failure of reserve design. Traditionally, reserve design was often parochial and piecemeal with little or no communication among conservation practitioners. Working in isolation, scientists, managers, and policy makers often pursued separate conservation interests consistent with the fundamental interests of each discipline. Unfortunately, separate and often conflicting agendas emerged that actually suppressed cooperation and encouraged competition among "individual party lines" to suit each special interest. For reserve design to succeed under the contemporary nonequilibrium viewpoint, not only is ecological knowledge and its application essential but interaction and cooperation among scientists, managers, and policy makers is also crucial (Rogers, this volume; Zedler, this volume).

To facilitate reserve design, a transdisciplinary, interaction matrix is proposed. The purpose of the matrix is to provide a framework that determines how different disciplines can engage in a cooperative relationship to jointly achieve an ecologically suitable reserve design.

As constructed, the transdisciplinary, interaction matrix is symmetrical (Table 19.2). The participating disciplines are listed as both row and column headings. Each cell of the matrix represents a pairwise interaction (feedback) between disciplines as contributors (rows) or recipients (columns) of key information. Because the interaction matrix is symmetrical, the cells along the principal diagonal represent the distinct conventions characteristic of each discipline. Other cells of the matrix at intersections between different disciplines represent reciprocal interactions. Thus, for disciplines to be truly cooperative, the information content of cells of the primary diagonal must relate to the total suite of transdisciplinary information contained in other intersecting cells of the same row or column. For example, specific ecological knowledge and understanding that result from scientific research, provide the ecological models and key indicator variables required of management and also provide the issue clarification and scientific rigor and oversight needed by policy. Likewise, conservation actions by management, coincident with reserve design, provide an evaluation of ecosystem integrity or ecosystem distress for policy and simultaneously feed back research needs or priorities to science. Finally, a feasibility analysis for reserve design is conducted by policy makers, establishing programmatic goals and conservation strategies for both scientific research and adaptive management practices.

Table 19.2. A generalized, transdisciplinary, interaction matrix among science, management, and policy.

		Recipients		
		Science	*Management*	*Policy*
C o n t r i b u t o r s	**S c i e n c e**	EXPLANATION Exploration, characterization, and interpretation: descriptive and quantitative models, statistical analyses	Ecological models: response rules, causal structure, range and thresholds of key indicators of ecological integrity or threats	Scientific rigor: oversight and issue clarification
	M a n a g e m e n t	Framework of research needs and priorities	PREDICTION Manipulation and monitoring: create maintain, or restore integrity; eliminate, avoid, or mitigate threats	Performance evaluation: measures of satisfaction or utility, available technology
	P o l i c y	Program support, relevancy, practicality, mission guidelines	Program support, scope, management guidelines	GOALS/STANDARDS Feasibility, programs and strategies: address social, economical, and political norms

The matrix format emphasizes the information content of interactions among cooperating disciplines as contributors or recipients in a joint conservation endeavor. The cells of the matrix along the principal diagonal identify the conventions adherent to each discipline. The remaining cells of the matrix describe the reciprocal interactions among disciplines.

Case History: A Bioreserve Design for the Connecticut River Tidelands

Reserve design is one of the most fundamental tools conservationists have to protect, maintain, or enhance the integrity of ecosystem function and heterogeneity or patchiness, and ultimately, biological diversity. An application of contemporary conservation theory based on the nonequilibrium paradigm is presented as a case study describing a bioreserve design for the Connecticut River Tidelands ecosystem.

Science Quantifies the Ecological Dimensions of the Connecticut River
Tidelands Ecosystem as the Basis and Justification for a Bioreserve Design

The Connecticut River is the largest river system in New England. It stretches 660 km along its main stem from northern New Hampshire to Long Island Sound, draining approximately 29,100 km². Located at the lower reaches of the river, within the influence of tides, is a continuous series of salt, brackish, and freshwater tidal wetlands known as the "Tidelands of the Connecticut River." Considered exemplary in New England, the Tidelands of the Connecticut River have been designated a "Last Great Place" by The Nature Conservancy and deemed as a wetland ecosystem of international importance under the RAMSAR convention.

At the landscape scale, the dominant ecological controls of the lower, Connecticut River watershed are physiographic processes. Geologic control imposed by resistant metamorphic rock (Bell, 1985; Rodgers, 1985) and the lack of a well-developed coastal plain are reflected by the small size and limited extent of the tidal wetlands. The Tidelands occupy isolated embayments and islands associated with pendant bar systems, as well as drowned coves and valley marshes of tributaries. In many places, sediment accumulation relative to sea level rise has enabled tidal wetland development to keep pace with submergence (Gayes and Bokuniewicz, 1991; Patton and Horne, 1991). Hydrologic processes are the chief definitive agents of the Tidelands ecosystem. The role of hydrology is especially pronounced due to two distinct flooding regimes: the rigors associated with seasonal river discharge of uncertain magnitude and timing, and the unceasing ebb and flood of the saline, oceanic tides. Hydrologic processes are the key factors shaping the Tidelands habitat and dictating appropriate ecosystem responses, for example, the composition, abundance, turnover, and diversity of organisms. Hence, the regionalized landscape perspective emphasizes the open, dynamic, and heterogeneous character of the Connecticut River Tidelands ecosystem. These ecological dimensions of the Connecticut River Tidelands ecosystem provide the basis and justification for a bioreserve design. See Rogers (this volume) for an example of operationalizing conservation in river landscapes of arid areas.

Science Provides Management with Ecological Models and Key Indicators of
Ecosystem Integrity or Distress to Supervise the Bioreserve Design

Modeling of various ecosystem aspects of the Connecticut River Tidelands is used by management to guide conservation practice and supervise reserve design. For example, an empirical model that relates gradients of plant communities to gradients of hydrologic parameters (i.e., salinity, inundation, and current flow) is used to identify and monitor core reserve sites. In situations where good quantitative information describing ecosystems is lacking or unavailable to managers, key indicators are used (Woodley et al., 1993). The criteria for choosing appropriate indicators is discussed by Keddy et al. (1993). Sensitivity, efficiency, and above all, ecological meaningfulness are the most important indicator qualifications for conservation purposes. Rare species, limited by autecology, habitat, or geographical range, are generally indicative of ecological integrity. Keddy et al. (1993) issue a cautionary note: the lack of rare species does not always imply a loss of integrity. Potential indicator species include several vascular plants and animals designated as "Connecticut's Endangered Species, Threatened Species, and Species of Special Concern" (The Connecticut Endangered Species Act, Chapter 495, General Statutes). Other indicators used in the Tidelands program include natural communities and seminatural cover types.

Conversely, monitoring for ecological distress is usually threat specific where cause and effect are known (Woodley, 1993). Listed in decreasing order of perceived severity are habitat conversion, nuisance species, water pollution, human disturbance, sea level rise, and physical barriers. For example, invasive weeds, like Purple loosestrife *(Lythrum salicaria)* pose a threat by displacing more desirable plants. Exotic plants, invasive or not, are symptomatic of much larger problems (Keddy et al., 1993) associated with human activity. The potential impact of various stressors, and likely conservation actions figure prominently in reserve design.

Restoration opportunities are a significant means to restore ecological integrity in a settled landscape such as southern New England, where previous conservation efforts were not possible, had failed, or were ignored. Restoring historical tidal flows and removing invasive species are common maintenance practices.

Science Provides Policy with Recommendations for a Bioreserve Design and Clarifies Ecological Issues

The aim of the bioreserve design is to preserve the ecological integrity of ecosystem processes that maintain the exemplary tidal marsh communities and associated species of the Connecticut River Tidelands ecosystem through regional land conservation. Perceived as an open, dynamic, and heterogeneous landscape, the Connecticut River Tidelands ecosystem warrants a reserve design that incorporates ideas of representativity, redundancy, substitutability, and connectivity in the landscape. Representativity is typifying the full range of ecological processes and patterns present, that is, a regionalized cross section of sorts. Redundancy means representing the same process or ecological component repetitively. Substitutability means representing one process or ecological component in place of another. Akin to redundancy, substitutability also refers to accommodating novel changes that may occur in the future. Connectivity refers to an adequate proximity of space or time to allow for the transfer of energy, exchange of material, or movements of organisms. To incorporate these ideas, a bioreserve design is advocated for the Connecticut River Tidelands that utilizes a watershed approach and an integrated network of sites.

The ecological dimensions of the Tidelands ecosystem are determined by the shape and hydrography of the regional watershed. Basin-wide ecological stratification of the heterogeneous landscape mosaic proceeds by drainage order or some other regionalized ecological land classification scheme. Superimposed on the drainage basin is an integrated reserve consisting of a network of multiple core sites integrated within a series of concentric zones acting as corridors or successive buffers to the core sites. The level of protection afforded in each zone is mainly based on the hydrologic relationship of the strictly protected dominant core sites to any perceivable threats or constraints due to patterns of human settlement and conflicting land uses within the basin.

Multiple core sites are necessary to ensure adequate landscape representativity and the redundancy necessary to secure the patchiness characteristic of nonequilibrium systems (Pickett, 1980; Wiens, this volume) for metapopulation turnover within patches (Gilpin and Hanski, 1991) or rescue among patches (Hanski, this volume), as a hedge against losses due to natural or accidental disturbances (Pickett and Thompson, 1978; Peters et al., this volume), and ultimately for habitat shifts due to sea level rise (Patton and Horne, 1991). Corridors are designated as a hedge against insularization to allow for the movements, migrations, and dispersal of mobile organisms (Noss and Cooperrider, 1994) and to allow natural phenomena, like flooding, to proceed without constraint. Other buffer areas function primarily to shield the core areas from exposure to adverse natural events or human impacts. Buffers also generate an edge with adjacent land to allow organisms unconstrained access into and out of contiguous areas. The intended function of buffers is to keep core areas open by creating a permeable boundary that operates as a filter (Schonewald-Cox and Bayless, 1986). The matrix zone is a mixed use zone of natural and seminatural patches of land enclosed within the regional basin divide. It is designated to define the external limits of the managed areas beyond the core areas but still potentially within a diminishing gradient of influence. The bioreserve design concept is a viable conservation strategy essential in settled landscapes such as southern New England, where the acquisition of large, continuous tracts of conservation land is just not possible.

Management Is Responsible for Establishing the Bioreserve Design that Will Effectively Preserve and Manage Samples of Natural and Seminatural Landscapes that Are Representative of a Larger, Regionalized Landscape and yet Still Maintain the Ecological Integrity of Ecosystem Function, and Heterogeneity, and Ultimately, Biological Diversity

A bioreserve design was proposed and located in the lower Connecticut River watershed that consists of a network of individual core sites bound by a series of concentric zones acting as corridors or successive buffers (Fig. 19.1). The level of conservation effort in each zone is based on the direct relationship of the zone to the dominant core and any perceivable threats or constraints due to

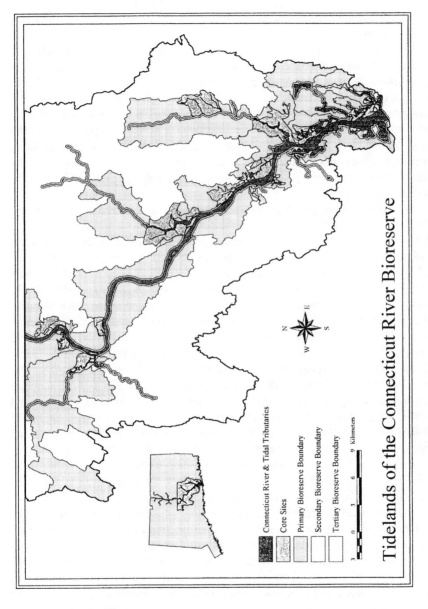

Tidelands of the Connecticut River Bioreserve

Connecticut River & Tidal Tributaries
Core Sites
Primary Bioreserve Boundary
Secondary Bioreserve Boundary
Tertiary Bioreserve Boundary

N
W   E
S

0   3   6   9 Kilometers

Figure 19.1. Watershed-based, core-dominant bioreserve for the Connecticut River Tidelands ecosystem.

conflicting land uses. The site design is itself dynamic and will incorporate new information as it becomes available both for individual marshes and for the larger watershed system.

There are presently 17 core sites within the Tidelands region. Core sites are ecologically significant sites targeted for conservation. Thirteen of these core sites are tidal wetlands representing salt, brackish, and freshwater tidal wetlands. Core sites are not exclusively tidal wetlands. Core sites often include adjacent uplands. One site is a large forested tract. When possible, core sites are delimited by natural boundaries taking into consideration the probable impacts af adjacent land use. Because this is a water-based system, secondary drainages are often included in the core site boundaries. Within these core sites, The Nature Conservancy is actively pursuing land protection and implementing research and monitoring programs.

The primary bioreserve boundary serves as a physical corridor to preserve the integrity of the hydrologic processes maintaining the tidal marsh system. The 100-year floodway is used to define this corridor such that all wetlands and water courses are connecting core areas, including tidal wetlands not already designated as core sites, for example, the main river channel, tidal distributaries off the main stem, and tributaries leading into the main stem with their associated wetlands. This boundary is inclusive of all river processes including seasonal flooding events. Where the 100-year floodway does not exist adjacent to the river, a minimal, 100-m buffer was used along river escarpments. Occasionally, core sites are located on nontidal tributaries of the Connecticut River above the extent of the 100-year floodway. In such cases, the buffer is positioned to supplement the floodway. This purpose of this latter buffer is to protect against nonpoint sources of pollution. National Wetlands Inventory (NWI) wetlands (Metzler and Tiner, 1992) plus a 100-m buffer to these NWI wetlands were used as part of the primary ecosystem boundary. This part of the boundary extends upriver along primary and secondary tributaries up to and including the first major wetland. The rationale for this boundary is that a major wetland should buffer areas downstream from most nonpoint pollution sources. There are areas in which the core site and primary ecosystem boundaries overlap.

The secondary bioreserve boundary includes buffers for the tidal marshes against adverse impacts and provides organisms with relatively unconstrained access to and egress from the core areas and corridors. The boundaries of this buffer are the Connecticut River mainstem and major tributary drainage basins.

The tertiary bioreserve boundary defines the external limits of the Connecticut River Tidelands ecosystem boundary. It defines a matrix zone that includes all natural, seminatural, and developed lands within the Connecticut River drainage basin.

Scientific monitoring programs are used within this system to detect incipient environmental and ecological change as distinct from "normal" or historical (background) levels. Monitoring key indicators of ecological integrity (for example, rare species or high-quality natural communities) and ecological distress (for example, invasive/exotic species, water quality, sedimentation rates) are necessary.

Adaptive management must correspond with scientific monitoring and scientific research. In the context of preserves as open, dynamic, and heterogeneous ecological systems, the inevitable consequences of environmental and ecological changes usually do not threaten ecological integrity. In a settled landscape such as the Connecticut River Tidelands, however, intervention into the ecology of the preserve system is inevitable to secure conservation interests from specific threats or at least to maintain ecological integrity. Under such complex circumstances, management decisions must be information-intensive and adaptive to changing circumstances. Within the Tidelands system, examples of adaptive management include fencing off the area used by breeding piping plovers (*Charadrius melodus*), a federally listed bird, along a barrier beach at the mouth of the Connecticut River. Fencing off nesting areas has greatly improved the breeding success of birds at this site by protecting the birds from predators and human impacts. Another example of adaptive management is the active removal of invasive species either through marsh restoration projects or labor-intensive hand removal.

Management Provides Scientists with Feedback Indicating Research Needs Appropriate for the Bioreserve Design

Continuing ecological concerns often warrant further studies that incorporate themes useful to management, for example, baseline hydrologic conditions, ecological land classification, ecology of key indicator species, global climate change, sea level rise, biological pests, water diversion, ditching and dredging, pollution and water quality, shoreline modification, and watershed development.

Management Provides Policy Makers with an Evaluation of the Bioreserve Design and Other Measures of Performance

The level of conservation commitment provided by the bioreserve design is periodically evaluated by measuring whether ecosystem processes or components are adequately represented. Landscape studies depicting changes in land cover/land use summarize the status and trends of the cumulative proportion of the landscape committed to conservation. The suitability of lands for conservation varies directly with land use constraints.

Conservation statistics are compiled by the results of attitudinal surveys, key indicator variable studies, reports of management successes (for example, regional land integration), suitability of buffers, and the like.

Policy Implements Conservation Strategies that Address the Significance of a Bioreserve Design

The goal of the Tidelands program is to protect the tidal marsh and riverine ecosystem of the Connecticut River including the exemplary tidal marsh communities, the riverine communities, their associated globally and state-rare species, diminished or declining species, water quality, and the integrity of the ecological processes that maintain the system (Tidelands Strategic Plan—1993). This is not to say that the entire watershed need not be taken into account. Rather, the watershed, and in particular, upland buffers to the wetlands, are considered in relation to the ecological processes maintaining the Tidelands ecosystem.

The Tidelands Strategic Plan—1993 was written by The Nature Conservancy staff to cover a five-year period from 1993 to 1998. Included within this plan are conservation strategies within the following areas: land protection, biological monitoring and management, conservation biology research, water quality monitoring and pollution control, ecological restoration, government relations, and outreach. All strategies are related to the ecological systems that the Tidelands program is trying to conserve, and the stressors affecting those systems. Staff from all program areas within the Connecticut Chapter of The Nature Conservancy are involved in the development and implementation of these strategies with a Tidelands program director coordinating the chapter's efforts. In addition to the areas covered by the conservation strategies, public relations and fundraising are integral components of the Tidelands program and are necessary to its existence.

Feasibility is strictly defined in terms of "measures of success" for each strategy. These measures represent the actions to be taken over the five year period of the plan, and are reviewed and revised annually by The Nature Conservacy staff and the chapter's Board of Trustees. New information from research and management is evaluated at this time and incorporated into new or evolving conservation strategies.

A brief description of how each conservation strategy relates to bioreserve design follows.

Land Protection Strategy.

To date, the Connecticut Chapter of The Nature Conservancy owns land outright and under conservation easement within the Tidelands area. Land conservation through fee and easement gifts and purchases, management agreements, and voluntary protection agreements are a significant component of the Tidelands program, particularly within the core sites. Within the secondary and tertiary bioreserve boundary, partnerships are a critical component of land conservation as limited Conservancy resources will be focused within the core sites. Groups such as local land trusts and the State of Connecticut are already conducting land protection efforts within these areas in accord with their organizations' specific goals. Landowner education is a critical component of land protection efforts throughout the bioreserve.

As part of the Tidelands program, the chapter has also been refining its site design process for the core sites. Many of these site designs have been expanded to ensure adequate buffering of potential threats. In addition, internship projects and research are providing valuable new information in determining which tracts to pursue for conservation actions. By taking into account factors that will influence the vulnerability of a particular parcel to change from its natural state, tracts within a core site may be prioritized for conservation action.

Biological Monitoring and Management Strategy.

The purpose of biological monitoring and management is to ensure the long-term integrity of the Tidelands ecosystem by focusing on key processes as well as species and natural communities. This involves the development of core site conservation plans including species or natural community specific goals, objectives and actions, and the development of a systemwide strategy to address invasive species particularly as they impact key species and natural communities.

Conservation Biology Research Strategy.

Research priorities are continually reviewed and revised as new information becomes available. These priorities are developed with input from The Nature Conservancy staff (at state, regional, and national levels), chapter trustees, state biologists, and scientists conducting research within the Tidelands system or who are knowledgeable about the system. Funding to conduct research priorities is critical and comes from private individuals, foundations, industry, and state and federal grants.

Water Quality Monitoring and Pollution Control Strategy.

The relationship between water quality and the health of the Tidelands' aquatic natural communities and related species is a critical component of ecosystem integrity. A plan is currently under development, largely through the work of interns and volunteers, that will indicate the major water pollutants and their sources within different parts of the system site design, based on an analysis of land uses near to and affecting core sites and the results of water quality monitoring by other groups or agencies.

Biological Restoration Strategy.

Restoration efforts may direct reestablishing the integrity of the Tidelands ecosystem by focusing locally on key species and natural communities. Several restoration projects are currently underway within the Tidelands system involving multiple partners. For each project, careful consideration is given to the long-term feasibility and impacts of any actions. Both species and natural community levels of restoration efforts may affect the Tidelands site design over the long term.

Government Relations Strategy.

Work with town, state, and federal agencies has been and will continue to be an important partnership for the Conservancy in many aspects of conservation efforts. The State of Connecticut has protected many tracts of land within the Connecticut River watershed and has recently joined the Conservancy in protecting 207 acres within one of the core sites—the largest tract of land within a core site formerly owned by a single, private individual. The Nature Conservancy and the Connecticut Department of Environmental Protection are also working cooperatively in other areas such as biological research, restoration, and outreach.

The U.S. Fish and Wildlife Service has established the Silvio Conte Fish and Wildlife Refuge, which includes the entire Connecticut River watershed. Refuge administrators and biologists are currently working closely with Conservancy and state staffs, multiple conservation groups, and local residents in determining the plan of action for this refuge. In Connecticut, The Nature Conservancy would like to work closely with Conte staff particularly in the areas of land protection, biological monitoring and management, and outreach.

Outreach Strategy.

A Tidelands outreach strategy was developed to both increase public awareness of the Tidelands program and to help mitigate the known threats to the system through education and public awareness. For each threat, outreach objectives and actions are described with specified target audiences. Depending on the threat and related outreach actions, target audiences may occur only within the core sites or may include residents/users of the entire watershed.

Policy Lends Programmatic Support to Management and Science as an Incentive to Continue Conservation Bioreserve Design Work

The Nature Conservancy, Connecticut Chapter, has built a partnership with the academic community, local researchers, and other partners such as local land trusts to conduct basic and applied research. Two grants programs, the Tidal Wetlands Research Grant Program and the H. Allen Mali Research Fund, are an integral part of the chapter's Science and Stewardship Program through which funding is provided for high-quality and relevant conservation issues that advance the mission of The Nature Conservancy. Proposals are evaluated for scientific merit and relevance to The Nature Conservancy's conservation goals. These goals for the Connecticut Chapter are determined by staff with input from the chapter's Science and Stewardship Committee, the Board of Trustees, the Eastern Regional Office, and National Office of The Nature Conservancy. Research goals relevant to the Tidelands site design include characterization of significant landscapes, natural communities and species, and threats to these systems. Scientific merit of research proposals is determined by peer review and relevance to Conservancy goals by mainly in-house review.

The Tidal Wetlands Research Grant Program promotes basic scientific research leading to an understanding of physical, ecological, and evolutionary processes that are directly applicable to conservation issues within the tidally influenced portions of the Connecticut and Housatonic Rivers. In keeping with the new conservation paradigm, the objective of this program is to advance applied research on ecosystem processes and context that relate to rare species, threatened natural communities and habitats, as well as invasive species. Funded research studies should increase the understanding of the physical or biological processes of these tidal systems and/or threats to these systems or the processes that maintain them. Results from such studies may then be applied to creating or improving site designs.

Stewardship goals are addressed through the H. Allen Mali Research Fund. This grant program supports scientific studies on lands of The Nature Conservancy or on topics relevant to conservation biology in Connecticut. Investigations related to the maintenance or enhancement of species or natural communities of concern within the state are encouraged. Examples of selected priorities relevant to the Tidelands site design include establishment of effective means of controlling invasive species; establishment of protocol for long-term monitoring of key indicators of ecosystem integrity or distress, such as selected natural communities; and inventories for selected invertebrate groups, particularly insects and mollusks.

The Connecticut Chapter's Internship Program is another venue through which management and monitoring needs are addressed. Students work on many aspects of the Tidelands program including management plans for core marsh sites. These plans provide information on site-specific threats that may be used in improving site designs. Through partnerships with local land trusts, individual land trusts provide internship funding for projects relevant to both the Tidelands program and the land trust. For example, the Essex Land Conservation Trust provided funding for two interns who were supervised by Conservancy staff. These interns characterized the vegetation of the Essex shoreline, a town within the Tidelands area and within the secondary bioreserve boundary. The document produced by the interns provided management recommendations for the conservation of the shoreline that can be used by The Nature Conservancy staff, land trust members, and town commissions.

Acknowledgments

We gratefully acknowledge Dr. Steward T. A. Pickett for his assistance and insights, which form the foundation of this manuscript. This work was partly funded by the Conservation Biology Research Program of The Nature Conservancy's Connecticut Chapter (contract CTFO 051091) awarded to Nels Barrett.

20

Ecosystem Processes and the New Conservation Theory

John J. Ewel

Summary

Heterogeneity and ecosystem processes are closely linked, and together they have implications for conservation. At the scale of communities, life form and temporal patterns may exert stronger control over some ecosystem processes (such as fluxes of energy and materials) than do species. Where conservation involves restoration, every effort should be made to include the original community's full complement of life forms, phenologies, and seasonality of fire occurrence and water flows.

The natural heterogeneity of substrates, and the variation in ecosystem processes that this entails, is sometimes masked by a single species that spreads over vast areas. Where dominance by a single species has been facilitated by the actions of people and domesticated ungulates, recuperation of the former diversity in ecosystem processes can only be achieved by restoration of biotic diversity. On the other hand, where single-species dominance is a natural phenomenon, both pattern and process are likely to be relatively homogeneous; the imposition of heterogeneity (and diversity) under these circumstances may be an artificial conservation strategy.

Notwithstanding the many conservation benefits of moderate disturbance and the heterogeneity this creates, there are two circumstances when it can work against the steward. One of these occurs when opportunities for regeneration permit invasions by nonindigenous species. The second involves nutrient loss from exposed soil, a process that is most marked in large patches and in wet climates. Heterogeneity in moderate doses serves conservation well, but in excess it can imperil conservation of natural resources.

Introduction

Conservation efforts tend to be directed toward three objectives: preservation of species and their attendant genetic variation; protection (or, where needed, restoration) of ecosystem patterns that reflect a landscape's biological and environmental diversity; and maintenance of the processes that occur within ecosystems. The first two are straightforward—conserve species and conserve communities. The third is most often regarded as a no-cost benefit that logically results from the first two: save the species and save the communities, and the services provided by the ecosystem will be forthcoming.

It is the conservation of species and complexes of populations that have been especially well served by recognition of the roles of heterogeneity—gaps, patches, periodic disturbance, episodic devastation, and the like. The reasoning is straightforward and well known, and it has been rigorously tested in many systems: (1) patchiness in pattern creates heterogeneity in resource availability, (2) heterogeneity in resource availability provides an array of opportunities for colonization and survival, and (3) the existence of multiple opportunities fosters diversity, thereby accomplishing the principal objective of conservation biology.

Heterogeneity is not merely a series of tree-fall gaps in vast expanses of old-growth forest, or logs and boulders strewn across stream beds. It occurs at many scales, ranging from the differential nutritional value of leaves on a tree to mosaics of land and sea in archipelagos. This heterogeneity is reflected in ecosystem processes, both those that manifest themselves off-site, such as fluxes of materials and energy, and those, such as the reciprocal influences of organisms and environment, that occur at smaller scales.

Heterogeneity and Ecosystem Processes

Beneath the broad umbrella of global patterns, it is at the community level where conservation biologists tend to practice their trade. This is because communities are assemblages of populations (the main level of interest), and preservation of the parts is dictated by maintenance of the whole. At the community level, heterogeneity manifests itself in four ways: species composition, physiognomy, seasonality, and substrate. Each of these has different implications for ecosystem processes.

Species Composition

There are many familiar and well-documented examples of the impacts of species on ecosystem processes: *mor* profiles develop in the soil beneath conifers, *mull* beneath broad-leaved trees; browsing ungulates tip the balance between woodland and savanna; nitrogen-fixing species hasten succession; and sizes of predatory fishes and their prey affect the structure of benthic vegetation. Perhaps even more remarkable is the fact that disappearances of some species have had effects that were far less dramatic than one might have guessed beforehand: the forests of the Appalachians still process energy and protect soil, despite the loss of the chestnut; the world's oceans still churn, driving global weather, despite the collapse of sea turtle and whale populations; and swamps of the Caribbean and southeastern United States support detrital food chains and slow the seaward flux of phosphorus, despite the loss of parakeets and parrots.

Is there any difference between heterogeneous mixes of species and simpler communities, with respect to ecosystem processes? The answer depends on what is meant by ecosystem processes. If the term refers to fluxes of energy and materials, then the answer is that the mix of species may not matter very much; a species-depauperate forest on a remote tropical island can afford as much protection against soil erosion as does a hyper-diverse continental rain forest. If, on the other hand, ecosystem processes refers to the internal workings of the system—trophic interactions, symbioses, pollination, within-system recycling— then the makeup of the community exerts a tremendous influence, and processes within diverse communities are invariably more complex than those in simple communities.

What about differences between equally heterogeneous communities of markedly different composition? Here John Harper's analogy with timepieces is apropos: spring-and-gear-driven watches bear no internal resemblance to today's electronic resonators, yet each accurately conveys the hour. One ecological example comes from experiments in which ecosystem processes were compared between species-rich tropical successional communities and communities constructed of an equally rich but different mix of species (Ewel [1986a] and references cited therein). The two proved similar with respect to herbivory, nutrient retention, exploitation of soil by roots, primary productivity, abundance of insects by feeding guild, and nematode abundance and species composition. There were differences, to be sure, but the surprise finding throughout was the remarkable ability of one unique combination of species to duplicate the functional attributes of another.

Would two such strikingly different ecosystems have similar impacts on the surrounding landscape? Perhaps not. To invoke Harper's watch once again, the mechanical model requires periodic repair, whereas the other mandates a constant throughput of heavy metals. By the same token, species composition determines, in large part, the hospitability of a forest to migratory birds. There may be many species combinations that lead to similar internal workings, but substitution at the species level could dramatically affect interactions between systems. This has never been tested experimentally.

Physiognomy

The physical structure of an ecosystem is a function of its species composition. Nevertheless, when viewed from even a modest distance, the detailed attributes of individual species disappear, whereas a characteristic, overall appearance of the community remains. In the case of forests, this physiognomic signature is a combination of plant architecture and size. The greater the diversity and range of architectural models, and the greater the range of age classes within species, the more complex the physiognomy. Thus, physiognomy is dictated in large part by two kinds of diversity—life forms and age classes.

Different life forms influence ecosystem processes in many ways (Ewel and Bigelow, 1996). Shrubs tend to access different nutrient supplies than trees, epiphytes intercept and redistribute water, fallen palm fronds crush seedlings and create moist microhabitat for amphibians, vines bind tree crowns, and so forth. Some life forms routinely affect off-site ecosystem processes. Flammable grasses and resin-rich shrubs, for example, lead to volatilization and redistribution of nitrogen, sulfur, and carbon across the landscape; replacement of sedge-dominated marsh (much of Florida's Everglades, for example) by closed-canopy forest changes both the magnitude and the seasonality of water throughflow. A conservation program that adds or removes a species may have only modest impact on an ecosystem's workings, but one that affects the combination of life forms is almost certain to affect ecosystem processes. Thus, from the perspective of ecosystem function, a first and foremost conservation guideline might be to preserve (or restore) the life-form composition.

Age and size tend to be correlated in higher plant species (with plenty of exceptions, such as perennial grasses). Gram for gram, young plants sequester mineral nutrients and carbon faster than do older plants. The amounts stored in the biomass of communities composed of old individuals tend to be higher than the amounts in communities or patches dominated by young plants, but the net rate of accrual is higher in the latter. Thus, when conservation objectives call for capture and immobilization (creation of significant carbon sinks, for example), the plan should call for youth; when retention of materials previously amassed is called for (retention of cations, for example), then a forest dominated by large, old trees will do the job best.

Seasonality

Plant growth, and the change in vegetation physiognomy that accompanies it, is not the only temporal change of significance to ecosystem processes. The pulse of seasons is accompanied by acceleration and deceleration of ecosystem processes, some of which are driven by changes in solar radiation, others by rainfall, and still others by temperature.

These abiotic causes tend to manifest themselves most conspicuously through their influence on plant phenology. Some relationships between phenology and biotic processes such as food choices of herbivores are reasonably well documented, even in parts of the tropics once regarded as relatively aseasonal (van Shaik et al., 1993). Examples of the influence of plant phenology on processes usually thought of as primarily abiotic, such as nutrient leaching, are not as common. In one recent example from the humid tropics comparing simple communities dominated by one or two life forms (trees alone or trees plus large, perennial monocots), my coworkers and I found that rainstorms in the short dry season had very different effects on the two communities. The trees lost their leaves with the onset of drought, and when a downpour came it was accompanied by significant leaching beneath the trees-only stands. Where the trees were under-planted with monocots, however, leaching losses were minuscule. The difference in the two was undoubtedly related to uptake: the leaf-free trees did not capture soil nutrients mobilized by the rainwater, whereas uptake by the monocots, which are evergreen, continued.

A less subtle example, one that is common today throughout the world's grasslands, concerns season of burning. Lightning ignitions were once most common at the onset of the rainy season, when vegetation was still dry and the year's first convection storms appeared. Human-mediated ignitions, in contrast, are concentrated at two very different seasons: most wildfires tend to occur in the midst of the dry season, when ignition is effortless, whereas most prescribed burns are done in the cool season, when wind patterns are most predictable and fire temperatures are moderate. Neither of these mimics the environment in which grassland species presumably evolved. The outcome is an inevitable shift in dominance, and in some cases it may lead to extinction.

Not all seasonal pulses in ecosystem processes are due to plants. The droppings of migratory passerines may be inconsequential, but the redistribution of biogeo-chemical wealth by nesting seabirds is clearly significant. Thus, seabird conservation is certain to have impacts that extend well beyond the species targeted for protection.

A comprehensive conservation approach must pay due attention to temporal changes in ecosystems, such as phenology of components, the timing of pulses (fire and oscillations of water flows, for example), and migrations, both local and long range. One that neglects the seasonal march of ecosystem processes jeopardizes the sustainability of the resource.

Substrate

The growing medium is often dictated by geology, and it exerts undeniable influence on species' distributions. One alga is found on silt bottom, another on fallen logs; this tree grows in swamps, that one in well-drained sands; and this herb tolerates high concentrations of nickel, whereas its neighbor does not. Nevertheless, species distributions can, under some circumstances, transcend boundaries imposed by substrate, and management actions can have important implications for this process.

One example is *Imperata cylindrica*, a grass that, thanks to logging, agriculture, and fire, now dominates a broad range of once-forested soil types throughout the southeast-Asian tropics. It is flammable, well dispersed, and effective at holding sites against incursions by woody colonists. A return to dominance by a greater diversity of species, species whose more specialized habitat requirements reflect the heterogeneity of the soils, probably requires fire prevention and establishment of larger-stature plants that can out-compete the grass.

A second example, counter to the first, concerns the *Gilbertiodendron*-dominated forests in west-central tropical Africa (Hart, 1990). This leguminous tree, like *Imperata*, dominates vegetation across a broad range of soil types, but there is no strong evidence that its distribution reflects a widespread episode of past destruction or an overriding environmental factor, such as recurring fire. Clearly, there are other species well suited to the various soils on which it occurs. Why, then, the widespread dominance by such a habitat generalist? No one knows for sure. One possible explanation is that *Gilbertiodendron* is exemplary of Connell and Slatyer's (1977) inhibition model of succession whereby, in the absence of local disturbance, long-lived organisms achieve dominance. Perhaps the rest of the world is subjected to more frequent disturbance than is this part of Africa, which is free of wind storms, geologically stable (and very old), and not routinely subjected to fires. Does this pose a dilemma for conservation biologists? The imposition of disturbance would add heterogeneity, and plant species that specialize on particular soils might dominate some patches of landscape, thereby increasing overall diversity. On the other hand, the natural trend in these forests has been toward decreased diversity—the right-hand side of the diversity-disturbance parabola (Connell, 1978).

The interaction between substrates and species is not unilateral, for just as substrates can dictate species' distributions, species can exert marked effects on substrates (Stone, 1975). When they do, they influence ecosystem processes through their impacts on soil. Yellow poplar, for example, leads to calcium enrichment (Kalisz, 1986); leaf-cutting ants, beetles, and gophers churn soil, affecting both its physical and its chemical properties (Alvarado et al., 1981; Kalisz and Stone, 1984); and the development of successional communities everywhere is hastened and sometimes redirected by nitrogen-fixing trees and shrubs, both natives and aliens (Reiners, 1981; Vitousek and Walker, 1989). In ways like these and many others a potentially homogeneous substrate is made heterogeneous by organisms.

Dangers of Heterogeneity to Conservation

The same chain of reasoning that makes heterogeneity such a popular ally can, in some circumstances, work against the cause of conservation. When patches permit colonization by unwanted species, or when they facilitate irreversible depletion of nutrients (e.g., Walker et al., 1981), then they are to be avoided.

Patches Attract Problems

Modest disturbance does free resources, facilitating regeneration of species having a broad array of life-history traits. When the potential colonists include vast numbers of nonindigenous species, however, creation of opportunities for colonization is not in the best interests of conservation. Consider the case of the Hawaiian Islands, at 3500 km from the nearest donor continent the most remote archipelago on Earth. Here the native flora is outnumbered by human-introduced aliens, many of which are more effective than the native species at taking advantage of newly available sites for regeneration. This leads to shifts of dominance away from communities composed primarily of native species toward communities dominated by nonindigenous species.

What can conservation biologists do to impede the shift from native to alien vegetation? The most common strategy, and one whose importance cannot be denied, is to attack the aliens, thereby reducing the threat they pose to native ecosystems. The disadvantages of this approach are twofold. First, in places like Hawaii that already support hundreds of nonindigenous species, the battle can never be won. If the threat of invasion by one species is contained, another stands ready to take its place. Because native systems cannot be sustained without immense human subsidy, conservationists must commit to eternal warfare on exotics—a depressing prospect.

The second disadvantage to this approach is that it often attacks the symptoms rather than the disease. Some invasions by alien species are facilitated by changes in the environment, as when dense populations of nonindigenous earthworms attract introduced pigs, which dramatically modify the opportunities for plant regeneration by churning the surface soil and by preferentially dispersing alien plants, or when water is channeled and shunted from the land, creating hydrologic conditions that differ greatly from those under which the native biota evolved (e.g., Ewel, 1986b). In such cases the aliens may be better adapted to the newly created environmental conditions, and it is almost inevitable that they will triumph in head-to-head competition with natives. The best option under these circumstances may be to attempt to restore the initial environmental conditions: elimination of (alien) earthworms may be an essential precursor to the elimination of (alien) pigs, which may in turn be required before (alien) guavas can be controlled, just as restoration of surface flow may be needed before native plants can prosper throughout the Everglades once again.

Suppose that aliens are already present in such diversity and profusion that they cannot be excluded, or that conservation biologists resist adopting the engineering mindset needed to restore abiotic conditions? There are two options. One is to guard against opportunity for colonization in the native communities. This is extremely difficult because most agents of disturbance and heterogeneity such as wind storms, senescence, and landslides are uncontrollable. Nevertheless, some positive actions can be implemented. One example is prevention of human-created gaps created by harvests of single trees from closed-canopy forests; such small-scale harvesting is commonly regarded as ecologically innocuous, yet in some cases, it can pave the way for invading species. Another action, on a larger scale, is consolidation of land holdings, thereby reducing the ratio of edge to area and reducing the threat of invasion.

The second option is to live with nonindigenous species. That is not to say that such a decision should be reached only by relaxation of effort, by giving up the fight, as it were. Rather, we may have to tolerate some aliens in some situations. Those we accept should be ones that offer the least threat to native biological diversity and exert the least dramatic change on ecosystem processes. By accommodating aliens, we may be able to conserve most diversity at the species level and still retain ecosystem services such as gas exchange, fixation of carbon and nitrogen, and soil protection. It is diversity at the community level, that is, loss of community types, that is sacrificed by such accommodation.

Patches Leak

To sustain biological richness, the abiotic features of the ecosystem must be retained, and in the case of terrestrial ecosystems the most vulnerable abiotic factor is soil fertility. To jeopardize soil nutrient status is to tinker with the pendulum of plant competition and influence the forage quality of plant parts. Furthermore, undesirable losses of nutrients from terrestrial ecosystems do not disappear; they often show up as equally undesirable nutrient enrichment of aquatic systems.

Especially in warm, humid climates, where leaching is reduced by transpiration, the presence of actively growing vegetation can mean the difference between net retention and loss of nutrients. Recovery, of greenery if not species composition, is faster in wet climates than in dry, so gaps in a rain forest are inevitably shorter-lived than gaps in a semiarid woodland. In wet climates, it is the sustained excess of rainfall over evapotranspiration that drives nutrient loss from the soil (Bruijnzeel, 1989), a process that can be more accentuated in gaps than under vegetation (Parker, 1985). In semiarid regions, on the other hand, it is episodic deluges—freak storms or the rainy seasons triggered by El Niño-southern oscillation—that accentuate losses from vegetation-free gaps.

Even within the same bioclimate zone, not all patches leak equally; nutrient loss tends to increase with two factors, patch size and duration. Size and duration of vegetation-free patches influence rates of nutrient loss in parallel ways (Fig. 20.1). In small patches, or in short-lived patches, nutrient loss is probably very small. For small patches, this is because the patch is underlain by living roots from surrounding plants, which continue to take up nutrients, whereas for short-lived patches it is because soil organic matter, both living and nonliving, has the capacity to absorb a substantial fraction of the nutrients normally freed by gap formation. Large patches and long-lived patches, on the other hand, incur much greater rates of loss: large gaps because their core area exceeds the reach of roots from surrounding vegetation, old gaps because organic matter decomposes, and the inputs required to sustain soil heterotrophs are gone.

If the step-model functions depicted in Figure 20.1 are approximately correct, then rates of nutrient loss per unit area are likely to be parallel, if not equivalent. Nevertheless, the total losses (amount per area or per time, in the case of patch size and duration, respectively) will differ because patch area increases as the square of the radius (if circular). Thus, big holes in terrestrial communities pose proportionally greater threats to soil fertility than do those that are long-lasting.

What are the messages for conservation biology? First, losses of soil fertility are often far more subtle than erosion, and nutrient accrual that took place over millennia can be lost in a very short time, especially in regions of high rainfall. The best defense against such loss is vigorous, healthy vegetation. There is a critical size of gap—probably a size that is unique to each combination of soil, vegetation, and climate—below which nutrient losses are likely to be negligibly small. The mound of the gopher and the single-tree windthrow, for example, will not jeopardize soil quality in grassland and forest regeneration, respectively. Likewise, short spurts of soil exposure are unlikely to lead to significant nutrient loss. The breakdown of soil as a living system takes place on scales of months

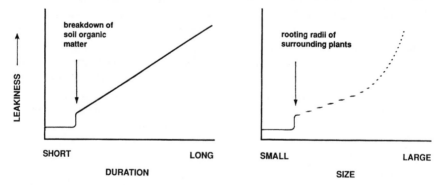

Figure 20.1. Hypothesized impacts of gap duration [left] and size [right] on nutrient loss. Losses are extremely small under both short-lived and small gaps; they increase faster as a function of size than as a function of duration because of the exponential relationship between radius and area.

to years, not days or weeks, and most soils, even without plants, have substantial nutrient-retention capability. The dangers to site quality—and to conservation—are accentuated with increasing precipitation, likelihood of episodic rains, patch size, and patch duration. The first two are beyond our control, but they can be used to identify zones and seasons of high risk. The latter two can be manipulated, and it behooves all conservation biologists to keep the abiotic side of ecosystem processes firmly in mind as we implement stewardship and restoration actions.

Acknowledgments

The ideas in this chapter were developed in part as a result of research supported by National Science Foundation grant DEB 9318403. I thank S. Tartowski and S. T. A. Pickett for reviews.

21

Measurement Scales and Ecosystem Management

Doria R. Gordon, Louis Provencher, and Jeffrey L. Hardesty

Summary

Ecosystem management involves managing relatively large areas with a regional perspective within a quasi-experimental framework. Monitoring the effects of management decisions in the context of quantified management and monitoring objectives is necessary to evaluate the extent to which those objectives are being met. The temporal and spatial scales at which environmental perturbations influence components of the natural system need to be defined in conjunction with the dispersion of those components in the landscape. These data then provide a model of the dynamic mosaic of landscape components that should be mimicked by management directed at maintaining or restoring ecosystem function and processes. Understanding the functional hierarchy of the components is critical to determining the variables in which to monitor response. Within this hierarchy, the responses of larger scale variables to perturbation should be slower than those at smaller scales that operate with faster dynamics. Incorporating the spatial and temporal dynamics of species and communities of concern within the model of environmental perturbations should help identify both the scales at which perturbation or management operates and the smaller scale at which monitoring will detect changes before slowly responding variables are significantly altered. As a result, monitoring should focus both on variables that are dynamically slower and more rapid than the scale at which management is applied. Variables that respond slowly to perturbation should be monitored at high precision over large spatial areas to detect change over short time frames. Monitoring carefully selected indicators of these slow variables is critical because they may provide the most powerful management tools. More rapidly responding variables that are functionally linked to the species of concern should be identified and monitored at lower effort. These variables are expected to have lower impact on the management target, but provide an early warning of change in the system. Finally, explicit evaluation of spatial scale effects within sampling designs for experimental tests of management techniques, or comparative evaluation of management units using spatial analysis, should elucidate the scales at which future monitoring will identify significant changes in the variables examined.

Introduction

Ecosystem management (EM) has been defined by the Ecological Society of America as "management driven by explicit goals, executed by policies, protocols, and practices, and made adaptable by monitoring and research based on our best understanding of the ecological interactions and processes necessary to sustain ecosystem composition, structure, and function" (Christensen et al., 1996). This concept (in various forms) has been adopted as a primary mandate by at least 18 federal land-managing agencies (Congressional Research Service, 1994). Some states, like Florida, also have formally adopted an EM approach. Although the definition and implementation varies with the agency, in all cases EM involves more than just the biological function of managed ecosystems. A review of the literature documented 10 primary components of EM that range from ecological to institutional and social (Grumbine, 1994).

A full discussion of evaluating success within an EM context would include measurements across a range of disciplines, from institutional variables like interaction and cooperation among adjacent land managers and institutional flexibility, to economic variables, to ecological ones (Christensen, this volume). Integral to all aspects, however, is adaptive management: involving continuing measurement of the system responses to management and modifications of that management if the measurements so indicate. For the purposes of this chapter, however, only the ecological responses of large systems in a human management context are considered.

Clearly there are numerous approaches to examining the effects of temporal and spatial scales on meeting ecosystem management or large landscape management goals. We address the influence of scale on ecosystem management efforts at two levels: (1) determination and monitoring of landscape level objectives for the size and dispersion of communities and seral stages and (2) the scale(s) at which to monitor population and community level objectives. Although the concepts we introduce should be generally applicable, throughout this chapter we use examples from our work on upland longleaf pine communities at Eglin Air Force Base (Eglin) in northwest Florida. This base contains the largest remaining stand of longleaf pine uplands, a system that once extended across the southeast, covering over 222,000 km^2 (55 million acres) (Ware et al., 1993). Eglin manages 992 km^2 (245,000 ac) of this habitat, of varying condition and management history within a 1,870 km^2 (463,000 ac) area.

Landscape Mosaic Determination

One of the most significant changes that ecosystem management imposes on traditional land management at the landscape level is the emphasis on dynamic function and process. As such, the concepts of nonequilibrium dynamics and heterogeneity are explicitly incorporated into this management framework, following the theoretical trends in ecology and conservation science (Pickett et al., 1992; Christensen, this volume; Fiedler et al., this volume). Traditional management has been relatively inflexible and static, with a single target condition (e.g., prior to European colonization, sustainable harvest of one or more products repeatedly from the same stand on a given rotation). Patch dynamics were largely ignored and ecotones rigidly maintained in space. However, we increasingly understand that the function and, therefore, conservation of systems is at least partly dependent on retaining their dynamism in space and time (Gardner et al., 1992). Managing for a dynamic landscape is challenging for the most experienced of land managers.

One approach to incorporating this dynamism into management is provided by the body of theory that suggests that a few factors, operating at different temporal and spatial scales, are driving pattern and process in most natural landscapes (Holling, 1992). If we can identify those factors and the interaction of their effects over a number of scales in specific geographic locations, this might provide a tool for managers to identify objectives that mimic natural landscape structure and to evaluate the effectiveness of the management imposed.

In Figure 21.1 we identify both the spatial and temporal scales at which major atmospheric and soil-disturbing processes might have operated in northwest Florida. As has been documented (Turner, 1989; Holling, 1992; Bourgeron and Jensen, 1994), the overlap of these ellipses indicates that processes frequently interact in time and space. The shading of the ellipses indicates the relative impact that each type of perturbation might have on vegetation. For example, hurricanes and animal mounds both have high potential to uproot and kill plants (but on very different spatial scales); in contrast, many of the species in Florida are relatively fire-tolerant and can resprout from roots if aboveground biomass is killed. For managers, perhaps the most relevant information is that the time frame of interest is from a month to a century, so that their management can be conducted in a biologically significant time frame, and that while the spatial scale of interest ranges from 0.01 m^2 to 1000 km^2, management primarily will be imposed on scales >10,000 m^2 (1 ha).

In many places, land and sea processes affect variables such as wind and humidity, which influence the frequency and area of a perturbation likely to occur at a particular location (Foster, 1988; Turner and Dale, 1991). Understanding these variables should enable us to narrow the ranges on Figure 21.1 for a particular location. For example, if the probabilities of stand-altering processes are combined with the predominant weather patterns at Eglin, geographically explicit probabili-

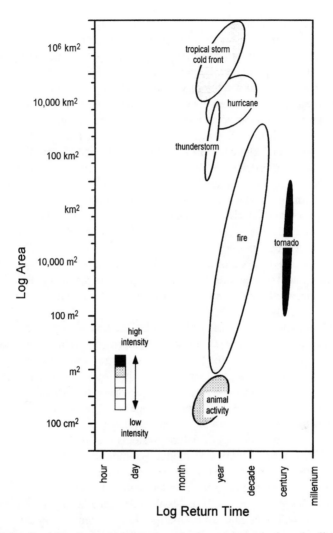

Figure 21.1. Spatial and temporal domains of primary atmospheric and soil-disturbing processes in northwest Florida longleaf pine ecosystems. Shading of the ellipses indicates the intensity of process effects on the vegetation.

ties of events impacting vegetation in particular areas of the base can be generated. We predict that storm activity will be greatest where on- and off-shore winds meet along the east-west ridge across Eglin; natural fires most frequently ignite south of the intersection, where less precipitation falls but lightning strikes are still relatively common. Further, the distribution of ravines across the base reveals a pattern of natural firebreaks that likely historically influenced the areas of fires ignited. These patterns increase our ability to establish probabilities for both the frequency and areal extent of fires in specific areas of the base.

A cellular automata or other spatially explicit model (Turner and Dale, 1991; Gardner et al., 1992) can then be generated from these probabilities to produce a hypothesized mosaic of patches of various seral stage and structure. This model should both elucidate and integrate our understanding of system dynamics (Lee, 1993) and provide guidance to managers about the size and dispersion of management units in which fire, silviculture, or other activities are conducted (Pickett and White, 1985; Franklin and Forman, 1987; Hobbs and Huenneke, 1992). One method would be to distribute relatively high impact activities (e.g., silviculture) across the landscape to mimic the size, frequency, and dispersion of similar impacts. Alternatively, such activities might be clustered in areas where they will not impede processes that periodically operate at large spatial scales, perhaps because of other development or geographically protected sites (e.g., preferentially place activities not compatible with fire management in units surrounded by natural firebreaks). As specified in EM procedure, such decisions should be made in the context of the larger geographic region, not just within the administrative boundaries of the management area (see Christensen, this volume; Fiedler et al., this volume).

The only way to evaluate the utility of this method is to generate the spatial model, conduct management to mimic the patterns depicted by the model, and see if the structure and pattern on the ground converges toward model predictions. Tools such as remote sensing, Geographic Information Systems (GIS), and spatial statistics should be combined to monitor the pattern on the ground (Turner, 1989; Wiens, 1989b; Haining, 1990; Turner and Dale, 1991; Rossi et al., 1992; Barrett and Barrett, this volume). However, precisely because of interacting effects at different scales, the responses of natural systems are likely to be unpredictable (Levin, 1992; Holling, 1993), and model assumptions as well as management strategies need to be reevaluated if no convergence occurs.

Although this approach of examining the scales of natural perturbations to the system provides guidance for broad-scale management and monitoring, land managers also will have numerous small-scale management objectives focused within communities. Often the species or groups of species of management interest are dictated by mandate or by the larger management objectives for a property. At Eglin, populations of special concern include federally endangered species such as the red-cockaded woodpecker (RCW) and the longleaf pine habitat they require (Bean, this volume), game species such as deer and turkey, and understory communities essential for carrying the prescribed fire that now largely replaces the high-frequency lightning-ignited fires natural to this habitat. Because the relevant temporal and spatial scales are smaller in this context, we need to develop understanding of how and what to manipulate and monitor at these smaller scales.

Community and Population Monitoring

The responses of populations to phenomena at various scales are mediated by interactions among coupled and decoupled variables, which may be species or species and their abiotic environments (Turner, 1989; Holling et al., 1995). Interacting coupled variables (mass action) have dynamics that are directly and approximately proportionately related to one another (linear or nonlinear). Coupled variables therefore operate at the same scale. Examples are traditional predator/prey or competitive interactions.

Decoupled variables have dynamics that are only weakly interactive and almost independent, but still influence each other. A variable that is both decoupled and responding more slowly to perturbations (hereafter "slow variable") than another variable ("fast variable") can be viewed as the environmental background of the faster variable (O'Neill et al., 1986). Variables in a decoupled interaction operate at different scales. One well-known example is that of spruce budworms and spruce in northwestern North America. The foliage density is almost independent of budworm densities, which are controlled by avian and small mammal predators. The rapid predator/prey interactions are decoupled from the slow accumulation of foliar biomass over time (Ludwig et al., 1978; Holling, 1986). Interestingly, after some threshold of foliar density is reached, both the coupling and the interaction speeds shift: at high foliar density, predators can no longer easily locate prey budworms. As a result, the density-dependent predator-prey interaction becomes decoupled, and the interaction of the two species slows. Budworm herbivory then rapidly defoliates the spruce. Similarly, in aquatic systems, trophic interactions decouple nutrient concentrations and lake ecosystem productivity (Carpenter and Kitchell, 1993).

One of the potentially most useful implications of decoupled interactions is that the amplitude or relative variability of the responses is dependent on the temporal or spatial scale. As proposed by O'Neill et al. (1986), O'Neill (1989), Holling et al. (1995), and others, the effects of perturbations transmitted between variables increase from larger to smaller scale variables, and decrease from smaller to larger (Fig. 21.2).

This theory suggests that the dynamics of populations or variables of interest might be best influenced by a management activity that directly impacts a resource operating at a larger spatial or temporal scale than the variable. (The potential role of faster-scaled variables as indicators of management success is discussed below.) We can use a graph of generation time of the species with individual size for plants or home range size for animals (Fig. 21.3) to identify the spatially larger and dynamically slower ecological processes (Fig. 21.1) that managers can imitate or modify to maximally impact species of concern. Key processes are determined by first imagining crosshairs centered on a species. The outward extension of crosshairs will "touch" space-time areas occupied by ecological

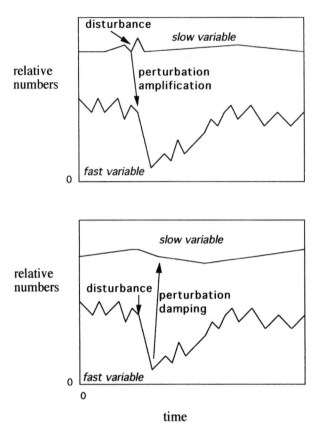

Figure 21.2.   Perturbation propagation between slow and fast variables. (Upper) Perturbation amplification: a perturbation of the slow variable has increasing effects on smaller scale, faster variables. (Lower) Perturbation damping: a perturbation of the fast variable has decreasing effects on larger scale, slower variables. The arrows indicate the causal direction of a perturbation and the sections of the curve where the variable is affected directly or indirectly by the perturbation.

processes that have larger spatial scales or longer return times than the species of concern. Management should strive to identify and use the properties of these slow processes and variables.

For example, RCWs are limited by the number of living older longleaf pine trees that have been infected by the fungus that causes heartrot and renders the pine heartwood soft enough for cavity excavation; therefore, either acceleration of the slowly colonizing heartrot fungus, or creation of artificial cavities will enhance RCW population size (aging of the pines and infection by the fungus have much slower dynamics than do RCW populations) (Walters et al., 1992). Drilling artificial cavities accelerates provision of a resource that only slowly

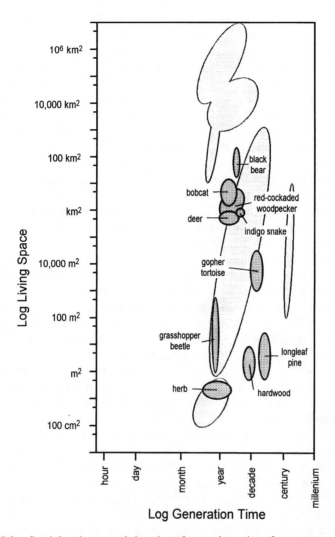

Figure 21.3. Spatial and temporal domains of several species of management concern in longleaf pine systems at Eglin Air Force Base superimposed on the general spatial and temporal domains of natural processes from Figure 21.1. Spatial range is determined by the home range areas of male and female animals, or space occupied by sessile organisms. Temporal range is determined by age at first reproduction.

naturally accrues. Increasing the change in this slow variable should have greater impacts on the number of RCW groups than would increasing change in a faster variable.

The relationship between scale and response speed and amplitude also has implications for monitoring programs. We suggest that monitoring will be most effective if decoupled interactions responding both more slowly and more rapidly than the population of interest are identified and also monitored. However, different sampling designs are necessary to detect changes in variables with different dynamics. Because slow variables have low variance over large areas, one would either have to monitor for a long time or intensively in several blocks over a large area to have enough statistical power to detect a biologically significant change. We generally have no choice in terms of time, especially if we are concerned about the risk to a more rapidly changing variable; instead we need a high-power sampling design that encompasses and enables estimation of variance across a large area. Determination of the scale that results in low within-block variance may require incorporation of spatial sampling techniques (e.g., Haining, 1990:189–191; Rossi et al., 1992).

Faster variables, because of higher variation across smaller scales and greater response amplitude to perturbation, can be monitored with lower effort over smaller areas. As a result, monitoring at a scale lower and more dynamic than the population or variable of interest should provide an early warning of perturbations to the system. Examples from the literature include local blue-green algae responses to elevated phosphorus concentrations in aquatic systems, as compared to detectable changes in average phosphorus concentration in the water (Edmondson, 1991), and responses to changes in habitat quality reflected in the asymmetric expression of sexually selected traits by birds, mammals, and insects. The degree of asymmetry in sexually selected ornamental traits such as length of both tail feathers in birds and canines in gorillas increases with environmental stress at significantly lower stress levels than do naturally selected traits (Hill, 1995).

Identification of the appropriate decoupled variables at scales around the variable of interest is clearly complex, and will involve carefully designed research projects. We end this chapter with an example of how these scale considerations can be identified and should be useful for management decision making on the ground. This work is a collaborative effort between The Nature Conservancy, University of Florida, Tall Timbers Research, Inc., and Eglin Air Force Base.

In many places, as at Eglin, we do not yet understand the dynamics of the system, much less the variables that respond faster or slower to perturbation. One approach to these questions is to examine the effects of measurement scale on components of the system that should operate at different scales. We have collected soil, vegetation, and invertebrate data at two spatial scales in several stands of fire-suppressed and frequently burned old-growth longleaf pine stands. This design enables us to test whether the mean and relative variability differs if variables are collected over 10 m as compared to 50 m distances across a larger area. The distance between sampling plots will be called the sampling step. As an example, we have determined the mean and coefficient of variation (CV) of longleaf pine seedling density. Data have been analyzed using a split-plot analysis of variance where sampling step is tested at the subplot level.

Figure 21.4 shows data from twenty-four 200-ac plots of fire-suppressed and

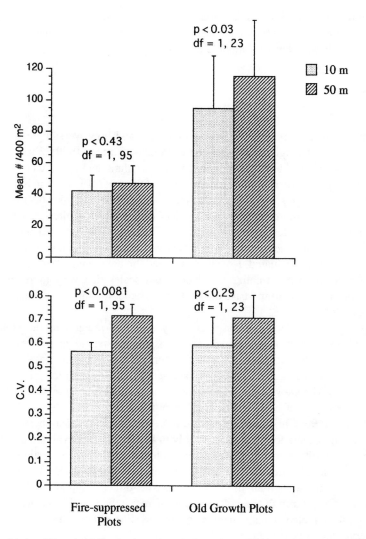

Figure 21.4. (Upper) Mean (± 1 s.e.) and (lower) coefficient of variation (CV) for longleaf pine seedling density in subplots separated by 10 or 50 m in fire-suppressed and old-growth plots. The p-values represent the effect of sampling step scale on either mean or CV in each habitat type.

six 200-ac plots of old-growth longleaf pine stands. Longleaf pine seedling densities were generally higher when sampled at 50 m than at 10 m, although only significantly different in the old-growth plots ($p < 0.03$). As a result, site condition or history appears to influence the effect of scale. Although no clear trend appeared for the CV in old-growth sites, a significantly greater amount of variability ($p < 0.0081$) was detected at the 50-m sampling step in fire-suppressed plots. Sampling step effects have also been detected for planthopper density, but not for hardwood density (unpublished data).

Conclusions

The importance of scale in terms of both what and how we sample in natural systems is increasingly understood by ecologists (Turner, 1989; Wiens, 1989b; Levin, 1992). The challenges of managing large natural and modified systems for multiple objectives necessitates identification of the most efficient and informative monitoring strategies. We suggest that these will include variables operating with both slower and more rapid dynamics than the variable of management focus. Selection of variables to measure should become easier as measurements at different scales are incorporated into more research efforts and the explicit contribution of scale is evaluated. However, we need the development of new methods that enable us to renormalize the data across scales that vary more than two orders of magnitude to increase our ability to predict the interactions among and responses of variables.

Operation of natural processes at many scales is integral to the continued survival and evolution of species. As a result, understanding of these scales, management of landscapes to allow their natural function, and methods to evaluate whether those management objectives are being met are all critical needs of land managers. EM requires that management methods be implemented as tests of weak hypotheses of how best to meet specific objectives (Walters and Holling, 1990; Lee, 1993; Christensen, this volume). We hope that researchers will work with land managers to most effectively both implement and evaluate those tests. In turn, the ability of managers to implement long-term treatments within management units should provide researchers the opportunity to conduct longer-term, larger-scale research than might otherwise be directly supported by funding agencies. Only through the collaboration of researchers and managers will the cross-scale directed management mandated by EM policies and necessary for conservation be possible.

Acknowledgments

We thank the Longleaf Pine Restoration Project team for data collection and K. Thomas for assistance with figures. D. Maehr, R. Labisky, P. Moler, and M. Sunquist provided data on animal species' home ranges and generation times. This work was sponsored by The Nature Conservancy and the AFDTC/EMNW, Air Force Materiel Command, U.S. Air Force, under cooperative agreement number F08635-94-2-0001. The views and conclusions contained herein are those of the authors. The U.S. government is authorized to reproduce and distribute reprints for governmental purposes notwithstanding any copyright notation thereon. Lance Gunderson and an anonymous reviewer provided useful comments on this manuscript.

22

Biogeographic Approaches and the New Conservation Biology

Daniel Simberloff

Summary

The equilibrium theory of island biogeography was a foundation of the "new conservation biology" of the 1970s. An influential series of refuge design "rules" was ostensibly based on the theory. Some of the rules, which advocate certain configurations for refuges, have been shown not to derive from the theory, and all of them are questionable on a variety of grounds, especially their focus on species richness and their failure to account for important factors. The equilibrium theory paradigm in conservation has been replaced by a metapopulation paradigm, but the equilibrium theory has left a crucial legacy: the search for causes of small population extinction. The metapopulation paradigm, like the equilibrium theory, may not accurately depict much of nature.

Biogeographic research also influences conservation through its emphasis on species' ranges. Both gap analysis and the study of the spread of nonindigenous species require areographic detail. Even historical biogeography has a role in conservation in such matters as which species are truly indigenous, the extent to which species' fates are intertwined, and the ways global climate change will affect species' ranges. Species' genetic structure is a growing research focus in both biogeography and conservation.

Introduction

Biogeography—the study of how, why, and when species came to be arranged as they are—overlaps broadly with ecology. Historically, biogeographers tended to study regional patterns, whereas ecologists focused more on local details of spatial arrangement, but this demarcation was never absolute. The distinction between biogeography and ecology was blurred even more by the tremendous interest generated by the dynamic equilibrium theory of island biogeography. Within biogeography, a dichotomy between ecological and historical biogeography has to some extent been bridged by adoption of similar procedures by both branches (Simberloff, 1983) and by common interest in events such as Quaternary range changes. More recently, biogeography has come to encompass the study of geographic distributions of alleles and genotypes.

Because one must know where a species or community exists to recognize changes in it and to manage it, it is no surprise that biogeographers have long contributed to conservation biology. What is more interesting is how academic developments in biogeography have affected the evolution of conservation biology.

The Equilibrium Theory

Until the mid-1970s, conservation biology was not a distinct science. Rather, it was part of ecology and wildlife biology and largely consisted of research on the autecology and habitats of species of particular concern. In 1974–75, the application of the dynamic equilibrium theory of island biogeography (MacArthur and Wilson, 1967) to the design of nature preserves helped inaugurate a new conservation biology (Simberloff, 1988) as well as a new kind of biogeography (Simberloff, 1983; Myers and Giller, 1988). The equilibrium theory envisioned species as divided into very distinct populations with little interaction, with many extinguished in the short term, but with sufficient movement among sites that local extinctions are often temporary. The number of species at each site is thus a dynamic equilibrium between local extinction of existing species and immigration of new ones. The key feature of this conception is the relative isolation of the component populations: reproduction is far more important than invasion in maintaining populations. For both biogeography and conservation biology, this view of nature dictates the study of the exact causes of population extinction. However, the basic statistic of the theory—the number of species on an island—is a community-level statistic, and conservationists thus focused on species richness as the key concern.

The original proposal of the equilibrium theory (MacArthur and Wilson, 1967) envisioned "island" as a metaphor for any insular habitat—a lake, a forest surrounded by fields, and the like. Thus it is hardly surprising that habitat islands were prominent in the equilibrium theory literature. And, because nature reserves are usually habitat islands, it is small wonder that the equilibrium theory was applied to conservation. Several papers (references in Simberloff, [1988]) suggested a series of rules, ostensibly based on the theory, for the design of reserves. When this set of recommendations was adopted in World Conservation Strategy (I.U.C.N., 1980), the MacArthur-Wilson theory became enshrined as the key basis for conservation planning (Fig. 22.1).

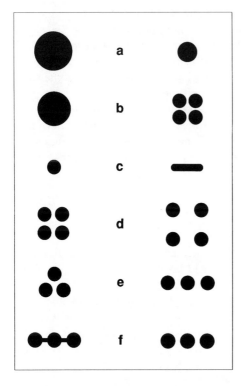

Figure 22.1. Rules of refuge design, in which each configuration on the left is seen as superior to that on the right (after I.U.C.N., 1980).

As was inevitable when a single overarching and simple theory is applied to a complex, varied set of problems, a reaction set in from two directions. First, the theory was increasingly questioned as a model for nature. Early criticisms (e.g., Lack, 1976) had been summarily rejected (e.g., Grant, 1977; Ricklefs, 1977), primarily because the key complaint— that crucial detailed natural history and autecology were ignored—was simply too boring (Simberloff, 1994a). Later similar criticisms (Gilbert, 1980; Williamson, 1981, 1989a; Jehl and Parkes, 1983) were buttressed by too many data on population disappearance to be so easily dismissed. These critics argued that most of nature is not structured as the equilibrium theory envisions and that local population extinction is infrequent. Most empirical "support" for the theory consisted not of observations of species turnover but of demonstrations of a monotonic species-area relationship (Simberloff, 1974). Because the species-area effect can be predicted on grounds other than the equilibrium theory, particularly a relationship between habitat diversity and area (Connor and McCoy, 1979; Williamson, 1988), this support is far from convincing.

Second, most of the rules of refuge design (Fig. 22.1) were found not to be consequences of the equilibrium theory (references in Simberloff, [1988]). Rule a (Fig. 22.1) just restates the species-area relationship, which can be predicted on other grounds as just observed. Rule b, on the relative species richness of a single large refuge and several small ones (the SLOSS issue), is not dictated by the theory. Neither is rule c, about refuge shape, or rule e. The theory does predict rule f, that refuges linked by corridors will contain more species at equilibrium, because immigration rates will be raised. However, the predicted species richness gain may be slight and the cost of corridors prohibitive; further, corridors could also cause species loss (Simberloff and Cox, 1987; Simberloff et al., 1992). Rule d, which might be predicted by the theory if other assumptions are made, turned out to lead to another way of framing conservation questions that is discussed later.

Aside from the fact that most of the rules do not derive from the theory, one might question whether the underlying terms of the theory are even appropriate for conservation. The theory deals in numbers of species, not their identities. Perhaps saving the most species regionally is an appropriate conservation goal, but even this goal need not mandate that the best tactic for each refuge network is to maximize richness. One could imagine a refuge system that saves many common species, all in no need of protection, while another system saves many fewer species, but a high fraction of these depend utterly on this system. Surely the latter is the desired configuration.

Even if the maximization of richness in each refuge network were not the goal of conservationists, and even if most rules do not derive from the equilibrium theory, is it not conceivable that the rules are good rules anyway, that they might turn out empirically to achieve the ends conservationists seek? Although Soulé and Simberloff (1986) viewed them as a dead issue and not useful in conservation planning, arguments about two of them persist. The SLOSS issue seems immortal (e.g., McNeill and Fairweather, 1993), and corridors are widely advocated (e.g., Bennett, 1991) even as they are hotly debated and their advocates acknowledge that various complaints are sometimes valid (e.g., Hobbs, 1992). At this point, I recommend that the SLOSS issue be put to rest for the main reason given by Soulé and Simberloff (1986)—the different configurations never have the same habitat—that corridors be viewed independently of the equilibrium theory, and that the latter be seen as an historically interesting development rather than a management guideline.

Although the MacArthur-Wilson paradigm no longer dominates ecology (Williamson, 1989a), it has provided a theoretical framework in which to consider a number of patterns and phenomena (Haila and Järvinen, 1982). Thus, even if conservation biologists today rarely cite the equilibrium theory as a guide for specific actions, it has left an important legacy: the search for patterns and causes of extinction of small populations. Many workers who noted the existence of a species-area relationship were agnostic about the causes of extinction. MacArthur and Wilson (1967) posited demographic stochasticity as the cause. However, many other explanations are possible for an increased extinction probability in small, isolated populations, and population viability analysis (PVA; Ralls and

Taylor, this volume) seeks an understanding of their modes of action and relative importance. PVA is a direct outgrowth of the equilibrium theory, but it is not explicitly biogeographic. Metapopulation theory, on the other hand, is a biogeographic theory and partly a sequel to equilibrium theory; it is adumbrated in rule d (Fig. 22.1) for refuge design.

Metapopulation Theory

The reasoning behind rule d is that, if sites are close enough to one another, intersite recruitment may forestall local extinction or redress it quickly. Brown and Kodric-Brown (1977), casting the distribution of an insect population among flowers in an equilibrium theory context, termed this phenomenon the "rescue effect." However, as noted, the MacArthur-Wilson theory envisioned local populations as so isolated that their maintenance is primarily a consequence of endogenous reproduction. Thus, although this fact was not noted at the time, the rescue effect really constituted a paradigm shift or, at the very least, a blurring of the equilibrium theory paradigm. If many species' populations are really structured as these flower-insects are, the world is not accurately represented by the MacArthur-Wilson theory.

Independently, Levins (1969, 1970) had developed a concept of a species' metapopulation, an ensemble of local populations, all serving as source areas for invasion of other sites, the whole thereby resistant to extinction even as local populations frequently disappear (Fig. 22.2[a]; cf. Hanski, this volume). The influence of the equilibrium theory on metapopulation theory is questionable (Hanski and Simberloff, 1996); metapopulation models have antecedents well before the equilibrium theory (e.g., Andrewartha and Birch, 1954). It may be relevant that Levins (1969, 1970) was at least as concerned with how natural selection would operate in a metapopulation as with resistance to extinction.

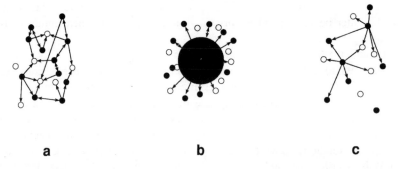

a   b   c

Figure 22.2.   Three different types of metapopulations: (a) Levins model, (b) Boorman-Levitt model, and (c) Pulliam model.

In any event, Levins's papers inspired others (e.g., Richter-Dyn and Goel, 1972) to construct models exploring whether metapopulation structure can aid the persistence of a species. Of course any theory concerned with how extinction is stemmed was relevant to the burgeoning new conservation biology. The weaknesses in the MacArthur-Wilson equilibrium paradigm previously described combined with exciting theoretical results of metapopulation modelers (reviewed by Hanski and Gilpin [1991]) to culminate in the replacement of the MacArthur-Wilson paradigm by the metapopulation paradigm as the governing worldview in conservation biology (Hanski and Simberloff, 1996). Although this paradigm shift was occasionally noted explicitly (e.g., Merriam, 1991), it was largely imperceptible and tacit, all the more surprising because it was so abrupt and complete. It constituted a change in focus from a community level—the number of species at a site—to a species level. This change is ironic, given the wide current advocacy of conservation at the community and ecosystem levels.

The shift was generated by more than scientific results. The equilibrium theory and its emphasis on the species-area relationship devalued small sites (Simberloff, 1982a). So did the first general law of the new conservation biology dealing with genetics, the 50-500 law (Simberloff, 1988). Small sites could be justified on ad hoc bases—a particular site happened to contain a surprisingly large number of species, a set of endemic species (Nott and Pimm, this volume), a particular threatened species, or was held in such affection that it was worth saving as a public relations gesture. But, on the whole, small sites were under assault.

The metapopulation paradigm rescued small sites by attributing new significance to at least some of them. Even if particular species are absent from a site, these absences might be temporary and a network of such sites crucial to these species' continuing existence. Lande (1988a) proposed on theoretical grounds that, at demographic equilibrium, a population does not occupy all available suitable sites. The Furbish lousewort (*Pedicularis furbishiae*) provides a well-known example in which local extinction rates are high and a supply of suitable sites, including empty ones, is critical (Menges, 1990).

Boorman and Levitt (1973) defined a different type of metapopulation, with a large, central population as the source of propagules and smaller peripheral populations that may occasionally disappear, to be re-established by immigration from the central population (Fig. 22.2[b]). But the central population is immortal, and propagules from peripheral populations are irrelevant to its persistence. In this conception of a metapopulation, small sites are insignificant, so it is not surprising that the metapopulation model that has attracted conservation attention is the Levins model (e.g., Lande, 1988a).

Pulliam (1988) also proposed a "source-sink" metapopulation, but one that might appear much like the Levins model (Fig. 22.2[c]). One or a few sites would all serve as sources, and the majority as sinks, whereas the size of a site or its population does not indicate whether it is source or sink. Only extended observations of productivity, and of which individuals move where, can determine the sources. Thus, small sites may have conservation significance, but many sites—the sinks—will really be unimportant.

Data on rates of movement among sites and productivity of populations are much scarcer than data on which individuals are where, so it is difficult to determine whether any of these models pertains to much of nature. Harrison (1991a) and Harrison and Taylor (1996) believe the Levins model fits few cases and that more species conform to one of the source-sink models. Further, they find dispersal among patches in many species is so frequent that the system is a single population rather than a metapopulation, despite its patchy structure. And they suggest that many species that appear to be maintained by metapopulation dynamics are actually declining as the component populations disappear, largely because of habitat destruction.

I would go further—because almost all species are aggregated at some scale, a dwindling species will appear to be a metapopulation at some stage in its decline, so that one might often hypothesize a failure of metapopulation dynamics as causing the decline when in fact the metapopulation appearance results from the decline. The disappearance of Swiss populations of the scops owl (*Otus scops*) exemplifies this problem (references in Simberloff, [1994b]); most local aggregations disappeared because of habitat destruction, and the increasing sparseness of the distribution of aggregations is more likely a reflection of the destruction than a cause of the decline of the bird.

The challenge to conservation biologists is to generate many more empirical studies to know how frequently small populations and sites are likely to be important in regional persistence and if Levins-type metapopulation dynamics are crucial for certain types of species. For now, the key contribution of metapopulation theory to conservation is to force us to study dispersal rates and their importance.

Areography

Extinction at Range Margins

Understanding the role of metapopulation dynamics thus requires information from areography, the study of the geometry of species' ranges (Rapoport, 1982). For decades, areography had largely been reduced to the status of record-keeping, but it has lately been rejuvenated, in no small part because of its relevance to conservation. Thus, range changes, formerly published as notes in obscure taxon-oriented or regional journals, are now seen as data vital in assessing models of the spatial organization of species and crucial for conservation. It is not coincidental that range changes are now sometimes referred to as "extinctions" (e.g., Hengeveld, 1990).

Range boundaries are dynamic, and probably a large fraction of this dynamism consists of population extinction (Williamson, 1989b), although rarely are there enough data to determine what is a population or part of a metapopulation. Thus, typical published range maps are snapshots of a changing phenomenon. The exact reasons for the changes are rarely studied. Nor are the reasons why species' ranges terminate where they do, and answering such questions usually requires transplant experiments and laboratory work.

For example, Neilson and Wullstein (1983) sought explanations for the range limits of Gambel's oak (*Quercus gambelii*) in the American West. They found the northern limit governed by spring freezing and summer drought, which determine seedling mortality. Suitable microsites become increasingly rare to the north, until they are so sparse that they are rarely colonized. Variation in spring freezing and summer drought will generate short term changes in locations of peripheral individuals, so the range boundary will change, although this phenomenon was not studied. Nor did Neilson and Wullstein ask whether this species is organized as a metapopulation.

By contrast, a study of the British range limits of prickly lettuce (*Lactuca serriola*) (Prince et al., 1985) presents a very different picture. Aggregations are not sparser toward the range limit. There is an abrupt boundary with a dense aggregation on one side and no plants on the other, and no climatic variable correlates with this boundary. A large range expansion in 1977 was followed by several years of range contractions. The authors interpret these observations in terms of a metapopulation model in which a minuscule reduction in environmental suitability can generate a boundary much sharper than the underlying environmental gradients. However, data are lacking to determine whether prickly lettuce really is a metapopulation.

These data make it seem unlikely that the disappearance of peripheral aggregations in Gambel's oak or prickly lettuce, whether these are parts of metapopulations or not, is important to the regional conservation of these species. The loss of isolated aggregations in the scops owl is obviously important to conservation, but as a reflection of the devastating effects of anthropogenic habitat change rather than as the cause of an impending extinction. On the other hand, Hanski et al. (1995b) show for the endangered Glanville fritillary (*Melitaea cinxia*) that all occupied patches are small and subject to extinction and that interpatch movement is infrequent, as envisioned by the Levins model. They also argue that the disappearance of this butterfly from the Finnish mainland likely resulted from increasing sparseness of the network of sites.

The Spread of Nonindigenous Species

Nonindigenous species are second only to habitat destruction in harming native communities (Simberloff, 1995). For a number of animal species using various dispersal mechanisms, a sigmoidal plot of square root of occupied area versus time since introduction represents their spread, as was predicted by an epidemiological model (van den Bosch et al., 1992) in which the parameters of the curve can be estimated from accessible literature data. Such predictability would be useful in determining what amount and type of effort to devote to combatting an invasion, including whether to mount an expensive eradication program at the outset.

Long-Term Range Changes

Species ranges of the geological past have also taken on added conservation significance. For example, the U.S. National Park Service removes exotic species from national parks (Houston et al., 1994). This policy ran afoul of animal rights advocates (e.g., Anunsen and Anunsen, 1993), who challenged the removal of mountain goats (*Oreamnos americanus*) from Olympic National Park on the grounds that they might have been native there during the late Quaternary. This contention (Lyman, 1988, 1994) has been rebutted (Scheffer, 1993; Houston et al., 1994) by a careful consideration of fossil evidence and records of early historic explorers.

Quaternary range changes have increasingly concerned conservation biologists in at least two ways. First is the question of whether conservation can target individual species or whether whole communities inevitably rise and fall together because of obligatory interactions among their members. For both animals (Graham, 1986) and plants (Davis, 1986), ranges of species moved great distances during the Quaternary but at very different rates, so that new assemblages constantly arose and old ones disappeared, at least suggesting caution in depicting communities as longstanding coevolved entities. Second, the same range changes in response to past climatic changes suggest the sorts of range changes that might be anticipated as greenhouse gases induce global warming. In particular, the fragmented landscape may render movement of species much more difficult than previously, leading to suggestions of massive networks of north-south corridors (references in Simberloff et al. [1992]).

Gap Analysis

The evolution of Geographic Information System (GIS) technology combined with renewed interest in areography to suggest a different approach to conservation, gap analysis (Scott et al., 1987, 1993). By layering range maps, one can generate species richness isoclines. Areas of peak richness can then be compared to areas under various ownerships to determine such matters as the extent to which a reserve system includes sites of great richness, or who owns unprotected rich sites. However, a focus purely on species richness fails to recognize the criticism voiced above: we may not want to save sites with the most species; rather, we are concerned with species that are endemic (Nott and Pimm, this volume), threatened, unprotected, or rare. A GIS can produce isoclines of richness of rare or threatened species as easily as of total species (Scott et al., 1987), although the emphasis so far has been on the latter.

One can also map vegetational communities or ecosystems to see how many are contained in a refuge system. Here the problem is partly that there is no universally accepted classification of communities or ecosystems and partly that it is difficult to map them at a fine enough scale for conservation decisions. However, substantial recent progress has been made in both directions. For example, The Nature Conservancy has recently compiled, with the aid of its partners in Heritage Programs, other agencies, and academia, a classification of rare plant communities for the entire United States (Grossman et al., 1994) that should go far to standardize terminology and allow maps acceptable to all parties and usable in gap analyses. As for the difficulty of mapping, the use of Landsat Thematic Mapper imagery allows a rapid classification of much of the landscape into habitat types that, although fairly coarse, may still be useful for gap analysis (e.g., Kautz et al., 1993).

As an example of how gap analysis may be deployed at the ecosystem level, Crispin (1994) describes a project for the entire Great Lakes region in which managed area boundaries were superimposed on ecosystem maps to reveal gaps in protection. At the species level, Cox et al. (1994) used gap analysis to determine the distributions and current level of protection for several threatened vertebrates, such as the Florida panther (*Felis concolor coryi*), the mottled duck (*Anas fulvigula fulvigula*), and the black bear (*Ursus americanus*), and to recommend additions to state-protected lands.

Gap analysis will be a useful conservation tool, but there is a tendency to emphasize the technology and the flashy maps. The products of the GIS are only as good as the range maps, and, as has been noted, ranges change constantly and published maps are generally coarse, one-time caricatures. Further, the level of mapping detail needed for effective management of a species varies greatly with the species. For widely ranging animals like the Florida panther, the coarseness of existing habitat and range maps is not a problem. For more narrowly distributed habitat specialists, such as many invertebrates and plants, existing habitat maps are woefully inadequate. And the effort devoted to gathering detailed field data on the ranges of such species is far less than that devoted to manipulating the pitifully few data that exist.

Nestedness

A nested pattern exists for a group of species distributed among a set of sites if all species of each site are found on all sites containing more species. Several authors (references in Simberloff and Martin [1991]) have argued that nested patterns are important in refuge design. They contend that nestedness arises by selective extinction in various sites and that these extinctions occur in a fixed order, such that if a species goes extinct in a site, it will also disappear from all sites with fewer species. McDonald and Brown (1992) have thus predicted the extinction of montane mammals that will be caused by global warming from patterns of hypothesized post-Pleistocene extinctions accompanying postglacial warming.

In fact, most biotas that have been examined are nested, but there is generally little evidence that selective extinction is the cause, rather than selective invasion or habitat differences among sites (Simberloff and Martin, 1991). Many extinctions are unobserved but are assumed to have occurred based on questionable statistical procedures. This is certainly true for the montane mammal study (Skaggs and Boecklen, 1996), which is further compromised by incorrect data on past and present species' ranges. In any event, if all the data were available to calculate nestedness scores accurately, such a score would probably not be very useful in designing a refuge. From the same data, namely, which sites are occupied by each species, one would already have a better idea of the problems faced by each species on a particular proposed refuge (Simberloff and Martin, 1991).

Genetic Geography

The role of genetics in the "new conservation biology" was exclusively in the form of research on how inbreeding depression and drift threaten small populations. Now these forces are seen as generally less important than demographic and environmental stochasticity (Lande, 1988b, 1993; Holsinger and Vitt, this volume), but the advent of molecular techniques that can distinguish populations—particularly study of allozymes and mitochondrial DNA (mtDNA)—has generated new interest in genetics among conservationists. The key biogeographic component is consideration of how genetically distinct conspecific populations are, and the key question is what taxonomic level is worthy of conservationists' concern.

The dusky seaside sparrow (*Ammodramus maritimus nigrescens*) was endangered by habitat destruction (references in Rhymer and Simberloff [1996]). The U.S. Fish and Wildlife Service spent $5 million on habitat for this subspecies, but the decline continued until only six males remained. At this point, a legal donnybrook erupted over a proposal to interbreed these individuals with another subspecies, a project eventually carried out with private funds. The entire controversy was irrelevant, as demonstrated by an mtDNA analysis by Avise and Nelson (1989) showing that the dusky seaside sparrow was not genetically distinct from other Atlantic coastal populations and thus not worth the fuss. The genetic concern should be to preserve representatives of Atlantic populations on the one hand and Gulf populations on the other (Avise, 1994a).

Acknowledgments

Mary Ruckelshaus, Ilkka Hanski, and Mary Tebo made numerous helpful comments on a draft of this manuscript.

23

Conserving Interaction Biodiversity

John N. Thompson

Summary

Priorities for conservation of biodiversity have been based primarily on concerns over species and ecosystems: maintenance of species diversity and ecosystem functions, and retention of genetic variation within populations. But the links between species and ecosystems occur through the interspecific interactions that shape the organization of biological communities. We now know, as a result of research in evolutionary ecology over recent decades, that interspecific interactions can sometimes evolve quickly under changed environmental conditions. For example, restriction in gene flow among populations can lead to divergence among populations in the outcomes of interactions; increasing host density and increasing transmission rates of infectious parasites (as can occur in isolated reserves) has the potential to favor the evolution of higher levels of parasite virulence; and the loss of metapopulation, or broader geographic, structure in interactions may disrupt the coevolutionary process that maintains traits such as some resistance genes in plant populations. Hence, the continuing evolution of interactions as we change their spatial dynamics has important immediate effects on conservation efforts and priorities. Concerns over the evolution and conservation of "interaction biodiversity" cannot be something simply added on to the long list of other concerns in conservation biology. They must be an integral part of efforts to maintain evolutionarily viable conserved communities. To do so requires concerted research efforts to better understand the geographic scale at which interactions evolve among species, and the retention of some large unmodified landscapes in which such studies can be carried out.

Introduction

When Darwin introduced the concept of coevolution in the *Origin*, he did so using a thought experiment on reciprocal adaptation between flowers and bees.

He began with a population of a bee species coevolving with the flowers of a plant species through reciprocal morphological changes, and then imagined, if the bee population became extinct, how that plant population might come to coevolve with another local bee species. In retrospect, Darwin's thought experiment unwittingly captured one of the most difficult challenges facing coevolutionary studies today—how coevolving interactions will change in the face of escalating local extinction of species and invasions by new species or populations.

We often view the priorities for conservation from two ecological perspectives—maintenance of species diversity and ecosystem functions—and one evolutionary perspective—maintenance of the genetic diversity of populations. The major reason offered for maintaining genetic diversity is that it will allow populations to respond over evolutionary time to changed environmental conditions. The evolutionary concerns, however, often seem far off, a laudable and fully appreciated goal for the long-term maintenance of species diversity, but unlikely to have immediate effects on conservation efforts. For example, only 2 of the 10 most recent books in conservation biology acquired by my university's library this year include entries for "evolution" or "natural selection" in the index (although several of them do include the phrases genetic variation or genetic diversity). Yet the environments in which organisms live, and the selection pressures they are experiencing, are rapidly changing through habitat fragmentation, loss of species, and addition of alien species. Consequently, in this chapter I want to make two points: (1) the evolutionary concerns are just as immediate as the ecological concerns and (2) the conservation of interaction biodiversity over large landscapes is likely to be the most effective way to meet those concerns. By interaction biodiversity I mean the diversity of ways in which pairs and groups of species interact with one another within and among populations, and the diversity of outcomes that the interactions elicit.

Rapid Evolution of Interactions

Loss of genetic diversity in small, remnant populations through random genetic drift is the most commonly invoked evolutionary concern in conservation biology. This concern has spawned a great deal of refinement in recent years in the mathematical formulations and the empirical determination of effective population size (Lande, 1988b; Caughley, 1994; Nunney and Elam, 1994; Valle, 1995; Holsinger and Vitt, this volume). These studies have collectively suggested that the small size of many nature reserves is likely to result in the erosion of genetic variation in local populations of many taxa not connected through gene flow to other demes. Hence, the practical concerns in the conservation biology of particular species have become focused on maintaining, wherever possible, interconnected populations at levels that slow the rate of loss of genetic variation and decreasing the likelihood of extinction due to the vagaries of demographic processes.

That, however, is only part of the problem. The genetic variation retained in these populations is not just being held unchanged in Hardy-Weinberg equilibrium, or experiencing gradual loss through genetic drift. It is being reshaped through natural selection now, today, as populations respond to the increasingly rapidly changed conditions within virtually all biologically rich environments worldwide. We already know that interactions between species can sometimes evolve very quickly. For instance, populations of the checkerspot butterfly (*Euphydryas editha*) in California differ in the plant species that females prefer when laying their eggs, and preferences in at least one population have evolved in recent decades through human alteration of the landscape and introduction of an alien plant species (Singer et al., 1993). On the island of Hawaii, the population of the i'iwi, a honeycreeper with a long decurved bill used to extract nectar from flowers, has evolved in its bill morphology since the turn of century, apparently in response to changes in the availability of plant species through habitat fragmentation and extinction of one of its major competitors for nectar (Smith et al., 1995). And the introduction of myxoma virus to control European rabbits in Australia and Britain has resulted during this century in rapid coevolution of that interaction (Fenner and Myers, 1978; Ross and Sanders, 1984; Dwyer et al., 1990).

These three examples illustrate rapid evolution in interactions in response to the three different kinds of changed ecological conditions that have been of greatest concern in conservation biology: habitat fragmentation, extinction of species, and introduction (either deliberate or accidental) of new species. Similar examples of rapid evolution of interactions have been appearing with increasing regularity in the literature of ecology and evolutionary biology in recent decades (Thompson, 1994a). Together they suggest that the evolutionary problem to solve in conservation biology is not just retention of genetic variation but also how interactions are likely to evolve among species given different conservation decisions. Evolution is probabilistic, not deterministic, and hence it is not possible to predict precisely how interactions will evolve under any specific set of environmental changes. Progress in evolutionary ecology is, however, making it increasingly possible to understand the kinds of evolutionary changes that are likely to occur as environments change. At the very least, these studies suggest some immediate evolutionary considerations that need to be part of the ongoing debate about conservation priorities. I discuss three of them here.

Divergence in Interactions through Decreased Gene Flow among Populations

Severe restriction of gene flow among populations can lead to two results: rapid divergence among populations in the outcome of some interactions and extinction of interactions (and, therefore, sometimes species) that rely on metapopulation structure. The potential for rapid divergence comes from the observation that there is a distribution of outcomes in all interactions among species, and that the distribution can often vary greatly among environments (Thompson, 1988). For example, plant populations differ in how resistance genes against parasites are distributed among individuals (e.g., Burdon, 1994), and the efficacy of the resistance genes can depend on the local genetic composition of parasite population and the environment in which the interaction takes place, which influences can gene expression and parasite transmission rates. To the extent that distributed outcomes are genetically based, they are the raw material for the evolution of interactions, providing different selection pressures on different populations.

In the complete absence of gene flow, these changes can occur surprisingly quickly. In a large-scale recent laboratory study on the evolution of competition, the outcomes of interspecific competition evolved rapidly among two species in only eleven generations. Some of these evolved changes in outcome were due to selection imposed by the competitors on one another, but many of the changes in competitive ability also resulted from adaptation of the populations to the different physical environments in which the experiments were conducted (Joshi and Thompson, 1995, 1996). These laboratory experiments suffer from all the usual attendant problems that come from asking ecological questions in simplified environments. As with all well-controlled microcosm studies (Lawton, 1995), however, they provide an indication of the rate and direction of ecological change that is possible under some environmental conditions. In this case, they show that interactions can evolve rapidly under some ecological conditions.

What this means for conservation is that at least some, perhaps many, interactions in reserves cut off from one another are likely to evolve to different outcomes. Protocols for the maintenance of species and interactions that initially work for a group of reserves may quickly fail to work for some reserves as the interactions evolve to new outcomes in the face of restricted gene flow. Yet, my discussions with reserve managers and conservation biologists have led me to conclude that, if differences in interactions were observed among reserves, many would look initially for ecological causes rather than evolutionary change in the interacting species. That priority in the search for causes is based upon the assumption that interactions evolve very slowly.

Loss of Interactions through Loss of Metapopulation Structure

The second evolutionary consequence of reduced gene flow is loss of metapopulation structure. Recent studies have indicated that long-term ecological persistence of some interactions is increased through metapopulation structure (Hassell et al., 1991; May, 1994b). In some cases, longer persistence may be simply an ecological result: the local chaotic population dynamics of an interaction becomes stabilized over regional scales with the addition of metapopulation structure. In other cases, however, the interaction may persist because metapopulation structure stabilizes the evolutionary, not just the ecological, dynamics of the interaction.

One of the long-standing problems in coevolutionary studies is how reciprocal change in local antagonistic populations could withstand repeated bouts of the evolution of new adaptations and counteradaptations without one of the species going extinct. Time lags in response to new adaptations in the other species and differences in generation times between the species create the possibility that a rapidly spreading new mutation in one species will not be countered quickly enough in the other species to prevent extinction. But metapopulation structure can even out the playing field somewhat. A local population may lose and become extinct, but the interaction itself may be retained over a broader geographic scale through a combination of local adaptation and counteradaptation, gene flow, and recolonization.

Gene-for-gene coevolution between plants and pathogens may be one important form of coevolution that requires a metapopulation structure (Thompson and Burdon, 1992). Mathematical models of gene-for-gene coevolution that rely on frequency-dependent selection for maintaining resistance genes in plant populations and virulence genes in pathogen populations often become increasingly unstable as more ecological variables are added. Metapopulation structure allows for local instability of an evolving interaction by providing increased regional stability. It is part of a landscape view of ecology and evolution. Long-term field studies of the interaction between wild flax (*Linum marginale*) and flax rust (*Melampsora lini*) in natural habitats in Australia are indicating that local interactions between these two species evolve rapidly and are unstable, and that the species persist through metapopulation structure (Burdon and Jarosz, 1992; Burdon, 1994; Burdon and Thompson, 1995).

The interactions between wild flax and flax rust also show that the distribution of genotypes can change rapidly in a local natural population during the course of an epidemic (Burdon and Thompson, 1995). The problem of how to structure nature reserves in ways that minimize the spread of disease epidemics is becoming a major topic in conservation biology (May, 1988; Aguirre and Starkey, 1994; Dobson and Hudson, 1995). One argument for restricting gene flow among reserves is that it may prevent the spread of a disease. If, however, the maintenance of resistance alleles against some parasites requires a metapopulation structure of hosts and parasites, then the loss of that structure may slow the spread of any particular epidemic but rob a host species of its evolutionary future.

Effects of Increasing Local Population Density

Epidemics differ in degree of virulence. One of the most important advances in the study of evolving interactions during the past two decades has been the development of a firm theoretical basis for the evolution of virulence in interactions between hosts and parasites. Theory shows that interactions between parasites and hosts can coevolve quickly to new outcomes as density of host populations and transmission rates of parasites change. The pioneering work of Anderson and May (Anderson and May, 1982, 1991; May and Anderson, 1990) on the evolution of virulence, and subsequent mathematical and empirical work by them and others, has made it evident that increasing density of host populations and increasing transmission rates in horizontally transmitted (infectious) parasites favor higher levels of virulence (Herre, 1993; Bull, 1994; Ebert, 1994; Ewald, 1994).

Consequently, the maintenance of high local population density of hosts within isolated reserves is likely to favor the evolution of more virulent genotypes of horizontally transmitted parasites, all else being equal. There are, however, many empirically unexamined aspects of the evolutionary dynamics of virulence, including the effects of different forms of metapopulation structure. What is needed is retention of some large-scale ecological arenas in which to test the ideas.

Conserving the Geographic Mosaic of Interaction Biodiversity

It is unclear at just what geographic scale the persistence and evolution of most interactions takes place, but it is crucial that we find out. The study of conservation biology has moved from efforts to conserve single species to priorities based on more of a landscape view (Franklin, 1993b; Hansson et al., 1995; Pickett and Cadenasso, 1995). Much of that landscape view, however, has been developed based primarily on an ecological rather than evolutionary perspective, except for considerations of effective population size. Yet to be spliced to that view are the several decades of research in evolutionary ecology showing that populations differ in their adaptations and their degree of specialization to other species across landscapes at different scales.

The geographic mosaic theory of coevolution argues that the local adaptations and counteradaptations within local populations, and the differences across landscapes, are not just interesting and esoteric demonstrations of the power of natural selection to adapt populations to local conditions (Thompson, 1994a, 1994b). Rather, they are the raw material, and often only the raw material, for evolving interactions. In this view, most of the dynamics of coevolutionary change (and evolutionary change in interactions in general), and the persistence of evolving interactions, occur at geographic scales, above the level of local populations and below the level of the fixed traits found in comparisons of the phylogeny of species. The overall trajectory of the evolution of an interaction results from the reorganization of those local adaptations through gene flow, genetic drift, and population extinctions.

According to the geographic mosaic theory, therefore, there is a hierarchy to evolving interactions: from local populations to geographic structure and dynamics to the subset of adaptations that eventually become fixed in species. The evolution of interactions cannot be understood without analyses at all levels of the hierarchy. Yet, despite the fact that we have a long list of studies showing population differentiation in interactions, we have very few studies that have been designed to evaluate how geographic structure may ultimately shape the evolution of interactions. Study of the geographic structure of interactions is the needed link between analyses of single local populations, which make up the bulk of studies in evolutionary ecology, and the phylogenetic distribution of fixed traits, which make up the bulk of studies in trait evolution by systematists interested in biodiversity. If we are to establish biological priorities for conservation, we need a firm understanding of how the geographic mosaic of interactions shapes the evolution of interactions.

To gain that understanding we need more detailed analyses of the ways in which geographic structure influences the evolution of a variety of different kinds of interactions. Highly specialized and long-coevolved interactions—the kinds of interaction that are often presented to students simply as "neat, but highly unusual, stories"—are our touchstones for understanding the importance of the geographic mosaic. These interactions provide the next step up from laboratory microcosms in their relative simplicity for study. They are precious tools for both evolutionary biology and conservation biology. Through comparative analyses of these populations, we can ask whether the interactions are locally stable or rely on metapopulation or broader geographic structure for persistence. And we can ask how, and how rapidly, these interactions evolve with the local loss of species, invasions of new species, and habitat fragmentation.

For example, a group of us has been using the interactions between the prodoxid moth (*Greya politella*; a close relative of yucca moths) and its saxifragaceous host plants to evaluate the geographic scale at which interactions are evolving, and sometimes coevolving, between insects and plants. These moths passively pollinate their host plants as they oviposit into the flowers. In some populations, the mutualism is swamped by copollinators (Thompson and Pellmyr, 1992; Thompson, 1994a). We chose this particular set of interactions because it is geographically widespread, geographically variable, experimentally manipulable, and common in and around the major wilderness areas of western North America. The major wilderness areas provide the opportunity to evaluate the evolving geographic mosaic in these interactions in the least modified environments we have left, and they provide a contrast to the same interactions in more modified environments. As these studies are progressing, it is becoming increasingly evident that there is tremendous geographic complexity in these interactions, and the overall evolutionary trajectory of the relationships among these species is being shaped over broad geographic scales (Thompson, unpublished data).

This is the evolutionary reason why the few remaining large-scale geographic blocks of relatively unmodified landscapes are so precious. They harbor our last opportunities for studying the outcome of long-term evolution of interactions on a geographic scale, and they preserve the highly specialized and coevolved interactions that are being replaced elsewhere with weedy species or managed landscapes. No amount of money or efforts in restoration ecology can recapture the geographic mosaic of these long-term experiments in evolution.

Where?

Outside of the tropics and the far north and south of North America, Asia, and Antarctica, there are very few places that provide the kinds of landscapes that retain anything that closely resembles the past geographic mosaic of interactions. In North America, south of Alaska and the Far North, only two major regions remain that seem to be sufficiently intact to use as touchstones for understanding the evolved geographic structure of interactions. One of these is the wilderness core of central Idaho and surrounding areas. Although not pristine, however defined, it has been less fragmented over large geographic areas by human activity than any other region in the United States south of Alaska. About 4,000,000 ha of roadless, undeveloped public lands currently occur within the 213,505 km 2 of Idaho, with 22 percent of the "Northern Rockies ecoregion" under protection (Wright et al., 1994). Therefore, the potential for increased protection is still high. Many habitat types, however, in and around the wilderness core remain largely unprotected and are being fragmented at a tremendously fast pace. The currently conserved areas are skewed toward habitats in higher elevations, such as subalpine forests and parklands, and less than 6 percent of the existing reserves exceed 10,000 ha (Caicco et al., 1995).

The other area is the Canadian Rockies, for which Jasper, Banff, Kootenay, and Yoho National Parks form the core. British Columbia harbors about 60 percent of Canadian plant species, 75 percent of bryophytes, 85 percent of lichens, three quarters of Canadian bird and mammal species, 40 to 50 percent of the amphibian and reptile species, and a high (but undetermined) percentage of invertebrate species (Foster, 1993; Pojar, 1993). These species are distributed across the vast span of landscapes found in British Columbia, but the Canadian Rockies contributes importantly to this unusual richness of species. The Canadian Rockies have been much less fragmented than the American Rockies, but that is changing fast. Only about 13 percent of the Canadian Rockies are currently protected (Gadd, 1995).

It is obvious that we are not going to have many large reserves throughout the world, and what may eventually become at least moderately large reserves will be through restoration efforts. Restored environments are not the same as conserved environments. Although crucially important as a component of overall conservation efforts, restored environments lack the geographic structure of evolved interactions that we so desperately need for rigorous testing of hypotheses in the evolutionary ecology of conservation. And they cannot serve as guides for understanding how interactions evolve over the long term under different sets of ecological conditions and over landscapes at different scales. That, from an evolutionary perspective, is why the few large remaining relatively undisturbed areas are so critical to conservation biology.

Conclusions

The evolution of interspecific interactions is not something just to be added on to the long laundry list of concerns in conservation biology. It is part of the here and now dynamics of biological communities. If we appreciate that interactions can evolve rapidly, and that their evolution may often involve processes that occur at metapopulation or even broader geographic scales, it will influence how we establish biological priorities for conservation. Political priorities are different from biological priorities, but our responsibility is to make certain that the biological concerns and priorities are clear when we are asked. Evolutionary ecological studies are providing an important argument for the conservation of at least a few large conserved areas to use as reference sites for the importance of environmental heterogeneity, patchiness of habitats, and metapopulation and geographic structure in the persistence and evolution of interactions among species.

Acknowledgments

I am grateful for discussions on these issues with Olle Pellmyr and Steward Pickett. This work was supported by NSF grant DEB-9317424.

SECTION V

The Application of Conservation Ecology

Themes

*Richard S. Ostfeld, Steward T. A. Pickett,
Gene E. Likens, and Moshe Shachak*

Ecologists working on problems relevant to conservation experience tremendous pressures from within and without. The scientific knowledge produced by a shockingly small cadre of professionals is the basis for decisions and actions that affect whether species will persist, whether and by what means natural areas will be protected, and whether ecosystems will continue to provide the goods and services on which society depends. Scientists are perceived as friends or enemies of politically motivated groups, depending on the groups' agendas, their veracity, and their understanding of the science. Scientists are pressured by some groups to speak out more actively, and by others to remain quietly in the ivory tower of academia. The knowledge gained by scientific inquiry has attached to it some degree of uncertainty, but the societal problems to which that knowledge may be crucial often cannot wait until the uncertainty is reduced to inconsequential levels. Indeed, some of the uncertainty results from the probabilistic, stochastic, and multiplicity of ecological relationships. This inherent uncertainty makes science and scientists vulnerable to both intentional and inadvertent misrepresentation by those with political goals. Moreover, the negative connotations of doing applied, rather than fundamental, research continue to inhibit the appropriate use of scientific information in policy and management.

This section returns us to the real world of conservation practice, policy, and publicity. The chapters share in common the understanding that scientists and their science dwell in societies that desperately need guidance for prudent decisions. What roles can and should scientists play in this exchange? To address this question, contributions in this section distill some of the fundamental insights and information from prior sections into practical tools and standards for application. Some chapters provide more philosophical perspectives on how to make environmental decisions in the context of limitations to knowledge and alternative courses of action. Others discuss both the benefits and the risks of scientists interacting with the public. Finally, the tension between scientific endeavor and activism is openly discussed from several perspectives, with the outcome that activism is viewed as a legitimate, if perilous enterprise for scientists.

24

State-Dependent Decision Analysis for Conservation Biology

H. P. Possingham

Summary

If applied conservation biology is to be effective, it is essential we develop conservation theory within an explicit decision-making framework. Most conservation theory enables us to identify management options that are beneficial, but it fails to help us to choose between those options given limited resources. Although the tools for making decisions in a stochastic world are not simple, economists and management scientists already have an array of methods for making decisions in risky and uncertain circumstances. This chapter explores a specific example of the application of a decision theory tool to nature conservation. I use Markov decision theory to choose between management options for a threatened metapopulation. A presence-absence stochastic metapopulation model is coupled to Markov decision theory machinery to determine the optimal management strategy for a threatened metapopulation. An important attribute of this decision-making tool is that the best management strategy depends on the current state of the system being modeled. I explore other possibilities for the application of state-dependent decision theory in conservation biology.

Introduction

Many ecological theories and concepts supposedly help us in applied nature conservation: island biogeography theory, metapopulation theory, theories of coexistence, and the notion of a minimum viable population size. The relevance of these theories to managers is often unclear because they are not couched within a decision-making framework (Maguire, 1986). Many of these theories only inform us which management strategies are useful, but they do not enable us to choose between those strategies when resources are limited. For example, island biogeography tells us that big patches of suitable habitat are better than small patches, and corridors linking habitat patches are useful (Simberloff, this volume). However, island biogeography theory does not tell us how to trade off reserve size with connectedness; that is, should revegetation involve making corridors between patches or increasing the size of existing patches?

Where time and money are limited, we must be able to choose the best management options within the constraints of those resources. Logically then, we need to merge existing theories of population ecology with decision theory tools to help managers achieve the best result with the resources available to them. This union of decision theory and ecological theory will lead to a theory of applied conservation management. To date the best examples of this union involve using population viability analysis to rank different management options for threatened species (Ralls and Taylor, this volume).

In this chapter, I briefly review the idea of making decisions in applied nature conservation, I explain the difference between static and dynamic decision theory, then I consider a specific problem and finish with other possible applications of these ideas.

Static Decision-Making Problems

Maguire (1986) was the first author to emphasize the importance of decision theory to conservation biology. The case study by Maguire et al. (1987) is an excellent example of how a decision tree can be used to choose between management options for an endangered species. In their example, a range of options are listed and ranked according to their ability to minimize the extinction probability of the Sumatran rhino and their economic cost. Maguire (1991), Maguire and Servheen (1992) and Ralls and Starfield (1995) introduce further methods and considerations for making conservation decisions. All these seminal examples are what I refer to as "static" decision-making problems. In each case, the optimal strategy is a single decision, whether that be enacted at one point in time, or whether that one strategy is fixed for all time.

The static decision theory approach has been very useful in bridging the gap between conservation theory and management in an uncertain environment. However, this approach does not take into account the fact that a series of decisions need to be made. For example, consider the question of moving animals between captive and wild populations. This decision needs to be made continually and the number moved between populations will surely depend on the current state of each population as well as expectations of events and management actions in the future. The only way to properly deal with this sort of sequential decision-making process is to use Markov decision theory.

Dynamic State-Dependent Decision-Making Problems

Before considering the application of Markov decision theory to a particular problem, metapopulation management, we need to review the existing application of metapopulation theory to conservation.

Interest in modeling the dynamics of spatially structured populations is rapidly increasing. One class of ideas and models used to understand the dynamics of populations that are not spatially homogeneous is metapopulation theory (Hanski, 1991b; Verboom et al., 1991; Mangel and Tier, 1993b; Adler and Nuernberger, 1994; Hanski, 1994b; Day and Possingham, 1996; Hanski, this volume). Metapopulations are made up of a number of local populations. The simplest kind of metapopulation model follows the dynamics of the number of patches that are occupied, ignoring local population dynamics (Levins, 1969). In this case, the processes of local population extinction and colonization drive the dynamics of the metapopulation (Fig. 24.1). This is often referred to as a presence-absence metapopulation model and is the kind of metapopulation model used in this chapter (Richter-Dyn and Goel, 1972).

I illustrate optimal dynamic decision making by using Markov decision theory on an applied problem in nature conservation. In particular, I consider the optimal management of a metapopulation. In this example, the objective is to minimize the likelihood of extinction of the metapopulation, the standard objective in applied threatened species management (Burgman et al., 1993). Two management options for minimizing the likelihood of metapopulation extinction are to make more patches of suitable habitat or recolonize suitable empty patches. I show how Markov decision theory can be used to choose between these two management options and enable us to make optimal management decisions in an uncertain environment. We will find that the best decision at any time depends on the current state of the system, which in this case is the combination of the number of occupied and unoccupied patches.

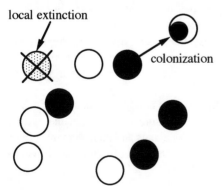

local extinction

colonization

Figure 24.1. The dynamics of a metapopulation are driven by local extinction and colonization.

Example of Metapopulation Management

There are two parts to solving a management problem using Markov decision theory. First we model the stochastic dynamics of the system, in this case, the dynamics of a metapopulation. Here I use the simplest stochastic metapopulation model possible—a presence/absence nonspatial model. Second there is the decision theory part, where the stochastic dynamic programming equations need to be formulated and rewards defined for the outcome of the management.

Assume that we are interested in a metapopulation where all the patches are identical and equally likely to colonize each other. This type of nonspatial presence-absence model has been used widely to explore aspects of metapopulation dynamics (Levins, 1969; Richter-Dyn and Goel, 1972). We also assume that patches are either occupied or unoccupied. A presence-absence model like this ignores the possibility that different patches may contain different numbers of individuals. However, where local population dynamics operate rapidly on the time scale of the extinction/colonization processes that determine the dynamics of metapopulations this is a reasonable simplification (Hanski and Thomas, 1994; Kareiva et al., this volume).

The details of the metapopulation model and the subsequent application of Markov decision theory are in Possingham (1996). For a particular set of parameters it is possible to numerically determine which of two strategies, recolonize one patch or make one new patch, gives the lowest long-term extinction probability for all of the possible states of the system. In the example in Figure 24.2, I assume there can be no more than 12 patches in the system. Figure 24.2(a) shows which management strategy is optimal for a particular set of parameters. A filled in circle indicates that the optimal strategy is to recolonize a patch, otherwise the optimal strategy is to make a new patch. The optimal strategy depends on the current state of the system and will change as the system moves from state to state.

Some of the results in Figure 24.2(a) can be quickly explained. When no more patches can be made the optimal strategy will always be to recolonize an empty patch. When all suitable patches are occupied (the leading diagonal of the results in Fig. 24.2[a]), then all the patches are occupied, recolonization is not possible, and the optimal strategy is to make a new patch. Between these extremes either strategy is possible. In general, the recolonization strategy is optimal when the number of occupied patches is small or the fraction of patches occupied is low.

Figures 24.2(b) and 24.2(c) show the response of the optimal strategy to a "worse" environment. If the extinction rate increases, recolonizing empty patches is only the best strategy when the number of occupied patches is very low compared to the number of unoccupied patches. Another way of making the environment worse for the metapopulation is to allow no natural colonizations. In the absence of management, this would mean that the metapopulation is doomed, however, because we can artificially colonize patches, the metapopulation has a reasonable chance of persisting. In this case the worsening environment

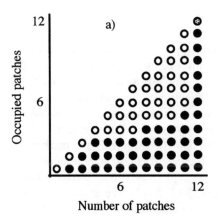

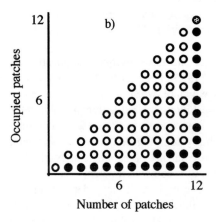

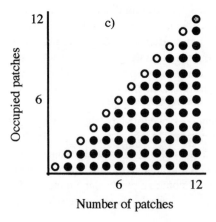

Figure 24.2. Optimal metapopulation management strategies for three different cases. Where a combination of number of patches and number of occupied patches has a filled in circle, the optimal strategy is to recolonize an empty patch, otherwise make a new patch. When 12 of the 12 possible patches are occupied neither strategy is possible: (a) the baseline case, (b) the rate of local extinctions is doubled, and (c) the colonization rate is zero.

causes recolonization to become the optimal strategy under all circumstances when it is possible, the reverse of what is optimal under an increasing extinction rate.

Application to Ecosystem Management

So far we have discussed a problem of single-species management where state-dependent dynamic decision theory can be useful in nature conservation—choosing between habitat creation and recolonization. Many other decisions for single-species management could be tackled in the same way. However, to emphasize the flexibility of this method let us briefly consider an example concerning the management of disturbances in an ecosystem.

Disturbances such as fire play an important role in maintaining biodiversity. Many species require disturbance, whereas others can only live in habitats that have not been disturbed for a long time. With these different requirements in mind, it would be prudent to manage disturbances so that no less than a minimum proportion of a region exists in each successional type. For example, let the vegetation in the area be classified into three successional states, early, mid, and late, and let the agent of disturbance be fire. Assume also that there is some background fire frequency, where each fire burns a random fraction of the landscape. In the absence of management we can model the stochastic dynamics of the amount of landscape in each successional state using a semi-Markov chain (Henderson and Wilkins, 1975; Moore, 1990). Suppose we have the resources to reduce or increase fire frequency and/or extent. A reasonable management objective might be to maximize the probability that every successional type covers at least 10 percent of the landscape. The state of the system is the current proportions of the successional types and, as before, Markov decision theory can be used to find the strategy that maximizes the management objective.

Discussion

A major shortcoming of Markov decision theory is the computational effort required for solving large and complex problems. The method does not lend itself well to problems where the number of possible states that the system can occupy is large. Nevertheless, by considering small tractable problems the method can be used to develop rules of thumb that may be robust. Given uncertainty and ignorance about many of the parameters and processes involved in the dynamics of ecological systems, the method described here is, at present, probably best thought of as a thinking tool, rather than a rigorous decision-making method for getting precise answers to specific conservation problems.

Often mathematical modeling can be seen as more of an educational process than a solution to a problem. I encourage others to integrate management tools into conservation theory, not only because it provides answers to specific problems but also because it forces us to explicitly list the available options and consider them objectively.

25

Expanding Scientific Research Programs to Address Conservation Challenges in Freshwater Ecosystems

Catherine M. Pringle

Summary

Here I highlight the scope of conservation needs and challenges in freshwater ecosystems and some of the different avenues by which we can expand our roles as scientists to actively address the complex environmental problems that face us. Losses in global aquatic biodiversity and associated ecosystem integrity are occurring at a rate unprecedented in geological history at all levels from ecosystems to species to genes. Shortages of freshwater and the increasing pollution of water bodies are affecting the life cycles of many species, including humans. Despite the severe environmental degradation of freshwater aquatic systems described here for North America, it is clear that without existing environmental legislation, things would be much worse. Countries "in transition" in Central and Eastern Europe and the former Soviet Union are facing regional degradation of water resources on a scale greater than what much of the Western world has experienced. However, many surface water quality problems resemble those that were familiar to the United States over two decades ago, before the Clean Water Act of 1972 was implemented, underlining the importance of scientific involvement in improving and protecting existing laws. As environmental problems become more pervasive, increasing opportunities exist for scientists to make connections between research activities and conservation. Two aquatic case studies from my own research are provided that illustrate how (1) basic research programs can be integrated with environmental outreach activities, (2) basic research can be redirected to make management applications for the protection of the biotic integrity of the systems that we study, and (3) addressing conservation objectives can lead to improvements in the quality of our science.

We all live downstream.—Unknown

Our faith in the power of science, and the knowledge and technology that it creates, may be an explicit roadblock to social action. The very benefits that science promises to deliver may be withheld from us because it is easier—politically, economically, socially, scientifically—to support more research than to change ourselves. The promise of science—a miracle cure—serves the politicians, who are always looking for patent medicine to sell to the public, and it serves the scientists, who understandably seek to preserve their special position in our culture. But it may not serve society as advertised. Indeed, the promise of science may be at the root of our problems.

—George E. Brown, Jr., Chairman of the House Committee on Science,
Space and Technology, 1992

Introduction

Brown's observation (above) is certainly provocative and many of us might say that it is overstated, but doesn't it contain more than a kernel of truth? Alternatively, if we disagree with his statement, shouldn't we actively participate in defining the role of science in society? It is clear that if we don't, Mr. Brown and others who share his opinion, and more extreme opinions, will continue to redefine it for us.

The global human population is expected to nearly double by the year 2050, from 5.7 billion in 1994 to about 10 billion (United Nations Population Division, 1992). Given the increasing rate of human alteration of local, regional, and global processes that accompanies this population growth (Turner et al., 1994), achieving a balance between human needs and environmental sustainability necessitates more than improving scientific theory and conceptual frameworks. For scientists concerned with environmental issues, a challenge for the future is to successfully combine ecological research with effective management, policy and/or environmental outreach, and education. As stated by Vitousek (1994), "We're the first generation with the tools to see how the Earth system is changed by human activity: at the same time we're the last with the opportunity to affect the course of many of those changes." This will require energy, innovation, and stepping outside of our traditional scientific roles.

Environmental Degradation in Freshwater Ecosystems

Since streams and rivers integrate processes occurring in the atmosphere and terrestrial environments that they drain, their consideration is integral to the conservation of biodiversity, ecosystem function, and integrity. Streams have been likened to blueprints of the terrestrial environment, reflecting the interaction of many different factors, including hydrologic modifications, changes in land use, point-source and non-point-source pollutants, groundwater contamination, acid rain and deposition, introduction and proliferation of exotic species, and climate change (e.g., Naiman et al., 1995).

Shortages of freshwater and the increasing pollution of water bodies are affecting the life cycles of many species, including humans, and are becoming limiting factors in the economic and social development of many countries throughout the world (e.g. Postel, 1992; Gleick, 1993).

Losses in global aquatic biodiversity and associated ecosystem integrity are occurring at a rate unprecedented in geological history at all levels from ecosystems to species to genes. Wilcove and Bean (1994) provide a comprehensive synthesis of these trends in North America, clearly illustrating the severity and magnitude of decline in the biological integrity of freshwater aquatic systems. The United States has lost over half of the wetlands that existed at the time of the American revolution. Less than a quarter of bottomland hardwood forests remain in the midwestern and southern areas of the United States (Wilcove and Bean, 1994). Of the 3.2 million miles of streams in the lower 48 states of the United States, only 2 percent are free flowing and relatively undeveloped. Currently, only 42 free-flowing rivers of over 125 or more miles in length exist. The other 98 percent of U.S. streams have been developed by dams, water diversion projects, and the like (Benke, 1990).

The degraded nature of freshwater systems is also reflected in the large number of aquatic species that are considered endangered. The Nature Conservancy has classified more than a third of the 800 North American fish species as rare, imperiled, criticially imperiled, or possibly extinct (Master, 1990). Comparable figures for unionid mussels and crayfish are 73 percent and 65 percent, respectively. Data compiled by the American Fisheries Society indicate that the rate of extinction of North American fishes has doubled over the course of the century (Wilcove et al., 1992). In addition, environmentally degraded systems are becoming increasingly vulnerable to invasion by exotic species. In many instances, species diversity is actually increasing as a result of invasion by exotic species, at the expense of environmental integrity. As an example, the Illinois River has lost 9 species of fish and gained 13 from introductions.

On a genetic level, populations of stream biota isolated in upstream reaches by dams and unsuitable habitat will inevitably suffer from reduced genetic variation. Also, in the United States, over 100 major salmon and steelhead populations or stocks on the west coast have been extirpated while an additional 214 face a high or moderate risk of extinction (Nehlsen et al., 1991). Replacement of depleted stocks with captive-reared fish has sacrificed the genetic diversity of many wild populations (e.g., Hilborn, 1992).

Losses in ecosystem products and services are accompanying decreases in the biodiversity and/or the integrity of aquatic systems. The loss of wetlands and their associated water filtration and buffering capacities has contributed to rapidly diminishing water quality and increasing incidence of floods. The Environmental Defense Fund and the World Wildlife Fund estimate that if half of the nation's remaining wetlands are destroyed, the cost to the nation to remove additional nitrogen from the water would be $38 billion (and nitrogen is just one of many nutrients that would have to be removed; Wilcove and Bean, 1994). The loss of wetlands in the Mississippi River drainage, in part a result of channelization and the construction of dams and levees, exacerbated the impact of the great flood of 1993, causing millions of dollars of damage (Allen, 1993).

Things Could Be Worse

Despite the severe environmental degradation of freshwater aquatic systems discussed above for North America, things could be worse. Countries "in transition" in Central and Eastern Europe and the former Soviet Union are facing regional degradation of their freshwater resources on a scale greater than what much of the Western world has experienced.

The basis for my comments on environmental degradation in Central and Eastern Europe derive from a series of international workshops that were jointly sponsored by the U.S. and Polish Academies of Science ("Preservation of Natural Diversity in Transboundary Protected Areas: Research and Management Options" and "Ecological Research and Environmental Protection" [Grodzinski et al., 1990]) and the U.S. and Romanian National Academies of Science ("Environmental Reconaissance of the Danube Delta" ([Griffin et al., 1991; Pringle, 1991a; Pringle et al., 1993b]).

The lack of sewage treatment plant facilities for domestic and industrial wastes throughout many areas of Central and Eastern Europe has resulted in riverine systems with high loads of human and chemical wastes. To provide one example of the scale of this problem: Budapest, Hungary, dumps an estimated three quarters of its sewage untreated (both domestic and industrial) directly into the Danube River (Pringle et al., 1993b). Both the Black and Caspian Seas, which receive the effluent of the many rivers draining Central Europe and the former Soviet Union are becoming increasingly anaerobic and toxic. The Black Sea was described as being "within an inch of falling into a coma" by specialists who attended the Ecoforum Peace Conference in Moscow in 1990 (Pringle et al., 1993b). Valuable fisheries have been wiped out: of the 26 species of fishes caught in the Black Sea by fisherman in the 1960s, only 5 remain.

Poland's Vistula River provides another compelling example of the severe regional degradation of aquatic ecosystems faced by European countries "in transition." The Vistula is the largest river in Poland and the second largest river emptying into the Baltic Sea. The river serves as a source of drinking and industrial water for most of the towns and industries located along it, and it also receives their sewage. While the Vistula is in fact the largest source of water in Poland, it is polluted almost from its source in the Carpathian Mountains by heavy industry (Kajak, 1992). Most of the upper reach of the river is acidified (Wrobel, 1989), caused by acid rain originating from heavy industry in Germany, the Czech Republic, and Poland. Pesticides and heavy metals are high and are often concentrated to dangerous levels along the food chain (Kajak, 1992).

Water quality and quantity problems throughout Central and Eastern Europe pose a serious threat to human health and economic development. The environmental situation has been referred to as a vicious circle, since environmental degradation has progressed to the extent of actually limiting development, and the lack of development is correspondingly limiting environmental protection (Marek and Kassenberg, 1990). For instance, the amount of river water of highest quality for drinking in Poland has dropped from 32 percent to less than 5 percent during the last 20 years. Postel (1992) reports that approximately three quarters of the nation's river water is too contaminated even for industrial use. Data from the Polish Inspectorate for Environmental Protection (Srodowiska, 1992), indicate that 82 percent of Polish rivers are not suitable for industrial use. Surface waters suitable for industrial purposes comprise only 14.5 percent and those suitable for agricultural use, a mere 3.3 percent. Pollution in surface waters is increasing due to continued contamination by industry and municipal sewage discharges as well as by agricultural sources. Water shortages are also limiting economic activity within Poland. As in the Western world, countries such as Poland are now discovering that they also have to address serious groundwater contamination problems in addition to present surface water problems. Indeed, problems of groundwater deterioration and depletion are emerging as an environmental crisis in many areas of the world and experts warn that this is just the "tip of the iceberg."

Many of the surface water quality problems in Central and Eastern Europe and parts of the former Soviet Union resemble those that were familiar to the United States over two decades ago (Hillbricht-Ilkowska, 1990; Cooper, 1990; Gromiec, 1990) before massive water cleanup and sewage treatment programs in response to the Federal Water Pollution Control Act of 1972. The Blackstone River ecosystem provides a case in point. The Blackstone arises in Worcester, Massachusetts, and flows 46 miles to Providence, Rhode Island. It was one of the first polluted rivers in the United States, receiving large volumes of sewage and textile mill effluent. In the 1930s, it was largely considered to be an industrial stream whose industrial importance came before any recreational advantage (Williams, 1995). The Clean Water Act transformed the Blackstone and it now supports thriving populations of waterfowl and fish including yellow perch, white perch, largemouth and smallmouth bass, black crappies, and hatchery-bred trout (Williams, 1995).

The Importance of Environmental Laws and Legislation: Improving and Protecting the Clean Water Act in the United States

As both scientists and citizens of the United States, we have enjoyed the benefits of environmental laws and legislation, despite their shortcomings. Over the past two decades, the Clean Water Act has reduced annual discharges of toxic chemicals and raw sewage into U.S. lakes and rivers by about a billion pounds and 900 million tons, respectively (Adler et al., 1993). It has cut the annual loss of wetlands from 450,000 acres to 290,000 acres or less. Because of this statute,

66 percent of our waters are now classified safe for swimming and fishing—up from 36 percent before the enactment. The U.S. Environmental Protection Agency and its sibling agencies in the states have forced 63,000 industrial and municipal polluters to treat their waste (Adler et al., 1993).

Revisions of the Clean Water Act, approved in House Committee on 6 April 1995, threaten to reverse this hard-won environmental progress. The revised act ignores a National Academy of Sciences report on wetlands delineation and redefines wetlands so that at least 60 percent of those currently protected under the Clean Water Act (e.g., large portions of the Everglades and most prairie potholes) would be accessible for development. A takings clause would provide compensation for alleged lost profits. Property owners would be entitled to compensation for any portion of their property lost or devalued (20 percent of fair market) due to wetlands regulation. Control of non-point-source pollution, which the Clean Water Act failed to address, would be abandoned as a national goal. Ammendments approved on 31 July would cut $362 million and $725 million, respectively, for enforcement of the Clean Water Act and the Safe Drinking Water Act.

We have learned many things about both the strengths and the weaknesses of the Clean Water Act over the last two decades (e.g., Adler et al., 1993), and we cannot afford to reverse this wave of progress. We need to build on the strengths of the act and improve its weaknesses (e.g., control of non-point-source pollution). Not only will our actions affect water quality and ecosystem integrity in the United States but they will invariably influence other nations that are looking to the United States, as they begin to tackle their own water resource problems.

The Expanding Role of Scientists in Conservation

Greater involvement in environmental and conservation issues by scientists is imperative to addressing the current environmental crisis. Our view of our role as scientists will determine, in part, how effectively we can meet this challenge. As stated by O'Brien (1993), "Asking certain questions means not asking other questions and this decision has implications for society, for the environment, and for the future. The decision to ask any scientific question, therefore, is necessarily a value-laden, social, political decision." How scientists frame their research questions often controls their answers (Schrader-Frechette, 1994b). Within the existing scientific framework, we can actively choose to frame research questions and programs to consider (1) principles of ecosystem science and landscape ecology (Meyer, this volume), (2) temporal and spatial variability over a range of different scales (Gordon et al. this volume), (3) consideration of humans as integral parts of the ecosystem—not separating them from the life-support systems on which they depend, and (4) multiple scientific disciplines when appropriate.

We can also choose to expand our involvement beyond the traditional role of scientific research to interface with management, policy, and environmental education. Progress in this direction has been hindered by numerous false dichotomies (such as maintaining scientific credibility versus involvement in conservation activities) when the focus should be on how to effectively apply quality science to environmental issues. A scientist can maintain credibility, while at the same time communicating with managers and policy makers (Zedler, this volume). We can consider alternatives in risk assessment (O'Brien, this volume) and socioeconomic and political realities when we frame research questions. A scientist's reputation is not diminished when he or she actively links basic research to environmental education. Although the role of scientists in environmental advocacy will continue to be debated within the scientific community, Karr (1993) makes the case that the scientific community should be calling for professional responsibility rather than debating advocacy. He argues that just as an engineer has a responsibility to point out a structural flaw in a bridge, professional ecologists would be irresponsible if they were silent when their scientific knowledge led them to conclude that the life support "bridges" on which society depends were being destroyed.

Cooperation Between the Scientific Community and Organizations Involved in Freshwater Conservation

As stated by Vitousek (1994) "Our colleagues outside the natural sciences, and outside academics, need our knowledge of how Earth is changing; we need theirs on what can be done about it. The effort we spend in finding non-scientists who will listen to what we have to say, the effort that goes into understanding their language and what they can teach us, is effort well spent."

Both scientists and conservation groups have become aware of the extent of the degradation of U.S. river systems and the necessity of making substantial changes in national policy. This common ground has provided the context for collaboration between these groups (e.g., Coyle, 1993; Doppelt et al., 1993; Anderson, 1993; Brouha, 1993; Pringle and Aumen, 1993; Richter, 1993; Woody, 1993; Duff, 1993). Also, a feature of recent policy developments in river conservation in the Western world is a broadening of views by scientists, conservationists, and managers (e.g., Newson, 1992; Dewberry and Pringle, 1994). All of these different groups are expanding their approach from a reductionist to a landscape perspective. The science of stream ecology has moved toward a stronger focus on entire catchments rather than stream segments. Conservation organizations have moved away from their preoccupation with streams based solely on recreation and aesthetics. This has further catalyzed collaboration between conservation groups and aquatic scientists (Pringle and Aumen, 1993; Dewberry and Pringle, 1994).

The public interest and support for improving water quality in the United

States is reflected by increased citizen action. Informed citizen action is being facilitated and guided by books such as *Rivers at Risk: The Concerned Citizens's Guide to Hydropower* (Echeverria et al., 1989), *How to Save a River* (Bolling, 1994), and *Lifelines: The Case for River Conservation* (Palmer, 1994).

An inspiring example is the Coalition to Restore Urban Waters (CRUW), which was formed in 1993 at a national meeting that included grassroots stream restoration groups from around the country and representatives from national environmental groups, state and federal agencies, and conservation corps (Mcdonald, 1995). This group is involved not only in restoring the biological integrity of rivers draining urban areas but is also working to bring disenfranchised neighborhoods into the process, partly to help plan and carry out projects and partly as a way of strengthening and empowering the community.

The Hydropower Reform Coalition provides another compelling example of citizen action. This group formed in 1992 to reform national hydropower policies and improve rivers altered by hydropower dams (Adler et al., 1993). Until 1993, hydro-relicensing was a relatively infrequent procedure that received little, if any, public attention. In 1993, 160 licenses affecting 237 dams expired. These dam relicensings are the beginning of a wave that will continue, with licenses for 650 dams expiring in the next 15 years. The Coalition is using this opportunity to restore river ecosystems through the relicensing process and to reform the way the Federal Energy Regulation Commission licenses all hydropower dams. Through the relicensing process, this organization has made much progress in river restoration through its focus on improving instream flows, restoring flows to dewatered bypass reaches, installation and/or improvement of fish passage facilities, protection of riparian habitat, planning for long-term dam maintenance or retirement, and riverwide planning and cumulative analysis.

Scientific information is crucial to the success of such freshwater protection efforts (e.g., Doppelt et al., 1993) and the recent groundswell in public interest, at both grassroots and national levels, is providing an avenue for scientists to effectively contribute their knowledge to freshwater conservation efforts. Increasingly, nongovernmental conservation organizations involved in reversing freshwater deterioration are requesting input from the scientific community. For instance, American Rivers has created a Scientific and Technical Advisory Committee to advise its Board of Directors. Likewise, the Pacific Rivers Council convenes special advisory panels of scientists to advise them on specific issues and stream ecologists have been important players in the formulation of its national rivers campaign (Doppelt, 1993; Doppelt et al., 1993). To facilitate cooperation between aquatic scientists and conservationists from nongovernmental organizations, a special symposium, "Current Issues in Freshwater Conservation" was convened by the North American Benthological Society (NABS) in 1992 (Pringle and Aumen, 1993). Presentations were made by representatives of major conservation organizations involved in freshwater conservation issues including American Rivers (Coyle, 1993), the Pacific Rivers Council (Doppelt,

1993), the Wilderness Society (Anderson, 1993), The Nature Conservancy (Richter, 1993), and the Sierra Club (Woody, 1993). Recommendations generated by the interchange of ideas at this symposium (Pringle et al., 1993a) that are being pursued and implemented include (1) development of a NABS conservation resource database that lists the names and addresses of NABS scientists willing to provide their expertise to conservation organizations, not as NABS members but as individuals—this database is currently serving as a clearinghouse to provide better linkage between scientists and conservation needs; (2) regular communication between designated liasons in the scientific and conservation communities on current environmental issues of mutual concern; and (3) environmental education/ outreach activities that include the creation of slide presentations (i.e., slides and script) on topics such as "How the benthic biologist (or concerned citizen) can have an impact upon instream flow assessments through hydro-relicensing" and "The ecological importance of stream riparian zones." These slide shows and others are made available to the scientific community, conservation organizations, and/or the general public through NABS.

Two Aquatic Case Studies

Although I don't have the answers, I do believe that we need to carefully think about our role as scientists in addressing environmental concerns. Toward this end, I provide two case studies from my own research program that illustrate different avenues by which we can expand our roles as scientists to actively address the complex environmental problems that face us. Between them, these case studies illustrate (1) how basic research programs can be integrated effectively with environmental outreach activities, (2) how basic research can be redirected to make management applications for the protection of the biotic integrity of the systems that we study, and (3) how addressing conservation objectives can lead to improvements in the quality of our science.

Case Study I: Development of an Environmental Education Program on Water Quality in Lowland Costa Rica

Over the past decade and a half, my colleagues, students, and I have been studying the ecology of lowland streams on Costa Rica's Caribbean slope (e.g., Pringle et al., 1986, Pringle, 1988; Pringle and Triska, 1991; Pringle, 1991b; Pringle et al., 1993d; Triska et al., 1993). One of the questions that we recently began to ask ourselves was, Given the knowledge and understanding of aquatic systems that we have acquired how can we apply this knowledge to address the severe water quality and quantity problems faced by the local community where we work? (Fig. 25.1).

The town of Puerto Viejo and adjacent areas in Sarapiqui province, located near our study site at La Selva Biological Station (owned and operated by the

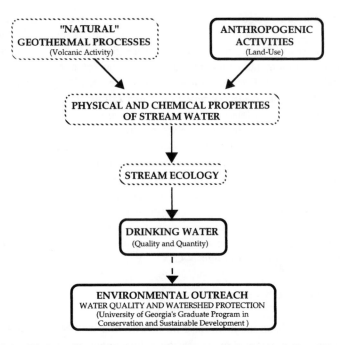

Figure 25.1. Diagram illustrating how our basic research in lowland Costa Rica has led to environmental outreach and education activities. The main focus of past and ongoing research activities (boxes delineated by dashed-lines) has been on ecological processes in streams and how they are affected by physical and chemical properties of stream waters, which are in turn affected by geothermal processes in the Earth's crust (i.e., volcanic activity). Human-induced changes in the landscape (e.g., deforestation, cattle grazing, banana plantations) interact with these "natural" geothermal processes to affect not only stream solute chemistry and ecology but also drinking water quality and quantity for lowland human communities. The sheer magnitude of deterioration of water resources in our study area led us to initiate environmental outreach activities on water quality and watershed protection (See Fig. 25.2).

Organization for Tropical Studies) are currently experiencing severe problems with drinking water quality and quantity. This is due, in part, to the fact that the population of Sarapiqui has grown from 3,000 to over 25,000 over the last three years due to immigration as a result of the recent and extensive development of banana plantations. Although the area annually receives more than 4 m of rainfall and is drained by numerous streams and rivers, local surface and ground waters previously tapped for potable water supplies are contaminated with fecal coliforms introduced by cattle.

In the case of Puerto Viejo de Sarapiqui (population approximately 10,000) and the nearby barrios, contamination of local surface waters and wells has resulted in the town's recent dependence on springs that originate approximately

8 km to the south in the lower foothills of the Central Mountain Range. The source of these springs is near the northern end of a national park (Parque Nacional Braulio Carrillo), the last intact tract of primary rainforest that spans extremes in elevation (from near sea level to 10,000) on Central America's Caribbean Slope. The intact forests of the park protect the water supply from anthropogenic contamination. In 1992, a 15-cm-diameter plastic tube was installed to transport water from the spring approximately 8 km to the town of Puerto Viejo where it is distributed to 95 percent of the downtown population. The quantity of potable water from this source is currently insufficient to serve other areas of Puerto Viejo and nearby barrios.

Although this new drinking water supply is currently uncontaminated by fecal coliforms, waters have a very high mineral content. Our studies indicate that this water is geothermally modified and can be classified as sodium-chloride-bicarbonate ($Na-Cl-HCO_3$) water (Pringle, 1991b; Pringle et al., 1993d), based on a classification of geothermal water types that is widely used by geochemists. Our past and ongoing research activities have focused on the spatial distribution and ecological effects of these waters in the landscape (Fig. 25.1).

Projected increases in the human population of Sarapiqui over the next few decades indicate that there will be increasing pressure on the relatively undisturbed watersheds of Parque Nacional Braulio Carrillo as a source of water for domestic and agricultural use. The development of a new graduate program in conservation biology and sustainable development by the University of Georgia's Institute of Ecology provided an opportunity to directly involve graduate students in an environmental outreach program in Costa Rica on water quality and watershed protection (Figs. 25.1 and 25.2). Two graduate students have taken a leadership role in developing an environmental outreach program "Water for Life" in the local community of Puerto Viejo as part of their masters theses projects. Rodney Vargas, a Costa Rican national, documented the history of municipal water resources in the Sarapiqui region. To communicate his findings to the adult community of Puerto Viejo, he developed environmental education materials (e.g., slide show, accompanying "script," and a watershed protection poster), describing past and ongoing problems with municipal water supplies of the area (Vargas, 1995). The environmental education program demonstrates the increasing dependence of lowland human communities on drinking water sources within forested watersheds of Parque Nacional Braulio Carrillo, emphasizing the importance of watershed protection in the forested highlands to protect drinking water resources.

In addition, Tina Laidlaw (1996) has developed an "Adopt-a-Stream" Program in Puerto Viejo (Fig. 25.2). The concept is based on the Georgia "Adopt-A-Stream" program and requires that local volunteers in the community commit to monthly monitoring of a local stream. Students in an upper-level agroecology class at the Puerto Viejo high school participate in the program and monitor a second-order stream passing directly through the downtown area. Students moni-

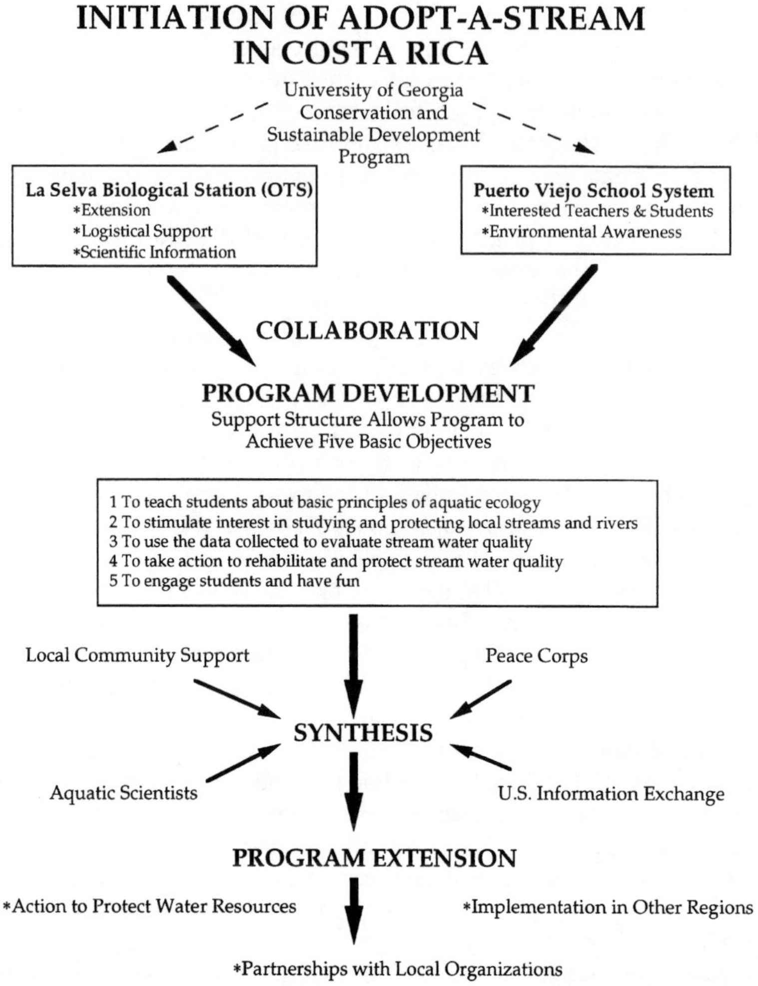

INITIATION OF ADOPT-A-STREAM IN COSTA RICA

University of Georgia
Conservation and
Sustainable Development
Program

La Selva Biological Station (OTS)
*Extension
*Logistical Support
*Scientific Information

Puerto Viejo School System
*Interested Teachers & Students
*Environmental Awareness

COLLABORATION

PROGRAM DEVELOPMENT
Support Structure Allows Program to
Achieve Five Basic Objectives

1 To teach students about basic principles of aquatic ecology
2 To stimulate interest in studying and protecting local streams and rivers
3 To use the data collected to evaluate stream water quality
4 To take action to rehabilitate and protect stream water quality
5 To engage students and have fun

Local Community Support

Peace Corps

SYNTHESIS

Aquatic Scientists

U.S. Information Exchange

PROGRAM EXTENSION

*Action to Protect Water Resources

*Implementation in Other Regions

*Partnerships with Local Organizations

Figure 25.2. Diagram illustrating the development of a volunteer stream monitoring program, "Adopt-a-Stream," in lowland Costa Rica. Aquatic ecologist, Tina Laidlaw, a graduate student in the University of Georgia's program in conservation and sustainable development, initiated a collaboration between the Organization for Tropical Studies' (OTS) La Selva Biological Station and the Puerto Viejo School System for the development and implementation of this educational stream monitoring program. In this program, high school students learn basic principles of aquatic ecology, evaluate stream water quality, and get involved in water quality and watershed protection efforts (from Laidlaw, 1996).

tor four stream components to evaluate water quality, including physical (channel alteration); habitat (diversity of bottom substrata); chemical (oxygen, nitrate, phosphate); and biological (fish, insects) parameters. Through their participation in the study, they have learned about the variety of organisms inhabiting the stream, pollution sources affecting water quality, and how they can work to improve and protect their streams and rivers. Based on the success of "Adopte Una Quebrada," plans are being developed to implement the program in several other local school systems.

The environmental education activities of both Vargas and Laidlaw have been highly successful in promoting public awareness about aquatic resources, while empowering citizens to become actively involved in water quality protection. The Organization for Tropical Studies' Education and Community Relations Program at La Selva Biological Station is currently administering the outreach activities initiated by these graduate students.

Case Study II: Research to Determine Management Strategies for Water Withdrawals from Streams Draining Puerto Rico's Caribbean National Forest

Puerto Rico's Caribbean National Forest (CNF) is the largest natural forest left in the Caribbean islands. The 11,269-ha forest is a major site for tropical research and it has been designated as a site for Long Term Ecological Research (LTER) by the National Science Foundation. Research is conducted in this highland area by scientists from the International Institute of Tropical Forestry, the University of Puerto Rico, and other universities throughout the United States.

Highland streams draining the CNF are characterized by a simple food chain typical of oceanic islands. The macrobiota of some tributaries is dominated by large numbers (20–30 individuals m-2) of freshwater atyid (i.e., *Atya* and *Xiphocaris* spp.) shrimps (Pringle et al., 1993). In other streams, where carnivorous fishes (i.e., *Agonostomus monticola* [Bancroft], *Awaous tajasica* [Lichtenstein], and *Anguilla rostrata* [LeSeur]) are present, freshwater atyid shrimps are often much less abundant. Our experimental research in streams dominated by atyid shrimps indicate that these relatively large invertebrates can have a dramatic effect on sedimentation, algal standing crop and community structure, and insect communities (e.g., Pringle et al., 1993c; Pringle and Blake, 1994; Pringle, 1996).

One of the factors that we did not consider in our basic research in the highlands was the impact of water abstraction from lower stream reaches on stream integrity and related ecosystem processes. In 1994, a water budget was developed for the CNF (Naumann, 1994). This budget indicates that 21 water intakes are operating within the CNF and 9 large intakes outside of the forest, resulting in significant stream dewatering. On an average day, over 50 percent of riverine water draining the forest is diverted into municipal water supplies (via water intakes) before it reaches the ocean. Several rivers have no water below their water intakes for much of the year.

This becomes extremely relevant when one considers the life cycle of shrimps and fishes that live in streams draining the CNF. Almost all of the stream macrobiota must spend some part of their life in the estuary to complete their life cycle. Their migrations of aquatic biota form a dynamic linkage between stream headwaters and their estuaries. In the case of amphidromous shrimps, newly hatched larvae migrate downstream and complete their larval stage in the estuary. On metamorphosis, the juveniles migrate upstream where they live as adults.

Stream dewatering is, of course, exacerbated during drought years. Since 1992, rainfall in Puerto Rico has been below average and on August 12, 1994, the Commonwealth was declared an agricultural disaster area by the U.S. federal government because of the drought. The Aqueduct and Sewage Authority imposed water rationing and areas of the capital city, San Juan, went without water for days at a time. This had serious negative effects on the economy (e.g., tourism, industry, etc.) and the pressure on the nearby CNF for water supplies is increasing.

The most recent water demand studies predict that between the years 1990 and 2040, the demand for water in the municipalities along the northern border of the CNF will increase from 28.3 million gallons per day (MGD) to 36.1 MGD (U.S. Army Corps of Engineers, 1993). Most of this increase in demand will occur before the year 2000. At present, all except one of the nine stream drainages within the CNF have dams and associated water withdrawals and a proposal is currently being considered to dam the last remaining undammed river (Rio Mameyes). The magnitude of current water withdrawals from the CNF is already in conflict with other important functions of the forest, including recreation, scientific research, and maintenance of the biointegrity of the island.

Given this dose of reality, my graduate students and I have consequently redirected our basic research to evaluate how dams and associated water withdrawals affect the migration of aquatic organisms to and from the estuary. For instance, we are assessing diel and seasonal migration patterns of freshwater shrimps so that the timing of water withdrawals from rivers can be adjusted so as to minimally interfere with the life cycle of these key stream organisms.

Studies conducted during the nondrought year of 1995 (Benstead et al. 1996; March et al., 1996) indicate that shrimps exhibit a strong pattern of diel periodicity in their migration patterns, with most larvae drifting during a few hours after dusk and little to no migration during the day. Water abstraction and damming in the lower stretches of one of the main river drainages within the CNF (Rio Espiritu Santo) significantly impacted shrimp recruitment due to direct mortality of over 50 percent of migrating larvae which were entrained into water intakes for municipal water supplies. During periods of low flow, below-dam discharge was insufficient to prevent saltwater intrusion leading to a sharp physiochemical discontinuity at the dam site. Furthermore, although the dam at the water intake did not appear to be a barrier to the upstream migration of returning juvenile shrimps, our observations indicated that it functioned as a predation gauntlet for the juvenile shrimps due to the accumulation of both freshwater and marine predaceous fishes below the dam. Our recommendations for mitigation include: (1) the elimination or reduction of water abstraction during post-dusk periods of peak larval shrimp migration (which would reduce shrimp mortality without significant decreases in water availability to human populations); and (2) the installation and regular maintenance of functional "fish/shrimp" ladders.

Redirection of our basic research toward more applied issues has also improved the quality of our science by forcing us to grapple with key questions that include (1) How has the spatio-temporal development of aquatic communities in streams draining Puerto Rico's Caribbean National Forest been affected by the interaction of both "natural" and anthropogenic factors? (2) How can we predict patterns of ecosystem function given the current and future massive water withdrawals that are planned for streams of this region? (3) What additional knowledge do we require about our study site's context in space and time to address these first two questions?

Concluding Remarks

In this chapter I have attempted to illustrate the scope of conservation needs and challenges in freshwater ecosystems and some of the different avenues by which we can expand our role as scientists to actively address the complex environmental problems that face us. The two case study examples that I provide from my own research program represent very small steps in the context of the magnitude and scale of environmental problems. Yet, if we as scientists do not use our available energies to bring scientific knowledge to bear on environmental problems—who will? As environmental problems become more severe and pervasive, there are increasing opportunities for us as scientists to make connections between our research programs and conservation.

Acknowledgments

The author gratefully acknowledges the National Science Foundation (grants BSR-87-17746 and BSR 91-07772), which supported research activities in Costa Rica and Puerto Rico, the Conservation Food and Health Foundation for providing funds for the development of the environmental education program on water quality in lowland Costa Rica, and the Organization for Tropical Studies for its role and support in the development and administration of this program. Special thanks are extended to J. Affolter, J. Benstead, T. Laidlaw, J. Match, R. Ostfeld, A. Ramirez, and S. Tartowski for their helpful comments and suggestions on the manuscript.

26

Standard Scientific Procedures for Implementing Ecosystem Management on Public Lands

Robert S. Peters, Donald M. Waller, Barry Noon, Steward T. A. Pickett, Dennis Murphy, Joel Cracraft, Ross Kiester, Walter Kuhlmann, Oliver Houck, and William J. Snape III

Summary

Modern ecological ideas, including fundamental principles of conservation biology, have not yet been fully embraced by federal land management agencies. Reasons include bureaucratic resistance to new ideas, influence by resource user groups who put maximum resource production above long-term ecological health of ecosystems, lack of external accountability by the agencies, and a misperception by agency personnel that conservation science is poorly developed. The fact that federal land management agencies often do not gather data or make management decisions using up-to-date science hinders their ability to ensure the long-term sustainability of both biological diversity and the specific resources, such as timber and fodder, for which the lands are managed.

For example, agency managers have often defended conventional patterns of timber harvest by arguing that the edge habitat they produce is beneficial to wildlife, even though edges are well known to have multiple deleterious effects on native, forest-interior species. To prevent such poorly informed management decisions, agencies must develop mutually compatible, explicit standards that determine how they will apply modern ecological science to land management. To facilitate this process, we propose a comprehensive set of standard procedures for implementing ecosystem management on public lands. These procedures incorporate our best contemporary understanding of population, community, and landscape dynamics to efficiently and systematically protect ecosystems and their constituent biodiversity.

Introduction

Management practices on federal lands have often harmed natural ecosystems, reducing not only biological diversity but even the resources—such as timber or grass production—for which the lands are primarily managed. For example, in many cases, fire suppression on national forests has changed the relative abundances of tree species, increased stand densities, or caused other structural and compositional changes that increase the probability of disease, fire, and loss of native species (Wagner and Kay, 1993; Anderson, H. M., 1994). Unwise grazing policies on Forest Service, Bureau of Land Management, and other federal lands contribute to the endangered or threatened status of many species, including Sonoran pronghorn (*Antilocapra americana sonoriensis*), Arizona willow (*Salix arizonica*), and desert tortoise (*Gopherus agassizii*) (National Wildlife Federation, 1994; USFWS, 1994a). The National Biological Service (1995) found that many U.S. natural ecosystems are seriously degraded or at risk of total loss. Many of these, such as Southern California coastal strand, Palouse prairie in the Northwest, and slash pine rockland in Florida have shrunk by more than 90 percent to become "critically endangered."

This unhappy state results from both scientific and institutional causes. Misunderstanding of ecological principles has led to management decisions that were often inimical to biological diversity. For example, predators once were seen as negative influences to be removed (Anderson, S. H., 1995), whereas habitat edges were seen as universal benefits to be increased (Reid and Miller, 1989; Woodley et al., 1993). Nor has there been adequate appreciation that ecosystems are highly interconnected and that management decisions in one place may affect ecological conditions on other lands. A classic example is how logging can produce silt that harms the reproduction of salmon far downstream (Pringle and Aumen, 1993). These erroneous assumptions have helped agencies yield to political pressures to increase production of timber and other resources at the expense of ecosystem sustainability. For example, by overlooking the essential role that top predators play in maintaining healthy ecosystems, agencies have been able to rationalize activities that extirpate them. Likewise, their erroneous conclusion that edge habitat is universally beneficial allowed them to rationalize excessive timber harvesting.

Federal Land Management Agencies Have Been Slow to Adopt Conservation Science

Despite advances in ecology and conservation biology during the past 20 years that have refuted many ecological myths (Parker, 1993) and provided comprehensive tools for sound ecological management (Christensen, this volume), and even though agency scientists have been doing some of the best ecological work (e.g., on modeling viable population sizes, Ralls and Taylor, this volume), agencies have been slow to base management policies on the new science. Scientists and conservation organizations have been educating, urging, and, in some cases, suing the agencies responsible for management to require them to collect adequate biological information and to act on it. For example, in 1992 the Sierra Club and other plaintiffs brought an unsuccessful suit against the U.S. Forest Service, when the agency refused to establish large blocks of mature forest that would be free of disturbances caused by logging, new roads, or wildlife openings. The suit complained that the Forest Service had rejected the proposal, and adopted a plan of ubiquitous fragmentation, without considering negative edge effects, island biogeography, and other concepts from conservation science (e.g., Noss, 1983; Meffe and Carroll, 1994).

Several factors have contributed to the historical unwillingness or inability of federal agencies to incorporate up-to-date conservation science in land management plans:

Misperceptions about the Quality of Conservation Science.

Resistance by the agencies is sometimes based on the incorrect conclusion that conservation science is too poorly developed to be dependable. People may draw this conclusion because they misunderstand the nature of science, mistaking scientific debate over particular points within the field for a generalized weakness (Pickett et al., 1994). For example, in the Sierra Club suit mentioned earlier, the judge concluded that the theory of island biogeography was too poorly substantiated to expect the Forest Service to use it in making decisions about how much habitat fragmentation to allow during timber production (Reynolds, 1994). He reached this conclusion partly because there is debate in the scientific literature over the degree to which the theory of island biogeography can explain patterns of species richness in habitat remnants, on mountain summits, and on other such terrestrial "islands" (Simberloff, this volume). The judge did not realize that scientific debate over the degree to which terrestrial habitat islands approximate real oceanic islands did not invalidate the basic conclusions, backed up by dozens of rigorous studies (e.g., White et al., 1990; Ås et al., 1992; Hansson, 1992a; Hansson, this volume), that small isolated habitat areas support fewer species than large ones, and that it is possible to make predictions about which species or groups of species will disappear first (e.g., Nott and Pimm, this volume). The inability of the theory to predict precisely how many species will disappear, given a particular logging pattern on a particular forest, in no way invalidates the general conclusion that fragmenting a contiguous forest, as was planned for the Chequamegon, will cause loss of biological diversity (Hansson et al., 1995).

Constituencies for Resource Use.

Arguably the most powerful force against incorporation of conservation science is pressure from resource-user groups. Historically, agencies have seen their primary responsibility to be the provision of sufficient resources for hunters, ranchers, and timber companies (Society of American Foresters, 1993), even if this comes at the cost of overstocking range lands and overcutting forests. Because conservation of biodiversity is a relatively new contender for agency concern and resources compared with traditional consumptive uses, there are frequently ideological, political, financial, or other constraints that predispose agencies to selectively interpret science.

Institutional Inertia.

Because the field of conservation biology is so new, with many of its primary tenets having been developed only within the past 20 years, it has not yet fully permeated government agencies (e.g., Hein, 1995). Conservation biology challenges individual agency professionals to think in new ways and asks the agencies themselves to develop and institutionalize new priorities.

Lack of Standardization.

No policy exists within and across agencies that sets standards for data sufficiency and types of studies required. Need exists for something analogous to the proposed interagency fire suppression policy (U.S. DOI and U.S. DOA, 1995) or the interagency standards for water quality assessment (Gurtz and Muir, 1994).

Lack of External Accountability.

American jurisprudence holds that courts will not substitute their judgment about how resources should be managed unless the agency's decision is without rational basis. In the words of the judge in the Sierra Club case, "in areas of scientific uncertainty the agency's choice of methodology is entitled to considerable deference" (Reynolds, 1994). This legal principle, called "agency discretion," means that agencies may make resource management decisions even if they are contrary to generally accepted principles of conservation science.

New Opportunities Exist for Helping Federal Agencies Manage Lands More Responsibly

Despite the political pressures that encourage retrogressive action by the agencies, the imperiled state of many ecosystems on federal lands has forced the agencies to recognize that change is necessary. Federal and state agencies are trying to modify their approaches to natural resource management to ensure long-term sustainability of natural ecosystems and biological diversity. This period of reassessment offers scientists a unique opportunity to help agencies develop policies to use sound science in plans for sustainable use of natural ecosystems.

Many agency efforts to revise resource policies center around "ecosystem management." Ecosystem management generally implies turning away from single-use, single-species management, and turning toward management designed to ensure the health of ecosystems. The goal is to use natural resources in a sustainable way that does not threaten the integrity of the natural ecosystems that provide the resources. Ecosystem management and related issues are discussed in several chapters in this volume, including Barrett and Barrett, Christensen, Ewel, Gordon et al., Meyer, Thompson, and Wiens, among others. Also see Alverson et al. (1994); Noss and Cooperrider (1994); and Crow et al. (1994).

Ecosystem management demands a much broader vision than have historical management practices. It requires that managers

- widen their focus from a few species of economic or other value to all the species within the ecosystem;
- focus on understanding and preserving not only species per se but also the interactions between species that collectively maintain the ecosystem and provide ecosystem services (e.g., Meyer, this volume);
- enlarge the management time frame to include long-term ecosystem processes, such as cycles of forest fires and succession to old growth;
- widen the planning process to encompass entire ecosystems, communities, and populations, including portions extending off-site onto adjacent nonfederal lands;
- pursue solutions to threats that originate off-site, such as nonnative pests and pollution;
- accommodate the needs of nontraditional users of federal lands, including nonconsumptive users; and
- use adaptive management.

To effectively manage ecosystems and other elements of biological diversity that extend across management boundaries, managers of ecologically linked land units must cooperate in planning and implementation of management procedures. Joint planning will depend on collection of scientific data that are mutually compatible among land units.

Suggested Standard Procedures for Biodiversity Assessments

To assist in the development of acceptable scientific procedures, in this section we recommend standardized procedures that we consider essential to any management planning program[1]. The procedures are organized as four main activities:

---

[1]This paper is the report of a workshop convened in July, 1994, by Defenders of Wildlife at the University of Wisconsin, Madison. Workshop attendees were conservation scientists and conservation law experts charged with developing a set of standard procedures that state and federal land managers could use to manage for biological diversity. The guidelines presented in this paper were developed at the workshop to help managers use up-to-date conservation science so that the impacts of timber harvesting, livestock grazing, and other types of resource extraction would not harm native ecosystems and species.

inventory and identification, evaluation of threats to diversity, management design, and monitoring and evaluating effects of management (Sections A, B, C, and D). Under these headings we identify and explain the importance of identifying communities and species at risk, choosing appropriate indicator species, measuring and mapping abundances, ranges, and other indicators of species and ecosystem health, calculating minimum viable population sizes for key species, calculating minimum dynamic areas for communities (defined later), and measuring edge effects. Agency procedures also should include methods for continued monitoring of ecosystem health so that short- and long-term negative impacts of management actions can be identified and rectified.

To be effective these procedures must be based on the best available science. Further, they should be standardized among agencies so that they can be consistently implemented across public lands. At present, agencies are frequently using divergent classification systems and data collection methods, as exemplified by the distinct ecosystem classification systems now being developed by the Forest Service and the Fish and Wildlife Service (CRS, 1994).

We realize that the standards offered below constitute an ambitious program. Because resources are limited, managers, scientists, and policy makers will have to jointly decide on which of the steps we recommend should receive highest priority. After priorities have been set, they should be reviewed regularly to make sure they are still appropriate given changing conditions and new information.

A. Inventory and Identification

As the first phase of their effort to assess the biotic status of their land base, managers[2] should assemble relevant and essential background information. Managers should

1. Prepare Maps of the Physical Environment.

Managers should prepare maps of topography, land forms, soils, climate and other factors that indicate the land's potential for supporting various types of biological communities. One important use of this information is to predict which kinds of natural vegetation could return to sites that are now substantially disturbed.

2. Prepare Maps of Biological Communities and Habitats.

Land managers need accurate and up-to-date information on the location and composition of the communities and habitats they oversee. Therefore

---

[2]When we use the term "managers" in this general sense we mean managers in conjunction with scientists and others involved in gathering and analyzing data and carrying out management.

- For each biotic community, managers should compile comprehensive species lists for as many taxonomic groups as possible, incorporating data from herbaria, museums, state natural heritage databases, and other sources. Species lists should include not only the conspicuous tree and vertebrate groups but also herbaceous plants, cryptogams, invertebrates, and fungi.

- Communities and successional stages should be mapped to fine scale (e.g., 100-m resolution) and their ecological condition assessed. A finely detailed classification scheme should be used that can distinguish among communities and successional stages that differ significantly in species or structural composition.

- In cases where present vegetation does not reflect potential vegetation, for example where grazing has changed the plant species mix from that which occurred historically, both present and potential vegetation should be mapped.

- Data should be gathered for ecologically related land both inside and outside the management unit.

- Data on both the biotic community and the physical environment should be entered into Geographic Information Systems (GIS) capable of displaying them as electronic maps.

- Methods of identifying, classifying, formatting and mapping data should be standardized within and among agencies.

3. Identify Communities, Species and Populations Known to Be at Risk.

Managers should flag for special concern communities and species that may be imperiled in their planning area. These include

- communities recognized by the National Biological Service (Noss et al., 1995), heritage programs, other federal or state agencies, scientific bodies, or private conservation organizations (Grossman et al., 1994; Noss and Peters, 1995) as rare, substantially reduced in area, seriously degraded, or in imminent peril of substantial loss regionally or throughout their ranges;

- species recognized by heritage programs, state or federal agencies, or others as threatened, endangered, or otherwise of special concern, or known or suspected to be declining in abundance across a significant fraction of their range, or known to be suffering a loss of range; and

- communities or species known to decline in the presence of human activity, introduced species, toxics, or other threats present in the planning area. These include species sensitive to habitat fragmentation or edge effects.

Agencies should collect all available data about these communities and species regarding their distribution, abundance, and responses to particular types of disturbance. In the case of species, assessment should include identification of specific habitat requirements. Where possible, distributions should be mapped.

4. Select, Inventory, and Monitor Indicator Species (ISs).

In addition to identifying species at risk, managers should identify other species that can serve as indicators of ecosystem structure and function and whose responses to changing conditions can be used to assess the success of management. Managers should include indicator species from all the following groups:

- species endemic to the area;
- species that require particular habitat or landscape features;
- species sensitive to environmental change;
- top carnivores or other species that require large blocks of habitat;
- keystone species that are known to have or suspected to have major influences on community structure or ecosystem function;
- species that are economically important, such as sport fish or game species;
- introduced species with substantial impact on natural communities; and
- species that exhibit metapopulation structure (e.g., Hanski, this volume; Wiens, this volume) or have ephemeral sub-populations that depend on disturbance (e.g., Hansson, this volume).

For these indicator species, agencies should collect and map where possible all available data regarding distribution, abundance, habitat needs, habitat specificity, and responses to particular types of disturbance.

5. Map the Historical Distributions and Characterize the Historical Compositions of Natural Communities.

Qualitative information and quantitative data should be gathered on distribution patterns from the pre-European settlement period both for the management unit and for ecologically related areas outside the unit. These data should be incorporated in GIS or otherwise mapped.

6. Characterize Current and Historical Disturbance Regimes.

Because disturbance regimes play a key role in determining patterns of biological diversity (Pickett and White, 1985; Huston, 1994), managers should identify and characterize patterns of disturbance from presettlement times to the present. As far as possible, information should include the frequency, intensity, areal extent, and distribution of each disturbance type. Current and recent anthropogenic disturbances, including grazing, timber harvest, and roads, should be characterized for comparison to the pre-European disturbance patterns that shaped the natural communities.

7. Determine How the Biological Communities Respond to Disturbances.

Managers should understand how disturbance patterns affect community structure and composition. For each major community type, managers should determine

- the immediate effects of the disturbances,
- the species composition and dynamics of communities that colonize after disturbance,
- the nature and composition of the intermediate successional communities that follow,
- the length of time necessary for "recovery" to old-growth conditions via succession,
- the predictability of successional sequences, and
- how disturbance size and intensity affect successional patterns.

Managers should pay particular attention to how indicator species and endangered and other "at risk" species respond to the type and intensity of both natural and anthropogenic disturbance, including rates and mechanisms of reestablishment.

8. Identify Area Needs and Edge Effects.

Managers need information on how habitat fragmentation and reduction in area affect the viability of natural communities. They should

- identify species and communities known to be sensitive to edge effects, limited habitat area, or genetic isolation. When data are not available for these species from the management area or its vicinity, managers should refer to published studies from other sites, particularly on closely related species.
- collect available data on how species diversity changes with area for major community and habitat types. These data are necessary to assess how decreases in acreage of a particular community may cause loss of species diversity within the community.

9. Identify Data Gaps and Priorities for Research.

Having compiled these data, managers should prepare a formal "Data Assessment Statement" describing the amount and quality of the data available to address preceding points 1 through 8. The statement should describe gaps in this basic knowledge and evaluate how this absence of information may limit a manager's ability to make informed land-planning decisions. Finally, the Data Assessment should identify high priorities for research based on both the significance of the gaps identified and on the relative costs of obtaining additional information. This research can be undertaken at the same time managers assess threats to diversity (Section B) and implement protective policies (Section C).

B. Evaluate Threats to Regional Diversity

Having assembled relevant information as described, managers should determine which immediate or potential long-term circumstances pose threats to natural ecosystems on their lands, and they should identify which communities, species, and other elements of native diversity have the most risk of being lost from the planning unit and from the region. In doing so, they should use the following analytical techniques, as well as their own judgment and expertise. The techniques described next fall into two categories—species and population approaches and community and landscape approaches (see Hansson, this volume). Both groups of techniques are required for a successful management program.

1. Species and Population Approaches

- Identify factors affecting viability and persistence of all indicator and "at risk" species. Managers (and scientific personnel) should use data collected in the inventory and identification phase (Section A) to identify factors affecting persistence of indicator species, endangered species and other species identified as being at risk (e.g., Ralls and Taylor, this volume; O'Brien, this volume). These factors include known or suspected biotic interactions, such as competition, disease, and predation, and associations with particular communities or habitats (e.g., Thompson, this volume).

- Estimate habitat needs for all species that are threatened, endangered, or otherwise at risk. Managers should attempt to predict minimal and optimal habitat requirements, taking into account known or suspected sensitivity to edge effects, area, and isolation effects. These factors should be analyzed both on the management unit and in surrounding, ecologically related areas.

- Perform quantitative and qualitative assessments of risk to vulnerable species. To minimize the probability of extinctions, managers should identify which factors most influence long-term survival of endangered and other vulnerable species on their management units and adjacent areas. When sufficient data exist, agencies should use quantitative population viability analysis (PVA) to estimate the risk of local extinction for individual species. Where appropriate and feasible, such models should be spatially explicit (e.g., Possingham, this volume) and should be based on the best available data on habitat needs, demography, and other aspects of sustainable population dynamics. In many cases, such PVA models will result in estimates of how much habitat is necessary for survival (e.g., Noon, McKelvey, and Murphy, this volume). Where there are too few data for PVA analysis, managers should identify threats and assess viability with less elaborate quantitative or qualitative population models. PVAs and or other methods should be used to assess how "at risk" species and indicator species are likely to respond to alternative management scenarios. These risk assessments should be routinely circulated for internal and external review and reevaluated every few years to incorporate improved data.

- Identify critical habitat for all threatened and endangered species present. To protect habitats that sustain rare and imperiled species and to meet the requirements of the Endangered Species Act, managers should use the data from inventory and identification, PVA, and the other risk analyses discussed to determine the minimum habitat necessary for long-term persistence of each endangered or other "at risk" species on the management unit or surrounding lands. These designations should include recommended management in and around critical habitat areas. Managers should also determine where the critical habitat needs of imperiled species overlap so they might be addressed simultaneously for maximum efficiency (National Research Council, 1995).

2. Community and Landscape Approaches

- Estimate the minimum dynamic area (MDA) for each major community type. Each community type historically developed partly in response to a particular disturbance regime. Thus the longleaf pine/wiregrass community developed and was sustained by frequent fires of a certain intensity and size. Conversely, eastern hardwood forests primarily developed where fires were relatively infrequent. Under modern conditions, two factors related to disturbance threaten the continued persistence of many community types. First, humans have changed many disturbance regimes, for example, increasing or decreasing fire frequency in different areas so that conditions no longer favor the original community. Second, many of the remaining areas of natural habitat are reduced in size and/ or fragmented. As patches of a community type become smaller, they typically become less able to recover from unfavorable disturbance events. For example, a remnant of eastern deciduous forest may be so small that, after having been subjected to repeated large fires, no mature growth remains. Whatever second growth does develop will be burned again before it reaches maturity. Therefore, for each community type on their land, managers must identify and maintain contiguous patches that are large enough to ensure their long-term persistence, given the prevailing and possible future disturbance regimes.

How does a manager determine how large a patch is large enough? Unfortunately, we still lack a generally applicable method for determining how large an area is necessary to maintain a particular community type, given the prevailing disturbance regime. Nonetheless, progress is being made. In 1978 Pickett and Thompson first suggested theoretical guidelines for calculating such an area, which they named the minimum dynamic area. The first practical attempt to calculate MDA for a specific community was undertaken by Givnish, Menges, and Schweitzer (1988), who applied the MDA approach to managing fire-dominated "pine bush" landscapes critical for the conservation of the Karner Blue Butterfly. They concluded that areas of 2,000 acres or more were necessary to sustain populations of the butterfly and lupine, its food plant. This application was simplified, however, because burning was controlled and therefore the extent of disturbance and its frequency was not random.

High priority should be given to extending this work so that we can apply the MDA concept to communities that experience random disturbance. In the meantime, at least one practical conclusion can be made, namely that the size of an MDA must be large where disturbance events are either large or common. In cases where disturbance events are especially large or frequent, MDAs may be larger than typical management units (e.g., a USFS Ranger District), and so will require planning beyond typical jurisdictional boundaries.

Once an adequate MDA model has been developed, the steps taken to gather data and utilize the model should include

- choosing the community types to be modeled;
- characterizing the historical and contemporary disturbance regimes for each community type, including frequencies, areas, and intensities of disturbance;
- assessing how each community type responds to the prevailing types of disturbance;
- estimating the minimal area needed to absorb the largest disturbance expected to occur within a 500- to 1,000-year period and still allow recovery via adjacent or nearby sources of colonizers (MDA analysis); and

- identify communities at risk. Managers should identify communities that are at risk of losing species or ecological processes. Identification will depend on ecological research that illuminates how individual species respond to particular threats, including fire suppression, excessive timber harvesting, overgrazing, pests, pathogens, exotic species, and extirpation of commensal or prey species (Section A).

  For many community types, some assessment of risk can be developed by characterizing the current disturbance regime and comparing it with the natural historic regime. For example, when the ecological community depends on frequent fires for health, as does longleaf pine savanna, a decrease in fire frequency from historical levels puts some constituent species at risk of extirpation. To the degree possible, information on disturbance and community response should be used in an MDA-type analysis to reveal which major communities occur in areas smaller than their MDA.

- Perform a landscape element completeness (gap) analysis. As a complement to identifying communities at risk, managers should assess the degree to which habitats and communities historically typical of the region are represented in the regional landscape (Usher, 1986; National Research Council, 1993; Klijn and Udo de Haes, 1994). In particular, managers should identify communities or successional stages that have become scarce in the region or that are not currently being managed to ensure their long-term viability. A thorough analysis would include creating distribution maps of all communities, estimating minimum dynamic area for these communities to predict long-term persistence, examinining known species-area relations, and determining habitat needs for sensitive species. Managers should also examine the extent to which community types are protected within existing reserve lands, such as national parks, wilderness areas, and Research Natural Areas.

  The National Biological Service (NBS), in cooperation with over 30 states, has already applied gap analyses to map the location of major vegetation types, ranges of terrestrial vertebrates, and other biotic elements and to identify those that are not adequately protected by the existing network of public and private conservation areas or by other conservation mechanisms (Vickerman and Smith, 1995). This approach should be widened to incorporate information on plant community successional stages and distributional and habitat need data for other taxonomic groups (particularly rare or endemic species).

C. Management Design—Formulate Specific Management Options to Address Recognized Threats

Managers should use the information gathered in Section B to devise effective strategies to conserve native communities and the species and processes that maintain them. These efforts should occur via parallel paths, assessing both the viability of particular indicator species and the long-term sustainability of the communities in which they occur. Although the analyses performed in Sections A and B rest primarily on scientific expertise and judgment, managers implementing the procedures outlined here must consider the wider set of issues surrounding multiple use. The procedures outlined here produce elements that can be integrated into the management alternatives required for long-range planning and the environmental impact process specified by the National Environmental Policy Act.

Priority for protection should be given to elements of biological diversity that are regionally, nationally, or globally rare and/or at risk of extinction. High-priority species, for example, would include those federally listed as threatened or endangered, whereas high-priority ecosystems would include those identified by the NBS (Noss et al., 1995) as "critically endangered." Species or communities that are nationally or globally common and secure would receive lower priority for protection even if they happen to be rare on the planning unit. Thus, managers should avoid management that replaces old-growth forest with second growth even if second growth is less common than old growth on the planning unit. Managers should be aware that the local increase in species diversity that often occurs following habitat fragmentation usually detracts from rather than adds to biological diversity on regional or global scales.

1. Develop Species Recovery Plans.

Federal agencies are already under obligation to devise recovery plans that ameliorate threats to threatened and endangered species and allow populations to recover (USFWS, 1994b). The success of these plans depends on the adequacy of the biological data, making it critical that the best data possible be collected and assessed. Successful recovery plans also depend on whether adequate habitat can be provided to ensure viability of the species—for many species, such as the Sonoran pronghorn (*Antilocapra americana sonoriensis*) (USFWS, 1994a) or the masked bobwhite (*Colinus virgianus ridgwayi*) (NWF, 1994), absence of suitable habitat is the greatest barrier to recovery.

2. Assess the Potential to Conserve or Restore Communities at Risk.

Managers should determine what opportunities exist within and beyond their land base to conserve or restore habitat to ensure perpetuation of each community type at risk. Such an analysis would consider landscape context, dynamic ecological processes, and the impacts of natural and human-induced disturbances, in addition to the physical suitability of potential restoration sites. An assessment should begin by asking

- "Is, or will, this community type be threatened in the region because it lacks some successional stages in sufficient abundance?"If this is true, managers should assess the potential for providing more of the missing successional stages by using active or passive techniques of community restoration. This might be accomplished on the management unit alone or in conjunction with nearby owners.

- "Is, or will, this community type be threatened in the region because it exists on an area smaller than its required MDA?"If less than the necessary MDA exists, managers should assess the potential to provide more area for this community type by identifying lands suitable for the community type that are presently dominated by other vegetation. If the present dominant vegetation belongs to community types that are common on a regional basis, managers should explore the feasibility of using restoration techniques to replace this dominant vegetation with the rare community type.

3. Develop Specific Management Practices to Sustain Biodiversity.

When species or communities are at risk of loss, managers should allocate sufficient area and make management changes to sustain those species and communities. Such management would likely incorporate planning decisions on the following two spatial scales.

- Regional scale.

Some communities and species that require extensive areas must be managed on the regional scale. Communities typically require large protected or managed areas if they historically developed in response to a regime of large disturbance events. Species requiring large areas include top carnivores and some other species with wide-ranging or otherwise dispersed populations. Large areas may be managed by

- letting natural disturbance regimes and other natural processes reassert themselves;
- generally letting natural regeneration of damaged communities occur rather than using active restoration techniques;
- allowing limited commodity production in some areas, provided evidence demonstrates that such activities will not threaten elements of diversity and provided that close monitoring occurs; and
- cooperating with other agencies and private land owners because the necessary large areas of habitat are likely to extend across management boundaries.

- Local scale.

Management can be successful on small areas provided that the desired communities and species have attributes that allow them to persist in small areas. Such management can include:

- generally using more active management than for larger reserves (e.g., using prescribed burning to maintain small patches of prairie or pine barrens);
- often using active restoration;
- generally excluding extensive or intensive commodity production; and
- protecting small reserves under suitable programs such as Research Natural Areas and Wild and Scenic Rivers. Such reserves are often created to conserve specific species or communities at risk.

Management on both local and regional scales requires planning to maximize landscape connectivity. Specifically, isolated patches of habitat must be connected by corridors of habitat that allow native plant and animal species to disperse from one patch to another. This will allow migrants to exchange genes among populations and colonists to reestablish extirpated populations. In addition, species will be able to travel along corridors and colonize new areas, thereby shifting their ranges to adapt to changes in climate or in the disturbance regime. Corridors can often be placed along rivers to incorporate and protect aquatic communities.

D. Monitoring the Effects of Management on Diversity

To complete the procedures described above, managers must closely monitor the biotic resources they oversee. Such monitoring is vital for assessing population, community, and ecosystem trends and to evaluate how management affects the trends. Proper monitoring can warn of impending species loss, community degradation, and loss of ecosystem function so that midcourse corrections can be made in management (Landres et al., 1988; Goldsmith, 1991; Crow et al., 1994). A careful monitoring design can, in conjunction with different management treatments, act as an experiment to determine the effects of the treatments on biological diverisity. For example, monitoring the abundance of native grasses and forbs on both heavily and lightly grazed plots provides information on the effects of grazing on grassland health.

Monitoring should have the following characteristics:

- Because resources are limited, managers should choose efficient and reliable indicators of biotic change and apply the best possible analytical techniques.
- Because accurate monitoring is particularly important for communities and species at risk, these should receive top priority.
- Monitoring should take place at several hierarchical scales to identify and diagnose new and emerging threats to diversity (Noss, 1990).
- Monitoring must be sustained over time to allow detection of slow or incremental change.

Because much has already been written about monitoring methods in the context of land management (e.g., Landres et al., 1988; Goldsmith, 1991; Crow et al., 1994) and because the NBS is currently developing guidelines applicable to all federal lands, we will not describe these techniques further.

Monitoring schemes should themselves be the subject of critical evaluation by the scientific community. We recommend that agencies publish their monitoring programs and distribute them for review. In addition, monitoring results should be systematically reviewed and regularly evaluated by scientists from academia, from local to federal levels of government, and from the private sector. It would be efficient to integrate these reviews and evaluations into the adaptive management cycle, with periodic updates of the activities described in Sections B, C, and D every 5 to 10 years.

27

Whatever It Takes for Conservation: The Case for Alternatives Analysis

Mary H. O'Brien

Summary

Conservation biologists are increasingly called on to contribute to ecological risk assessments but should insist on the preparation of broader alternatives assessments. Ecological risk assessments generally fail to examine a broad range of options for human behaviors and generally attempt to determine what cannot be determined: The assimilative capacity of the environment to withstand particular levels of human disturbance. Alternatives assessments, on the other hand, consider the pros and cons of diverse options available to humans to alter nonconserving behaviors. Although allowing conservation biologists to avoid defending any alternative as "safe," alternatives assessment involves examination of ecological and social benefits of particular activities as well as the hazards (i.e., "risk") of those activities. Although risk assessment generally employs experts to estimate the threshold for risks of a narrow range of unwise options, alternatives assessment is based on a broader range of options (including conservation and restoration), utilizes more diverse bodies of knowledge, and reveals the ecological and social hope offered by particular options.

Introduction

Decision making regarding conservation is necessarily a political undertaking, because it involves bringing about changes in specific human behaviors that are threatening or degrading ecosystem function, diversity, structure, or interrelationships. In other words, we would not have to undertake conservation efforts if humans did not behave in ways (and numbers) that are nonconserving.

Any proposed changes regarding restraints on or alterations of such human activities as burning, constructing, planting, herding, cutting down, digging up,

338 / Mary H. O'Brien

discharging, damming, draining, straightening, extracting, transporting, dumping, paving or reproducing, however, are often intensely resisted, because (1) we humans are habitual about our behavior patterns (called culture) and so it is generally hard for us to change or even imagine change; and (2) particular money, power, and institutional arrangements often ride on retention of business as usual, on *not* changing. In other words, we are swimming upstream to undertake and implement decision making that results in conservation.

It is within this context that we need to look at the current dominance that "risk" holds as an organizing principle for decision making, as in such processes as risk analysis, risk-based decision making, ecological risk assessment, and population viability analysis. After all, theoretically, "benefits" could just as well be the organizing principle, as in benefits-based decision making, ecological benefits assessment, or population maximization analysis. But instead we organize around risk. We in ecology are being handed an ecological risk assessment framework that is being transferred directly from toxics regulation in our country, which is built around single-chemical risk assessment.

I come to this topic from a time 14 years ago when I thought risk assessment was "just one tool," that it could be improved, and that it could be useful for environmental protection (see, e.g., London et al., 1985). In the intervening years I saw too much to remain with those positions.

As is evident from the 104th Congress, Newt Gingrich and his colleagues *love* risk assessment. They want to require an extended risk assessment process to precede any significant federal regulation related to the environment, and they want to allow any industry that would be affected by the regulation to sit on a "peer review" panel that would determine whether the federal government had done the risk assessment correctly. Needless to say, those who would be regulated will present competing risk assessments. Perhaps Mr. Gingrich's love (and corporate love) of risk assessment is reason enough to examine the context of risk assessment.

At least three major questions can be asked about our current involvement in or acceptance of risk as an organizing principle for conservation decision making:

1. Why is so much environmental decision making being organized around risk?
2. What are the consequences of this?
3. What alternatives do we have to organizing our environmental decision making around risk?

Why Are We Organized Around Risk?

I believe that risk holds the eminence it does within environmental decision making because risk is the perspective that will allow humans the most room to continue undertaking nonconserving, nonecologically beneficial activities. During risk-based decision making, certain questions predominate, such as, Exactly how risky *is* this activity? What is the evidence this activity causes harm? What level of this activity does *not* cause risk? What level of risk is acceptable?

These are assimilative capacity questions. They ask how much of our human activities the relevant ecosystems can absorb or rebound from. This turns into specific questions, such as, What is the smallest population that will allow an endangered species to persist 500 hundred years from now? How much water needs to be left instream to retain essential riparian functions? What amount of dioxin can be dumped into the Columbia River each day while retaining mink reproductive capability? What amount of large woody debris or number of snags must be left on each acre of a clearcut?

More fundamentally and to the point, these are also minimal-change questions, because they focus on assessing the impacts of habitual activities, of business as usual, and determining how much of such activities can continue. Generally, alternative ways of behaving are not in the picture.

Of course, you might say, "Risk questions are inevitable. It would be foolish to *not* assess risk. We *have* to analyze what damage may be caused by human activities (either destructive or supposedly restorative)." And you would be correct. As humans, we necessarily judge the risk of different options in every aspect of our lives: personal, occupational, local, national, and international. But think a moment about your own personal-life risk assessments: Don't you look at the risks of a number of *options*? And don't you look at the *benefits* as well as risks of each option? And haven't some of you implemented personal options that many other people never even considered or imagined?

I contend that most of the formal, institutionalized, risk analyses we undertake in environmental decision making are dangerous and nonconserving because (1) a reasonable range of alternative behaviors and processes is rarely considered and (2) the benefits of alternatives are not examined.

What Are the Consequences of Being Organized Around Risk?

One of the consequences of focusing on risks of a narrow range of alternatives is that we are called on to do something we cannot do with scientific integrity, and that is to determine safety; to determine the assimilative capacity of ecosystems or populations or species for damage, for example, from fragmentation or pollution. We know the difficulties of trying to base conservation decision making on such brinkmanship concepts as minimum viable population, maximum sustained yield, total maximum daily load, carrying capacity, and limits of acceptable change (see, e.g., Ludwig et al., 1993). The minimum population may turn out to have passed beneath some threshold or to have depended on habitat elements we didn't retain (Warren, 1992); the maximum yield may turn out to have been unsustainable in drought years or in currents affected by El Niño; the total maximum daily load of dioxin perhaps contributed to the decline of species we hadn't even considered. In other words, we don't know where the breaking point is between hazard and safety; the breaking point between insignificance and significance.

As Holsinger and Vitt (this volume) have noted with regard to individual rare species, for instance, we *do* know we should be trying to preserve as wide a range as possible of the species' geographic sites, because we *don't* know exact relationships between genetics and survival.

We are *certainly* able to answer the question, What are some of the potential hazards (i.e., risks) of this environment-impacting activity, or suite of activities? That is a legitimate risk assessment question. But answering that question would not address the question that is *really* being asked in most risk assessments of environment-impacting behaviors, namely how much of the behaviors to permit. (Ultimately, of course, environmental decision making is about deciding what activities, whether beneficial or stressful, will be undertaken, and at what intensity. When a risk assessment of an environment-impacting activity is being done, the decision at hand involves how much of the environment-impacting activity to allow.)

Simply discussing what we *can* know scientifically, namely some of the hazards caused by the activity, doesn't lend itself to *permitting* some specific level of the environment-impacting activity. Consequently, during risk analysis of a narrow range of options, we are almost always called upon to answer a question we cannot answer with scientific integrity, namely, How much of this activity (or suite of activities) is "safe," or "insignificant?"

We can't know the safety of particular, nonconserving activities because we don't know enough about patchiness, the diverse responses of a myriad of organisms, delayed responses, tortuously indirect effects, stochasticity, stunningly long time scales for some effects, or the cumulative nature of impacts of multiple human activities; in short, because we will never have enough information.

We could say that we will candidly acknowledge the uncertainties in our estimates of assimilative capacity, but in the world of public decision making, who pays attention to the caveats of uncertainties? Here is the standard, real-life scenario: Some entity (a corporation, a town, a nation) wants the go-ahead to continue nonconserving activities with the least modification or to initiate nonconserving activities to a maximum degree, and so it seeks a bottom line of permission. Someone of us (e.g., an ecologist in an environmental consulting firm) addresses the unanswerable questions of safety, sufficiency, or insignificance with estimates, which are perhaps accompanied by numerous caveats. On the basis of those estimates, a decision is made to permit some level or intensity of the activity, perhaps with some vague promises that adaptive management (i.e., modification, when seen as necessary) of the activity will ensue.

Then investments in equipment, transportation, storage, use, and extraction are made; permits are issued; budgets are secured for the activity; new habits and markets are built; and the monitoring slacks because unfavorable answers would mean having to change again. In other words, decisions will be made based on our pronouncements of safety or insignificance, minus the caveats regarding uncertainty.

Likewise, we could say we will add buffers to our estimates of minimum provisions of ecosystem function or structure for rare species, but from my 13 years of experience with toxics risk assessment and its consequences in the political world, that is a somewhat naive proposal, because decision makers will hear the word buffer, and regard it as just that: excess, unnecessary precaution; luxury, if we had unlimited funds; the wish list of biologists. When push comes to shove, any buffer will be chipped at or abandoned.

This is why this issue is so important. We *must* change the question being asked and refuse to be backed into a scientifically inviable corner.

A second consequence of risk assessment is that it focuses us on undesirable behaviors and we forget how things could be. I have often pictured risk assessment with this image: A woman is standing beside an icy mountain river, while four risk assessors stand behind her. The toxicologist says, "The water isn't even toxic, it's only icy. The woman should wade across the river." The hydrologist says, "I've seen rivers like this. This river probably isn't more than 4 feet deep at any point, and it probably doesn't have any whirlpools. She should cross the river." The cardiologist says, "She looks young, and she doesn't look hypothermic, so she probably won't suffer cardiac arrest. She should cross the river." The fourth risk assessor, an EPA bureaucrat, says, "Compared to ozone depletion and loss of biodiversity, her risk is nothing. She should cross."

The woman refuses to cross. "Why?" the risk assessors ask her. They show her their data, their models, their calculations. She's just a citizen, so they explain it all simply to her, with pictures. She repeats her refusal to enter the river. "Why?" they ask, exasperated.

The woman points upriver. "Because there's a bridge," she says.

That story illustrates the nature of risk assessment: Get us debating the risks of bad options, and maybe good options won't even come to mind. While ecological risk assessment focuses on minimizing risks, we forget about maximizing benefits. While it focuses on how much we can push the environment, we forget about how much we can change our own ways. While it asks us to figure out the resiliency of species, we forget to exercise the flexibility of human behavior. We are asked to tell the risk assessors whether the rock or the hard place is the worst place to be, and we work so hard to supply a careful, scientific answer to that question that we forget that there are other, better places to be.

The Alternative to Risk Assessment

What, then, is the alternative to risk assessment? The alternative is simple and it's what we do every day in our own personal decision making. The alternative is to compile a full range of alternatives, and then to examine the pros and cons, the benefits and risks, of each option.

We have a quintessential example of alternatives assessment in our National Environmental Policy Act (NEPA) regulations. These regulations, which govern the writing of environmental impact statements by U.S. federal agencies, require federal agencies that plan to undertake a project that may harm environmental health to consider all reasonable alternatives to that plan. The regulations accurately state the critical role played by public consideration of alternatives as being the "heart" of an environmental analysis. According to NEPA regulations, presenting reasonable alternatives to business as usual serves the purpose of "sharply defining the issues and providing a clear basis for choice among options by the decision maker and the public" (Council on Environmental Quality, 1986).

The fundamental process of NEPA, consideration of the risks and benefits of a broad range of reasonable alternatives, can be invoked in all environmental decision-making situations. All we have to do, when asked to participate in ecological analyses for any environmental decision making, is to insist that a full range of options be developed, and that both ecological benefits and harms be examined for each option. Social and economic benefits and harms will likewise generally enter into the analyses.

A key process in the development of the range of reasonable options, however, is broad participation: by practitioners of alternatives; by the interested public; and by groups such as Native Americans, who often have much to suggest regarding options.

With alternatives assessment, we will still examine risks, but this time we won't have to answer the unanswerable question, namely, How much of this unsafe activity is safe? Every one of the options may cause some harm; we don't have to claim safety. We merely have to say, "This option will likely bring these benefits, and might bring these harms. This other option will apparently bring very few ecological benefits, but might bring these harms. This option may bring substantial benefits, and so far we don't detect harms that might be caused."

As ecologists participating in alternatives assessment, we can't force anyone to choose what we think is the best option. But we *can* supply the best information we have about the pros and cons of each alternative. And, depending on the case, we may be able to eloquently and persuasively articulate the long-term benefits of particular options. It gives us the chance to articulate the benefits of a land ethic (Leopold, this volume) via one or more of the options.

Let me briefly describe an example of an alternatives assessment process in which Hells Canyon Preservation Council is currently involved. We are currently coordinating a project by the Hells Canyon CMP [Comprehensive Management Plan] Tracking Group, which consists of representatives (some of them scientists) from 15 environmental and hunting organizations and two tribes. We are writing an alternative (called the "Native Ecosystem Alternative") to be considered in a draft of the new Comprehensive Management Plan for the Hells Canyon National Recreation Area (NRA). The last Hells Canyon CMP was written in 1979. Our alternative must propose goals, objectives, and standards and guidelines for numerous aspects of the Hells Canyon NRA, such as access; recreation; aquatic conservation; Native American and Euro-American cultural sites; forest, riparian, and grassland habitat; geology; "unique biological habitat;" scenery; and soils. We are organizing our alternative around four principles:

1. Subordination of optional human activities to ecosystem protection (which is required on paper by the Hells Canyon NRA Act) and recovery.

2. Default decision making to no-impact or least-adverse-impact alternatives for human activities, whether management or recreational or commodity use.

3. Goals, objectives, and standards and guidelines that are measurable and based on ecosystem protection and maximum feasible recovery rather than on risk-based management for minimal wildlife and ecosystem values.

4. Monitoring schedules for human activities that could cause harm. Failure to meet the monitoring schedule for a given activity would result in cessation of that activity.

Forced by Forest Service regulations to subdivide the Hells Canyon NRA into management areas, we proposed three management areas based on ecology: forest, grassland, and riparian areas. In their alternatives, the Forest Service CMP Planning Team is carving the Hells Canyon NRA into management areas based on intensity of motorized and nonmotorized recreation and other human uses. A member of the Forest Service CMP Planning Team recently mentioned to our Tracking Group that the Forest Service now wishes that they, too, had based their planning around the "resource," as he calls the forest, grassland, and riparian ecosystems.

Our Tracking Group is not under the impression that our alternative will ultimately be chosen by the Forest Service. However, our alternative has already affected the alternatives the Forest Service is writing, and the Forest Service will be required to describe the environmental consequences (including environmental benefits) of our provisions in the Environmental Impact Statement (EIS). Our alternative widens the options being considered, it opens up a broader public debate, and it forces real choices.

That is one major value of alternatives assessment. The other major value, of course, lies in the fact that the likelihood of implementing environmentally beneficial alternatives can only be increased by an alternatives assessment process that brings such alternatives to the table.

All of us can probably think of other examples of decision-making processes that have involved alternatives assessment, that have involved analysis of the pros and cons of a broad range of options. Barrett and Barrett (this volume) have described The Nature Conservancy's approach to tidelands of the Connecticut River, which involves alternatives assessment, with various options being considered for the reserves, tributaries, surrounding towns, and the entire watershed. They describe the excitement of certain private landowners about living near a reserve, and of involving them in plans regarding endangered species on their property.

We ecologists, then, belong with alternatives assessment, not risk assessment. As conservation biologists, as conservation practitioners, we must move within the geography of hope, and refuse to be stuffed into a downward spiral of damage.

We have a key role to play in formulating alternatives; in analyzing potential benefits and harms of alternatives; in describing practical, feasible environmental visions of recovery and restoration; and in advocating for maximum attention to environmental needs and the potential for our human societies to change their bad habits.

Conclusion

There are reasons that so much environmental decision making in our country is currently organized around health and environmental risk rather than health and environmental benefits, around assimilative capacity concepts rather than the precautionary principle, around how much we can do *to* the environment rather than how much we can do *for* the environment, around how badly we can behave rather than how decently we can behave. But the reasons are not compelling, they have little to do with behaving as considerate members of the larger biotic community, and they spell disaster for ecosystems and for our future as an all-inclusive community.

We ecologists need to be conscious of how we are being used in environmental decision-making processes. Is our research, knowledge, time, or effort being sought for use in a risk assessment of bad options? Are we allowing ourselves to be sucked into debate on the wrong question? Are we ever flattered to be participating in a process from which the presence and wisdom of particular nonprofessional participants have been excluded? Are we agreeing to serve within unnecessarily narrow parameters?

We can help decision makers change the questions that are being asked. We can insist that a broad range of alternatives, including radically different alternatives, be examined. We can refuse to answer unanswerable questions of safety, and we can explain the grounds for our refusal. We can make sure practitioners of alternatives are part of the decision-making process.

As ecologists, we have enormous stakes in alternatives assessment processes. We can only describe the profound benefits and values of conservation-based behaviors if they are out on the table. We must face the reality that most current risk-based decision-making processes are depauperate, with critical options, key cards not even out on the table.

We must broaden the options considered, and then use our information and skills to make the case for the option that is ecologically and socially wisest. Much hinges on our determination to do so.

28

Conservation Activism:
A Proper Role for Academics?

Joy B. Zedler

Summary

This essay was stimulated by input from conference attendees who indicated that the conservation activist role is unusual, difficult, and inadvisable for academic ecologists, especially those who are just entering their careers. As an academic ecologist who actively advises on conservation issues at many levels, I hope to reassure young and established ecologists alike that it is easy, appropriate, and desirable to become involved in conservation efforts. Here, I define conservation activism as the direct transfer of ecological information to assist in developing conservation policy and legislation, designing nature reserves, providing testimony concerning land management, writing for the public, or educating the public, including elementary and high schools. Academics may also have opinions based on attitudes and beliefs that are unrelated to science; these are not included in the present discussion.

Role Models

Scientists can find many role models for conservation activism among our predecessors. Aldo Leopold (1949) wrote *A Sand County Almanac* not for his technical peers but for the public; he did much of the writing on his own time, sometimes rising at 3 A.M., sometimes in the evenings sitting by the fire (C. Leopold, personal communication, this conference). He found it difficult to put his words into poetic prose, but he persisted until he had what is now considered the basis of conservation ethics—a work that is revered by both academics and nonacademics. Leopold wrote for the layman and gained professional respect. It should be possible to do the same today, unless we have established a myth that conservation activism is not a proper role for academics.

Rationales and Rationalizations

At the conference, scientists not involved in conservation activism gave many reasons for avoiding the activist role:

1. "The system" offers no rewards for such work.
2. Nobody gives me brownie points for sticking my neck out.
3. I can't speak out until I have all the data.
4. When I talk to reporters about my work, they get it wrong and I lose face.
5. When I'm quoted or pictured in the press, my colleagues are jealous.
6. It's not my job; there should be a new breed of professionals who are paid to translate science for conservation efforts.

The first two comments suggest that scientists are prevented from taking an active role in conservation. Either the promotion-tenure system or pressure from peers is seen as a reason to keep quiet. Implied is the desire to be involved, but that some external factor prevents it. An alternative interpretation is that individuals want to avoid becoming involved, and these are convenient excuses.

The third comment, concerning inadequate data, indicates appropriate caution on the part of a scientist. The problem is that data sets are never complete; few ecologists ever have sufficient data to say without qualification what the situation is. Yet decision making should be informed even if the understanding is incomplete. Ecologists are free to indicate the shortcomings of their data when testifying or commenting in the conservation arena. Results that seem preliminary to the investigator may provide great insight to an audience who knows much, much less about the subject in question. Surely it is better to have a scientist provide guidance based on incomplete data than to have a decision maker draw conclusions without such input.

The remaining comments seem consistent with a lack of interest in, or dislike of, activism. I respect professionals who are honest about such attitudes. But there is a problem when academic ecologists create or maintain barriers, thereby deterring the activism of others. Attitudes that brownie points are needed and that "the system" must reward activism for it to be valuable are out of place. If we transmit to our peers the attitude that our knowledge of the environment can only be used to foster promotion and tenure, we perpetuate "the system." Most academics are very busy, yet we all undertake some activities by choice. If those activities involve letter writing to legislators, speaking to the public, writing for a magazine, or testifying at a public hearing, our peers may suddenly begin scrutinizing our free-time work. They may assume that we should be working on peer-reviewed publications whenever we address environmental issues, in violation of free speech. If peers judge conservation activism with scorn, rather

than admiration or tolerance, they are the perpetuators of "the system," not
its victims.

There are good reasons why professional ecologists should make their scientific
findings available to public conservation efforts. First and foremost, the information is needed to improve decision making. Second, many of our salaries are
paid by the public. Third, providing information makes the public aware of
ecological data and approaches, which improves people's understanding of the
natural world, and perhaps raises public esteem for ecology as a science. Fourth,
if *we* don't educate the public, the agency staff, and the politicians about the
science behind conservation, there may be no one else to help make informed
management decisions.

In my opinion, professional ecologists not only have a right to advise on
environmental decisions, we have a duty to do so. To become involved, I suggest
starting from familiar terrain, that is, with activities closely related to ongoing
research programs.

1. Seek funding from agencies that require an application of ecological research, for example, the U.S. Army Corps of Engineers, the Environmental
Protection Agency, the Highway Departments, and the like. In this case, there
is no barrier between research and its use in developing natural resource policies;
in fact, it will likely be required that data be interpreted and science-based
recommendations developed.

2. Work directly with agency staff to clarify ecological questions and determine
what type of information transfer will be most helpful. Staff may need an interactive model, a poster, verbal advice, a field demonstration, or a nontechnical report.
Sometimes just sending reprints is sufficient.

3. Write a simple summary of any applicable findings or the significance of
your findings and send it to agency staff along with citations and reprints. This
is a simple way to build on what we are paid to do.

4. Write up your research findings for a broader audience. Pass the text to
local papers, magazines, educators, visitor centers, and the like. This is harder
for persons trained in scientific writing, but the challenge is worthwhile.

5. Testify at hearings, even if what you have to say is brief. Scientific information and academic credentials will raise the level of discussion from opinion to
educated opinion.

6. Write reviews of projects; point out what's missing in an Environmental
Impact Statement (EIS) and why that information is essential to predict the
impacts. For example, cumulative impacts are rarely assessed adequately, yet an
EIS often concludes there will be none.

7. Offer your information to other activists when you see an application
for your knowledge. Respond when citizens request technical help for their

conservation activism. Realize that as scientists we have access to information that others do not.

8. Encourage your students to seek employment in positions where they will influence conservation decisions; maintain close contact so that they keep abreast of your research findings.

Risks

There are both costs and benefits to conservation activism, that is, risk of and reward for speaking out. There are certainly some risks. The worst of these is probably the loss of research funding if an opinion is expressed and someone with strong political connections dislikes it. I have lost research funding from two agencies that have influential political appointees at the helm, and such situations deserve caution. In both instances, however, I was able to reinstate the funding. Furthermore, I would take that risk again, if only to expose such corruption.

Some wonder whether it is possible to maintain one's scientific credibility if, for example, they are misquoted in the newspaper. The best defense is to qualify one's answers and to say what is not known. It is not necessary to draw conclusions beyond the data; listeners will likely do that themselves. There is also concern that one will be branded as an advocate. Again, this is not inevitable, especially if you consistently preface your remarks with references to your role as a scientist— "Speaking as a scientist,. . ." or "My objectively gathered data indicate that. . ." When speaking with the press, it is often necessary to remind reporters that you are not speaking out of emotion or advocacy; rather, you are providing data that is relevant to an issue. I always ask reporters to refer to me as a scientist or ecologist, rather than use the broader term, "environmentalist." It is important to educate reporters that scientists provide information based on data. Eventually, the public may come to understand that ecology provides the scientific basis for conservation.

There are many disappointments if one presents data in hopes of conserving a species or a piece of land, as one is often on the losing side. But the company is good, and informed activists who have access to your knowledge will be better armed for the next confrontation. It also requires sacrifice, but it need not be too costly in time or energy. There is much demand for technical advice in today's complex environmental issues, and activism can eat into one's professional time. The hardest part is saying no to a group who wants and needs advice when you know you could help. Some scientists solve this dilemma by becoming paid consultants. This is a solution, but it limits the list of information users to those who can pay. Giving free advice has a cost, but the personal rewards, and the freedom to advise whom you please, can certainly outweigh it.

Are there rewards? Absolutely. The activist community is enormously apprecia-tive. The public admires you, and decision makers show respect—even if they rule against the position your information supports. One gains appreciation from individuals whose opinions matter: kids, agency staff, decision makers, concerned citizens. The few times I have had a judge or congressman ask me to explain some aspect of wetland ecology—and then thank me for clarifying the issue— are memorable.

Opportunities

There are many opportunities for activism in today's world—with major chal-lenges to the Endangered Species Act, the Clean Water Act, the increasing role of states in regulation, and the need for state legislation to keep up with shifting responsibilities for resources. We learned from Michael Bean (this volume) that the Supreme Court's decision on guns near schools could carry over to environmental issues. From environmental impact analyses under NEPA and state regulations, there is a continuing need to predict impacts and evaluate the adequacy of cumulative impact assessment. It is clear to me that needs for technical expertise are expanding.

Every individual must decide how active to become. If you are already involved in bringing your ecological research to bear on conservation problems, please accept my admiration and appreciation. If your colleagues are not, perhaps you can make the difference by inviting their collaboration on a letter of comment concerning an issue of mutual concern. Some people have a hard time volunteering their expertise; they need to be asked.

If you are a professional ecologist who chooses not to become involved, I respect your decision. But if you blame that lack of involvement on "the system," I urge you to consider the impact of your own attitudes and actions on those around you. If you would like to be involved but believe the system is against it, identify the constraints and see if they can be changed. Deborah Jensen of The Nature Conservancy offered to come and speak to university administrators about the importance of getting involved with nongovernmental organizations and applying ecological science to conservation. Perhaps you can find ways to change the system.

No one needs to hinder the conservation activism of others. At the very least, make sure you are not part of the problem—that any negative attitude you possess does not propagate through the system. If you do not support the outside activities of your conservation activist colleagues, recognize that what professionals do on private time should not be a negative factor in promotion deliberations. Better yet, try a little activism on the side and experience the appreciation it brings; you may get hooked! It is a very satisfying reward to see one's research findings used to improve conservation policies and practices. Activism offers an opportunity to add value to your life.

Acknowledgments

I thank Dr. Mary O'Brien, Kathy Boyer, and the editors of this volume for reviewing the draft.

29

Getting Ecological Paradigms into the Political Debate: Or Will the Messenger Be Shot?

Graeme O'Neill and Peter Attiwill

Summary

There was never a more important time for ecologists to speak out about applied, socially relevant issues. Two forces make this difficult. First, applied issues quickly become the matter of politics, and ecology is often not strong enough to be used in independent arbitration, nor do many ecologists want to step into such an arena. The second follows from the first: the media puts emphasis on controversy in political issues, but since environment is somewhat peripheral to the main media thrusts of politics, economics, business, and arts, coverage of environmental issues tends to be shallow and nonanalytical. On the other hand, public-opinion surveys show a pressing demand for more and better coverage of science. If ecological science is to impose its authority on public environmental debate, ecologists must first work to ensure that the media are equipped with more and better-qualified environmental journalists and must then establish a productive interface with the media. There is an urgent need for a state-of-the-science document, directed at journalists and decision makers, summarizing the major ecological concepts that have been defined over the past 25 years.

Introduction

It was part of Peters' (1991) assessment of the state of ecological science that there is "a desperate need" for ecologists to publicize examples of good ecology so that they could participate in and contribute more effectively to applied, social issues such as the sustainability of forests and fisheries. This is easier said than done, and at the heart of the difficulty lies the topic of effective communication of ecological science. In this chapter, we develop some themes from our experiences as a professional science writer (O'Neill) and a forest ecologist (Attiwill) working in Australia.

Is Ecology a Weak Science?

Despite the fundamental importance of ecology, practicing ecologists are well aware of the weakness of their science (e.g., Peters, 1991; Weiner, 1995). Peters (1991:104) wrote that

> Ecology is dominated by complex and inadequately defined terms which confound the development of predictive theory. As a result, ecological classifications, ecological characteristics and ecological relationships may refer to phenomena that vary with each change in focus, scale or author, and ecologists are often not sure they are talking about the same thing. . . . The absence of such controls in ecology allows superficial association of concept with a diverse set of concordant observation, concept and theory such that an entire conceptual system may be erected and confirmed with observations.

This critique can be related to the way in which the media frame ecological issues and influence public debate and community opinion. It seems to us that the "fuzziness" of ecology as a science, as defined by Peters (1991), denies it the opportunity to be seen as an impartial arbitrator in public debate on environmental issues. As a consequence, the debate inevitably polarizes, with each side appropriating only so much science as it needs to sustain its position. When ecologists see their science used in this selective manner, and in a context for which it may not have been intended, they prefer to distance themselves from the public debate because of the risk that they may be seen as being aligned with one side or the other. Another interpretation is that environmental debates are political debates, and the boundaries between pure science and advocacy become blurred.

The Media and the Polarization of Environmental Issues

The media exacerbate this problem because they are attuned to controversy. Television in particular will almost always represent an issue in terms of diametrically opposed opinions, because it cannot afford the time to analyze any issue in depth. The ecologist who might wish to be seen as impartial is thus excluded. Even when ecologists do participate in public debate, the media will typically press them to declare, tacitly or explicitly, their support for one side or the other. The pressure to choose sides arises because the media, especially television, which have conditioned audiences to expect controversy, now cater for the public's preference to have issues represented in black and white. Controversial issues are presented as a form of theater; one's credibility or "believability" is determined as much or more by one's performance as by the information or views presented. Few scientists are comfortable or skilled in such a milieu.

The media generate and sustain the centrifugal forces that dominate public debate on environmental issues. Most media in Australia have abandoned the honorable role of independent observer because readers and audiences now expect to be told what to think about an issue; few people have the time or inclination to analyze an issue in depth and form a considered opinion—and those who do have an interest are rarely able to find the in-depth information they require in the popular press or on television. In the Australian print media, in-depth analysis of issues is restricted to the few broadsheet daily newspapers that employ full-time science writers, to a couple of tabloid daily newspapers, and to weekly news color magazines (e.g., *Time* Australia). An ominous trend is that newspapers—tabloid as well as broadsheet—generally are in decline and under economic pressure to move downmarket, with the result that substantial issues relating to the environment are receiving less attention. Quality coverage is even narrower on radio and television, being limited to programs on the government-sponsored radio and television networks.

Other forces in the media conspire to limit both the quantity and quality of coverage of environmental issues. The dominant culture in the media sees the world from a political-economic-business-arts perspective; the culture is self-perpetuating because editorial executives tend to clone in their own image. Scientific and environmental issues tend to be seen as peripheral to the main game; the vision is likely to be short term and parochial. Although few modern media outlets would consider appointing nonspecialists without academic qualifications to cover political, economic, or business news and issues, few would consider employing specialist, environmental journalists with formal qualifications in environmental science. Australia's few print-medium environmental journalists are mostly drawn from a background in general reporting. And because newspapers have a policy of rotating roundsmen to prevent them absorbing and reflecting the values of the interest groups they write about, they are likely to be only transient appointments.

As a result, coverage of environmental issues tends to be shallow and nonanalytical; depending on the particular outlet's political orientation, coverage may emphasize a "green" or "development" view. Through long experience of dealing with nonexpert journalists, both sides have become adept at manipulating media coverage and, thereby, public opinion. Sections of the conservation movement bring heavy pressure to bear on environmental journalists by promoting an extreme, nature-centered view of environmental decline and destruction (playing on "public gullibility and guilt," North, 1995); the "if you're not with us, you're against us" tactic is wielded with powerful effect. In opposition to such views, sections of the Australian mining and forestry industries have tended to downplay scientific issues and take a line that emphasizes national economic or employment issues. Although debate on forestry issues in the 1990s remains polarized, the mining industry is pursuing a more moderate approach, and many mining companies are now actively involved in research and promotion of environment-protection schemes.

It is in the nature of most environmental arguments that they arise before the ecosystem under threat or in dispute has been studied in any detail. In the hiatus between the initiation of the issue and formal scientific studies of an ecosystem's status and function, the contending parties can sustain their respective positions by reference to studies that may be of limited relevance to the ecosystem in question. Or worse, the contending parties can recite popular ecology, or "ecologisms." North (1995) follows the theme of our argument in proposing that ecologisms are generalities based on limited ecological information, and that they lead to the popular views that, on the one hand, undisturbed nature is stable, harmonious, and unchanging, whereas, on the other, disturbed nature is fragile. This is nothing new. For example, Barry Commoner, a leader of the emerging environmental movement in the 1960s and 1970s, wrote (1972) that "there are a number of generalizations that are already evident in what we know now about the ecosphere and that can be organized into a kind of informal set of 'laws of ecology'." Commoner's four laws of ecology are

- Everything is Connected to Everything Else
- Everything Must Go Somewhere
- Nature Knows Best
- There Is No Such Thing as a Free Lunch

Although these laws contain elements of truth that could be developed as testable hypotheses, the difficulty with them as they stand is that they contain no predictive power whatsoever. They are, in Peters' (1991) terms, tautologies in the sense that they accommodate every possibility. So, given these laws, it is then easy for Malcolm (1995), an environmental writer in an influential Australian daily paper, to outline the negatives of timber harvesting—removal of nutrients, introduction of weeds, soil erosion, silting and turbidity of streams, changed germination and growth conditions for plants, and reduced biological diversity favoring only a few, commercial plant species. The negatives have appeal and are easy to accommodate, because Nature Knows Best and Everything is Connected to Everything Else. But the ecologist must go farther. For example, is the removal of nutrients significant within measured rates of nutrient cycling and nutrient replenishment? Can we manage biological diversity within acceptable limits in the face of both endogenous and exogenous disturbance? The negatives can be stated with little knowledge of the literature, and little analysis and interpretation. The alternative argument—that sustained management of some proportion of our forest estate is ecologically feasible—requires detailed knowledge and analysis, which has little appeal for that part of the media that thrives on confrontation.

Attiwill's (1994b) long-term comparison of regeneration of mountain ash (*Eu-*

calyptus regnans, F. Muell.) forest after bushfire and after timber harvesting in the central highlands of Victoria, Australia, exemplifies the problem. Conservation groups rejected Attiwill's main conclusion that fire and disturbance by logging were to a large extent ecologically equivalent, and that mountain ash forest could be logged sustainably if logging practices are based on the ecology of natural wildfires. At the heart of their concern was the aim of the Australian Conservation Foundation to stop all timber harvesting in Australia's native forests, a policy that has been adopted by the emerging, green, political parties. They were also concerned that the forestry industry would seek to extend Attiwill's conclusions to sustain an argument that all tall eucalypt forests could be logged sustainably. The industry, like sections of the conservation movement, tends to be selective in citing findings from ecological research, thus denying the community a balanced view of conservation issues.

Attiwill's study was reported in *Time* magazine's Australian edition by a specialist journalist (O'Neill, 1993) with academic qualifications in environmental science. The article explained how the concept of patch dynamics could be integrated into logging practice to achieve sustainability. Conservationists sought to undermine Attiwill's credibility and his conclusions by attacking his scientific integrity and characterizing him as a lackey of the forestry industry. A similar attack was made on O'Neill, suggesting that he had been naive to report Attiwill's conclusions.

O'Neill's article was originally intended to explain the *ecological* concept of patch dynamics, and the role of natural disturbance in maintaining biological diversity, using a number of examples including mountain ash forest, rainforest, and coral reef ecosystems. Patch dynamics is not a new concept in ecological science, having emerged from pioneering studies of cyclone disturbance of rainforest and reef ecosystems in northeast Queensland in the late 1970s (Connell, 1978; Connell and Keough, 1985). Yet as a concept, it has never penetrated public debate on environmental issues in Australia; the green movement continues to emphasize outmoded, 80-year-old concepts such as fixed-trajectory succession toward a climax community (the very fundamentals of Clementsian ecology) and the stability and permanence of untouched or pristine ecosystems as stable associations of species through time. Conversely and almost as a matter of routine, it characterizes all ecosystems as 'fragile' in the face of human disturbance.

The persistence of such paradigms, long after ecologists have abandoned or modified them, attests to the crisis in ecological science that Peters (1991) describes, as much as to the imperfections of the modern media. Ecological science has failed almost completely to impose its authority on public environmental debate; where it should be leading debate, it plays a largely passive role or abdicates its responsibility on the grounds that the debate is so politicized or of such poor quality that good science would be sullied through any active participation. The result is a self-fulfilling prophecy.

Another reason why ecologists have failed to impose their authority on public environmental debate is that our peers see little need to go outside their confines, and indeed tend to denigrate ecologists who enter the debate. We tend to publish our science with a strategy of hope—hope that, having published, someone somewhere will see the relevance of it to something (Rogers, this volume). Adopting a strategy of promotion in place of one of hope is seen as aggression: "I see that so-and-so is chasing publicity again."

What Should We Do?

From the media's perspective, readers and audiences have no deep interest in environmental issues. Yet as Henningham (1988) has shown, the media grievously underestimate community interest in environmental news and issues, along with scientific and medical news and issues. Numerous public-opinion surveys (e.g., Lowe, 1995) have shown that the community consistently ranks these subjects ahead of all others, and gives a very low ranking to subjects such as politics and economics. Thus Henningham (1995) is optimistic about the future for scientific journalism in Australia.

How should ecologists get their messages across to scientific journalists? As long as ecology is perceived as a "fuzzy," malleable science lacking a strong conceptual framework, it risks being seen as weak and susceptible to manipulation, misrepresentation and politicisation. We therefore see an urgent need for a "state-of-the-science" document, directed at journalists and decision makers, summarizing the major ecological concepts to emerge during the past 25 years, and giving them relevance by citing the studies from which they arose or the modern issues to which they are relevant. Such concepts might include

- Quantifying environmental processes, particularly those identified from long-term studies. For example, the Hubbard Brook Ecosystem Study (Likens et al., 1977) began in the early 1960s and continues to be a yardstick for ecologists dealing with applied problems such as sustainable utilization and acid rain.

- The fundamental importance of disturbance and patch dynamics in determining ecosystem structure, function, and diversity (Pickett and White, 1985), particularly in forests (chapters in Pickett and White, 1985; Oliver and Larson, 1990; Attiwill, 1994a).

- The "overwhelmingly important" concept of diversity in ecosystems (Wilson, 1992). Although Everything is Connected to Everything Else is a nonverifiable truism, species are assembled within ecosystems neither in total disorder nor in total harmony and dependence. In Wilson's terms, there are big players and little players, and the biggest players of all—keystone species—influence biological diversity out of all proportion to their numbers.

- The concept of *K*- and *r*- selection (MacArthur and Wilson, 1967) of which it has been written that "it seems unlikely that there will ever be another such simple generalization with such explanatory power" (Begon et al., 1990). The red and gray kangaroos of Australia are typical *r*-selected populations that explode in numbers after drought or within the protected confines of national parks—especially where the national parks are islands within developed grazing lands. Yet we never hear this enormous explanatory power used in the mostly emotional public debate on whether or not and how we should manage and control populations of our national emblem.

- Quantifying or characterizing complexity in ecosystems and the relationship between complexity and robustness in terms of resistance and resilience. For example, the conservative response of nutrient cycling processes in forests following disturbance has been extensively studied (e.g., Likens et al., 1977; Jordan, 1987; Attiwill and Adams, 1993). These studies provide a definitive and quantitative basis for management to sustain long-term productivity.

We conclude that there was never a more important time for ecologists to speak out on issues that they know about, and we agree with Peters (1991) that there is a desperate need for ecologists to contribute effectively to applied social issues such as the sustainability of forests and fisheries. But who is to do this and how will consensus be reached on what is to be spoken about and how it is to be interpreted? Ecologists have no professional association (such as the medical profession has) that maintains standards and is a spokesperson for all of its members. The members of ecological societies cover a huge spectrum of interests—from hard science through ecologism to deep ecology. The last has major components of spirituality, and "mere ecological ideas, no matter how deep," may have little relevance to this spirituality (Seed, 1995). There is no doubt that most ecologists have experienced this dilemma between emotion and hard data generated by the ecosystems they have researched. But Seed (1995) goes farther, calling on us to communicate with plants and to enlist the power of plants to help us recover "our ecological Self." Seed's view is mystical, and the rise of mysticism has done much to confound and polarize the debate because mystical beliefs are undebatable and cannot be tested scientifically. They goad the opposition to use ridicule, so rational debate suffers.

North (1995) argued that, at least at the governmental level, sound science plus a component of the precautionary principle is increasingly driving policy in environmental issues in Britain. Whether professional ecologists participate in this process as individuals, or whether we can marshal our combined scientific force remains to be seen. In either case, success depends on our abilities to interact constructively with the media. Henningham's (1995) optimistic future for professional, scientific journalism sees this interaction as entirely positive; we should all be actively working toward its achievement.

SECTION VI

Synthesis and a Forward Look

Themes

Steward T. A. Pickett, Richard S. Ostfeld,
Moshe Shachak, and Gene E. Likens

This section employs two strategies in summarizing and bringing together the material and ideas presented in the previous chapters. First we attempt to capture some of the excitement and discourse that took place at the diverse meeting from which this book sprang. The tensions and exchange embodied in the meeting point not only to the areas of consensus, but also to the needs for additional data, further unification, and refinement of the fundamental scientific concepts and theory. A powerful agenda emerges as a result.

Lovejoy's chapter summarizes the presentations at the meeting and highlights some key comments that emerged from discussions after the talks. His insights reflect his wide experience both as a scientist and as one engaged in policy and conservation on the international scene. The immediacy of his pithy remarks points to the urgency of the task before us.

Talbot's chapter was stimulated by the closing plenary discussion at the conference. He condenses the insights from a wide ranging policy discussion and couches them in the ample context of his international policy experience in conservation. As an observer and participant in the evolution of the new ecological perspectives so important in this book, his remarks highlight the major problems to be solved in the unification of conservation science.

Harte, in a piece stimulated by the conference, takes a step back from the detail of the meeting to identify perhaps the most crucial scientific need to better prepare ecology to participate in the public discourse about conservation. He, too, has long and deep experience in inserting ecological understanding into the public sphere.

These three contributions illustrate some of the tensions that were felt in the meeting. In the rush and heat of discussion and the presentation of talks within short times, there is bound to be some inexact and inconsistent use of terms and concepts. The problems identified in this section can guide future conceptual refinement, point out the need for translation of concepts between the diverse perspectives of different ecological specialties, and promote empirical work that is the necessary partner to the best development and use of scientific ideas.

Another source of tension at the meeting was the disagreement about the nature of the paradigm shift, and whether or how to deal with that shift in the public discourse. Has ecological change been evolutionary or revolutionary? Has it been a stately progression of scientific knowledge or a radical shift in fundamental assumptions, scale of study, and dominant processes addressed? However one answers those questions, it is clear that the knowledge base supporting conservation practice has grown and been refined enormously. These three chapters identify key principles from the current state of ecological understanding, and recognize the needs for future research and interaction between managers and scientists. The current ecological understanding, based on openness of systems; multiplicity of dynamics, pathways, and end points (where such exist); and flux and adaptability of natural systems, still recognizes the limits within which ecological systems must remain to persist. The modern paradigm does not excuse human excesses relative to biodiversity, ecosystem function, and natural heterogeneity. Nor does it equate the unprecedented and extreme rates and levels of human patchiness and disturbance with those that have been a part of the evolutionary history of organisms and the ecological legacy of communities, populations, and landscapes, contrary claims in the popular press notwithstanding (e.g., Chase, 1995; Easterbrook, 1995; Mann and Plummer, 1995).

The second strategy we employ in the synthesis rests on a more leisurely approach, and seeks an explicit framework to unify the diverse kinds of knowledge and practice identified among the contributions to the book. Indeed, several of the chapters were sought after the conference to complement the contributions to the conference. The framework for synthesis we use attempts to unify paradigms, general theories, specific subtheories and models, and system manipulation. Manipulations can be done to generate new knowledge and, through management, to satisfy societal values. We extracted from the chapters the insights that assessed the ecological paradigm, the relationships between theory and models, and the connections between models and the practice of science or of conservation. Although we mention some of the same insights that Lovejoy drew from talks at the conference, we have an explicit and encompassing framework to test against them.

Finally, the synthesis section ends with a brief but provocative epilog by Joel Cohen. The world will be different in the future, based in part on the nature and success of conservation science. The epilog, by presenting one future vision of the world, reminds us of the importance of strengthening, unifying, and using ecological knowledge for conservation.

30

A Summary of the Sixth Cary Conference

Thomas E. Lovejoy

An initial confession: I don't like the word paradigm because it is too easily misinterpreted as containing an element of pretense. In any case, I agree with Rob Colwell that we are dealing more with evolution than revolution or paradigm shift in our science.

There are really two frameworks for conservation and ecology. On the one hand, people live and work within ecosystems; on the other, most of the time decisions are made within a societal context. I am forced to conclude that if the world works in one way and society in another, we are probably stuck with working at conservation and ecology in both ways unless the scales fall dramatically from humanity's eyes. In that context, the most interesting thing said in the course of the entire conference was Joel Cohen's response to Bill Robertson's question about why demographers by and large don't perceive a population problem: it was that demography and economics do not have a theory of scale—that they don't relate to the scale of the environment on which they depend. In contrast, those of us working in ecology and conservation are wedded to scale—multiple scales, both spatially and temporally over orders of magnitude. In addition Adam Smith's invisible hand operates with the tyranny of discount rates so that anything of long-term value is of little consequence in the calculus of market economics.

In passing, I am intrigued by Joel Cohen's illumination of the same pattern of human population growth over three orders of magnitude of scale, which brings to mind for no particular reason the similarity of patterns in John Wien's images of puesto in the Argentine and harvester ants in Colorado. This latter is probably more a comment on the visual nature of this social primate than anything else.

In any case, the discussion following Joel's talk about how many people do we want versus how many can the Earth support struck at the heart of the issue (Cohen, Epilogue, this volume). Sustainable development can occur at many levels of biodiversity from close to the natural to close to zero in the sense of Stuart Pimm's Hawaiian forest of alien species. At issue is quality of life (itself elusive of definition) and ecology is a keystone, if not *the* keystone, science to achieve that.

It is in fact the conservation imperative that forces us in the direction of ecosystem conservation and ecosystem management. The list of life forms in trouble is too long to be dealt with in any other way, and as we saw in Cathy Pringle's Puerto Rico example, and in Judy Meyer's Knolles Creek example, a particular conservation objective often cannot be achieved without considering the entire system. The discussion following Judy's talk about rivers as biological as well as water chemistry integrators reminds me of tropical limnologist Harold Sioli's dictum "The River is the Kidney of the Landscape."

Norm Christensen reminded us of how differently we think of the intersection of ecology and conservation today by reminding us of the outlook in 1916 at the time of the National Park Organic Act. In contrast to what seems from the perspective of today as a comfortable, deterministic closed-system view, we now deal with the dynamics of heterogeneity, disturbance, and nonequilibrium, as well as patchiness, and all over different scales. Nonetheless, there are return functions in natural and disturbed ecosystems, and our science is not sliding inexorably toward some stochastic black hole where entropy holds sway (perhaps as in the new chaos theory of *Dirt Bike Magazine*, cf. Bean, this volume). More than one speaker has reminded us that the new grows out of the old. There also have been multiple reminders to continue to expect surprises.

The discussion ranged from genes to landscapes without venturing far to sea. Kent Holsinger gave a lucid presentation of particular cases where conservation genetics will have a key role, but reassured us that by and large if a population is large enough to buffer against fluctuation and disturbance, the chances are it will do alright genetically. Kathy Ralls brought us up to date on population viability analysis, the roles it can play, and what it cannot do. Here as elsewhere through the meeting, the metaphor of engineering a bridge or a building for safety reminded us to be cautious about what we consider minimum (cf. Peters et al., this volume).

The population and community or ecosystem level provided us a bit of polarity. Dan Simberloff brought our attention back to the population level and Stuart Pimm reminded us not to confuse species richness with endemism. I, for one, would point to the value of also having representative ecosystems (cf. Barrett and Barrett, this volume) of different species richness so we can understand the relationships and functions at different levels of species diversity. I also hope we have not heard the end of the subject of species/area curves, as I believe we have yet to truly test their utility.

Dick Tracy reminded us that tiny populations may be functionally trivial in an ecosystem, but that shouldn't preclude concern about recovery of species in trouble to a fully participatory role. Of course, in many tropical communities where data is rare (as Nancy Knowlton pointed out) but species are abundant, most of the species in an ecosystem would be called rare and yet together are responsible for important functions. Barry Noon and Dennis Murphy—in one of a myriad of inspirations from the work of Russ Lande—provided us a way to begin thinking about generalized species/habitat models using species with different life histories, to which Ron Pulliam suggested adding different seral stages (e.g., Hansson, this volume; Peters et al., this volume). Doria Gordon gave us a stimulating perspective on how monitoring of coupled and decoupled variables can provide information for better ecosystem conservation. In one of the puzzlingly few references to the implications of climate change, Dan Simberloff reminded us how historically communities have disassembled and their species reassembled in different configurations (e.g., Peters and Darling, 1985). Our science must help us design landscapes that will permit, and not be obstacle courses for, dispersal. In the end, one has to conclude that although ecosystem conservation can subsume most species conservation, there will be situations that will require particular species conservation efforts: an obligate and continuing question in any ecosystem conservation effort should be, Are there any species requiring individual or special attention?

We, of course, had a feast of models and displayed our usual nervousness that one might, like *Imperator cylindrica*, take over from all others. Kareiva, Skelly, and Ruckelshaus talked of pseudo/spatial models and how presence/absence data may in some instances be sufficient, but Monica Turner pointed out that does not negate dependent variables, which require spatial data. Ilkka Hanski used fritillary data to show us an interesting version of the metapopulation model where a minimum of 10 patches seemed sufficient to have networkwide distribution. Nonetheless, caution was raised about simplistic adoption by the scientifically naive or ill-intended, noting among other things, possible effects of changes in the matrix. This resonated with Kathy Ralls's admonishment against using a single tool and the recurrent theme to build in a safety factor and not venture too close to a precarious minimum. John Lawton added color and information to our understanding of noise and how it relates to modeling. Theory and models cry out for testing.

Again and again—not just by Dan Simberloff—we were reminded of the importance of not becoming blind to the empirical. In Hawaii, disturbances invite exotic invasions; in Hansson's boreal forests they do not. Moshe Shachak's fascinating system of microphytic and macrophytic patches invites speculation whether after 10,000 years of human intervention, it is even possible to recognize whether exotics are in fact already present as functional parts of the system. Jack Ewel's experiments with trees and nitrogen leakage clearly showed the need to know about the actual species involved.

We were reminded a number of times to beware of time lags. Kareiva et al. found species-specific time lags on Mt. St. Helens. Kathy Ralls warned of them with respect to population viability analysis. Judy Meyer warned us not to accept today as natural or baseline, mentioning the "invisible present" (sensu Magnuson, 1990; for example, beaver in much of North America). This is an example of the recurring phenomenon of how our own ubiquitousness and interventions complicate the work of both science and conservation. Had Ron Pulliam been present for the latter part of the conference, he surely would have raised what he calls the "Tucson effect" in which each successive wave of settlers considered what they found on arrival to be both fine and baseline, while they also outnumbered previous arrivals, many of whom regretted the change.

We also discussed intensively how our work interacts with the rest of society, specifically raising questions about how much data is sufficient for decision making and how much for speaking out. I for one agree with Gretchen Long Glickman's assessment that questions should be raised sooner when the implications, as in the ozone hole or human-driven climate change, are vast. Maybe raising questions is a more comfortable way to deal with that responsibility? Certainly there is a serious question of, If scientists don't speak out, then who will?

We had a fair amount of debate about how to communicate to other segments of society mindful of Ron Pulliam's warning that we are often unaware of how often, or how, our information is "misused or used at all." Clearly we need to articulate benefits as well as risks and to show our motivation is to conserve opportunity not to create strait jackets for society (cf. Harte, this volume). Rob Colwell reminded us to be less sloppy in our language, for example, to use functional equivalency or similarity rather than redundancy. This in fact also helps us to think more clearly and avoid creating unfortunate opportunities for people looking to twist our results. In this respect, isn't it better to leave the "balance of nature" as something that has validity and meaning to the public at large rather than debate it as a scientific concept? Surely our time is better spent using our technical insights in technical ways as well as using them to improve, if and when we get the chance, the law? Mindful of Steward Pickett's mention of street meanings, don't we all clearly know what Secretary Babbitt meant when he wrote of being in equilibrium with the land? I think I subscribe to Dan's null hypothesis that science is working just fine—we have abundant evidence here that it is rich and dynamic.

The big question for most of the conference is where do people fit? Julianna Barrett showed it to us in the lower Connecticut River. Norm Christensen said we manage people not systems—actually I think we often manage systems by managing people. In any case, it is useful to think about people in the context of the ecosystem management approach that is manifestly making progress in South Florida (Harte, this volume) and southern California (Noon, McKelvey, and Murphy, this volume). If, as is clearly the case, the problems in those places arose from the aggregate of independent decision making, then clearly addressing such problems or avoiding them (Stuart Pimm's anticipatory conservation ecology: Nott and Pimm, this volume) derives from the aggregate of consultative decision making with a large voluntary component. It brings the people to the problem and the problem to the people. It makes an abstraction into reality. The principle is that an individual gives up a bit of influence over his or her particular piece of a mosaic in return for the chance to influence players with many other pieces that could affect the particular piece. What will drive what could otherwise be but a weak voluntarism? If the environmental problem is huge and undeniable, as in South Florida, it works. If it is desirable to avoid potential regulation as in southern California, it works. Nonetheless, there seems to be a real need, as pointed out by Michael Bean and others, to create positive rather than negative incentives and to harness creativity and market forces. Nonetheless, standards have to come from somewhere. Ecosystem management also gives scientists a chance to get their hands dirty with real-world conservation.

This process can help good results from getting lost, as in Kevin Roger's example. Partnerships become an imperative. There is an obvious need for metrics of success. And—this is not an obligatory bow to our hosts—adaptive manage-ment. We need to lift the ecosystem experiment from just being in the realm of science and insert this kind of science in resource management far and wide, public and private. Management as science and experiment. It has such elegant sense inherent in it. We might even find out why things happen.

I return to Joel Cohen to remind us not just of the complexity but also the scale of the challenge—to a growing human population that already uses energy equivalent to four times the energy fixed for food production (Cohen, Chapter 3, this volume). The growth of cities, with all their ills, has been a biodiversity godsend draining people from rural areas where biodiversity survives. But this is a short-term rescue and every new road constructed opens new opportunities for destruction. Every additional human requires some minimum quality of life and concommitant use of resources that cannot be denied. And the time: there is so little time.

We were reminded of the importance of restoration when the dignity, calm, and radiance of the Aldo Leopold, the man, were handed down to us by his son, Carl. May the green fire Aldo Leopold saw die in the wolf's eye be kindled in all of ours.

31

The Linkages between Ecology and Conservation Policy

Lee M. Talbot

Summary

This chapter focuses on the application of new conservation theory to the formulation and execution of environmental policy, and specifically, on the process of translating the theory into policy and management. The new paradigm in ecology renders most previous conservation theory obsolete, including much of the past basis for management of wildlife, fisheries, and protected areas. New conservation theory must be developed and translated into policy and action. Although additional knowledge is always needed, adequate ecological knowledge now exists if this knowledge is effectively used. However, such use faces strong countervailing pressures from the scientific and academic communities, among them the issues of pure versus applied science and advocacy versus passive observation. At the least, the scientists' role can involve communicating appropriate information to decision makers, managers, and the public. Such communication must be scientifically valid and should represent broad scientific consensus, particularly in view of the prevalence of disinformation in considerations of environmental issues. Effective communication needs to be carefully targeted and planned to address the specific needs and knowledge levels of the intended users. Dealing effectively with uncertainty is a central challenge. Key conservation issues are of such critical importance to human welfare that it is incumbent on ecologists to use their knowledge and information in a scientifically responsible way to further conservation goals.

Introduction

The "new paradigm" in ecology and the new conservation theory that stems from it are major themes of this book. The discussions have focused on the needs for a new paradigm and theory, their foundations and particularly their nature, that is, what are they. A major remaining issue is how to get the new conservation theory *used* and specifically, How can the new conservation theory be applied to the formulation and execution of environmental policy? The new paradigm in ecology makes most previous conservation theory obsolete and in the process shows why previous theory has had limited success. To achieve its potential improvements in conservation, however, the new ecological knowledge must be translated into policy, management, and in some cases, law. This chapter focuses on this process.

The first step is for scientists to provide the basic ecological knowledge and information that is necessary to create effective conservation theory (Ostfeld, Pickett, Shachak, and Likens, Chapter 1, this volume; Harte, this volume). This, then, must be translated into policy, that is, conservation principles, plans, or courses of action. To achieve this, the new conservation theory must be conveyed to those who make policy at the appropriate levels, local through international. In most cases, this process requires carefully targeted and planned communication by the scientists involved, along with public support and understanding. Once new conservation theory has become policy, however, there is still the need to convey it effectively to the managers who must carry out that policy, and to the educational system that produces future managers.

Conservation Theory and the New Paradigm in Ecology

The New Paradigm

In recent years there has been a major change in the understanding of ecosystems, which has been termed "the new paradigm in ecology." It should be emphasized, however, that the facts have been known by some ecologists and others for many years. What is "new" is the more widespread recognition and acceptance of this knowledge, which has come about only recently.

Formerly the dominant paradigm was that of an ecosystem that when mature was stable. It was closed, that is, not normally affected by external influences. It was regulated internally and behaved in a deterministic manner, and it existed in the absence of humans, indeed, in isolation from them. It was believed that if this stable condition were to be disturbed, the ecosystem would progress through a series of successional stages back to its original, stable, homeostatic state. In popular terminology this was the "balance of nature."

The "new paradigm" is of a much more open system, one that is in a constant state of flux, usually without long-term stability, and affected by a series of stochastic factors, many, such as global climate patterns, originating outside of the ecosystem itself. In this paradigm, an ecosystem may exhibit a variety of behaviors ranging from stability and cyclic fluctuation to unpredictability and chaos. As a result, it is probabilistic and multicausal rather than deterministic and homeostatic.

The current conceptual model recognizes that non-anthropogenic disturbances play a central role in determining ecosystem behavior, but that in addition, human disturbances often play an important, and frequently a dominant role in affecting the status of the system. It follows, then, that an ecosystem is characterized by uncertainty rather than predictability, and that it must be described as probabilistic rather than deterministic.

Application to Conservation Theory

The new paradigm in ecology has profound implications for conservation (Ewel, this volume; Hanski, this volume; Fiedler et al., this volume; Meyer, this volume). Most past and much present conservation theory is based on the old paradigm. As such it has never been wholly appropriate or, in many cases, effective. However, there are two caveats to this statement. First past conservation efforts, (e.g., the national parks movement) have achieved extremely important conservation results. And second, although the broader recognition and acceptance of the need for new conservation theory is very recent, it must be emphasized that for decades a few scientists, managers, and others concerned with conservation have recognized and, in some cases, acted on the dynamic and uncertain nature of nature.

Two broad examples illustrate the way that the old paradigm has served as the foundation for conservation theory and practice. The first involves wildlife and fisheries management and the second involves national parks and reserves.

Wildlife and Fisheries Management.

In wildlife and fisheries management the old paradigm's stable state, the "balance of nature," was the assumed stable state before hunting, fishing, or other types of management effort were started. This state was considered to be the fixed carrying capacity of the habitat for the target species, which was basic to the management concept of maximum sustainable yield (MSY). The MSY concept assumed that prior to fishing or hunting, the population levels of the target stocks were stable at the carrying capacity of the habitat, and if those levels were reduced and held at a level of roughly half the original population size, there would be a stable, maximum sustainable surplus of births over deaths available for harvest. The theory also held that if the harvest were stopped, the target population would recover back to its original stable level. We now know that real life rarely, if ever, follows this model and that the deterministic, single-species approach characterized by MSY is not appropriate (Holt and Talbot, 1978; Mangel et al., 1996).

National Parks and Protected Areas.

Underlying most conservation of areas (e.g., national parks and reserves) was the idea of a "balance of nature," which could be maintained if an adequate area was protected from human disturbance. In other words, the old paradigm led to the assumption that if an ecosystem was adequately protected, the ecosystem would persevere in a stable state and conservation would be achieved. Again, we now know that this model rarely reflects reality.

Although conservation efforts have achieved a great deal, the present status and trends of species, wild populations, and ecosystems worldwide are generally very negative (Talbot, 1994). The unprecedented increases in human populations with the consequent demands on resources and pressures on ecosystems (Cohen, Chapter 3, this volume), exacerbated by new technologies, create an imperative for much more effective conservation. Such conservation must be based on conservation theory that reflects the best current ecological understanding.

Science and Scientists

Information and Knowledge

There must be adequate scientific knowledge and information to provide the basis for developing new conservation theory. Further knowledge and information are always needed, particularly in the area of better understanding of the nature of disturbance regimes, their impacts on ecosystem dynamics, and how they affect biodiversity. However, this book demonstrates clearly that there is at present adequate ecological knowledge to develop new conservation theory, policy, and management, if this knowledge is effectively used.

The Scientists' Role

A continuing controversy exists within the scientific community over the proper role of a scientist. Should the scientist concentrate on "pure" science and avoid communication of scientific knowledge and information outside the normal scientific channels? Should the scientist focus on applications of science and become an activist on issues that are important for science? Or is there some middle ground (Harte, this volume; O'Neill and Attiwill, this volume; Zedler, this volume)?

The issues of pure versus applied science and advocacy versus passive observation have been contentious throughout the history of science. With ecologists, the problem is exacerbated by the popular use of the term "ecologist" to describe environmental activists. The predominate view at the conference from which this book derives, however, was that key conservation issues are of such critical importance to human welfare (as well as to the future of the science) that it is incumbent on ecologists to use their knowledge and information in a scientifically responsible way to further conservation goals.

At the least, such a role for scientists can involve communicating appropriate information and knowledge to decision makers and managers. The specific ways to do this vary with the level of policy and management, ranging from local through international. In general, these communications can involve testifying before legislative or other hearings; providing advice in a variety of ways; preparing written or oral comments or otherwise assisting with proposed policy, laws, and management regimes; working with legislative staffs at appropriate levels of government; collaborating and providing advice and information to nongovernmental organizations and other types of citizens' groups; and taking opportunities to share information with the public through presentations and popular articles. In addition, collaboration with other scientists, managers, and policy makers is also an essential part of the scientists' role (Pringle, this volume).

An Enabling Environment

Even if a scientist wishes to apply his or her skills to pressing environmental issues, however, there are strong countervailing pressures within the dominant culture of most universities and much of the scientific community. The system of academic and scientific rewards and incentives usually is oriented toward pure science and discourages or even penalizes its application. The problem is at least twofold: first there is a general perception that "if it is applied it must be bad science," and this is enforced by the rewards and incentives structure; and second, people who work on applied problems often either cannot get peer-reviewed papers (the currency of academic science) from the work, or they must move on before having done so. A related problem is the difficulty of obtaining funding for applied research. One result is that scientists and other academics who work on applied problems are often regarded and treated as second-class citizens.

There has been some recent progress on this issue. The field of conservation biology integrates traditional academic science into applied conservation issues, and some of the stigma of applied work is fading. Yet the problem remains a serious obstacle.

There is an increasingly urgent need to apply the best scientific/academic minds and information to the increasingly urgent resource conservation problems. It is time for the more traditional scientific and academic institutions to drop the outmoded prejudices and join the progressive institutions such as the Institute of Ecosystem Studies and George Mason University, which have fully integrated the societal needs for responsible application of good science into their structures of rewards and incentives.

The Issue of Consensus

If ecologists are to be effective in getting the new information used to develop theory and policy, there must be substantial consensus within the scientific community. However, such consensus is not the case at present and there are formidable obstacles to it. The first is that the new ecological paradigm represents a major change, and resistance to change is basic to human endeavor. One factor is an understandable and healthy scientific skepticism coupled with the scientists' need to see proof and be convinced objectively. This is basic to the scientific method. However, there is often an even more potent subjective factor. Scientists and other academics who have built their careers and reputations around the old paradigm often are fiercely resistent to change. Many have a professional and psychic investment in the status quo, and consequently they may see change as a threat or a repudiation of their own work. Although science is supposed to operate in an objective way, scientists individually may well operate quite subjectively. Consequently, achieving consensus on the new paradigm will not be easy.

Consensus is especially important in relation to new conservation theory. Conservation actions often involve conflicts between short-term, often personal, interests and long-term, often community, ones. Issues such as establishment of a park, protection of a wetland, reduction of pollution discharge, or reduction of fishing quotas to achieve sustainability all pit short-term economic interests against a longer-term public good. In such cases, those on both sides often seek to obtain scientific evidence to support their case. Although it is virtually always possible to find scientists willing to argue against a particular conservation action, the conservation side is much more likely to prevail if there is a broad consensus within the scientific community.

Disinformation

Disinformation is a related issue that represents a particularly potent threat to conservation in today's antienvironmental political climate. The new paradigm in ecology is particularly vulnerable to disinformation. Perhaps most significant is the issue of disturbance. Since ecosystems are now widely seen to be influenced by natural disturbance regimes, those who wish to counter conservation arguments (e.g., to allow additional vehicle impacts on fragile deserts, increased lumbering or hunting) now claim that ecosystems are resilient to such human-caused disturbance because it only mimics natural processes; or even, that such disturbance will be beneficial to ecological processes.

Disturbances that are caused by humans are often basically different from those that occur naturally. The area involved in anthropogenic disturbances, the timing, and the magnitude and type of disturbance and impacts throughout the system all can be fundamentally different from the natural disturbances. But the arguments to the contrary can be seductive to the public and even to scientists who do not understand the ecological complexities involved. The argument that clear-cutting only mimics natural fire disturbances is a good example of such a case where disinformation has been supported by some scientists.

A recent article in *Dirt Bike Magazine* further illustrates the problem. This article argued that nature is chaotic and, therefore, disturbances that are created by motorcycles (e.g., in the deserts of the southwest) only mimic nature.

Another dimension of the problem of disinformation derives from discussions of "functional redundancy." This terminology has been used in some scientific discourse to describe species whose functions within an ecosystem appear to be similar or even in some ways duplicative. In fact, no two species have ever been shown to be totally interchangeable ecologically. But the terminology has been picked up by those who wish to weaken the Endangered Species Act who have claimed that if species are "redundant" they are not needed, and nothing will be lost if they become extinct.

Again, although there may be some scientists who support these types of disinformation, valid information is much more likely to prevail if there is substantial consensus within the scientific community. However, consensus is important, but it is not enough. To combat disinformation and to promote conservation, scientists must be active and involved in conveying scientific principles and information to the public and the decision makers.

Standards

The problem of disinformation highlights the issue of standards in the ecological community. At least since the early 1970s the term "ecologist" has been applied to environmental advocates as well as to scientists of ecology. One result is that some of the public remains confused about what an ecologist really is. In the same period, environmental legislation has created a considerable demand for ecologists and environmental scientists. In response, these labels have been appropriated by some people who lack suitable scientific training and whose comprehension of ecology may be limited. This situation has led to increasing proposals for the establishment of standards within the ecological community, and some form of certification to identify truly qualified ecological scientists. Some efforts have been made, but success to date has been limited, and the issue remains.

The application of peer review is another part of this general set of issues. Peer review is a central part of normal scientific endeavor, either formally through the scientific publication process, or less formally through conference presentations and workshops. Much of the information used in policy debates, both with the public and decision makers, however, has not gone through such a peer-review process. A formal peer review would not be practical in most such cases, but there is still a need to ensure that the information coming from the scientific process is presented in a valid and useful way. One of the more effective ways to proceed is to develop effective partnerships or other working relationships between the scientists and others involved, especially nongovernmental organizations (NGOs) and government.

New Conservation Theory

The new paradigm in ecology renders the old conservation theories essentially obsolete and it creates the need for new theory. However, it does not, by itself, create such theory and this situation requires that new theory be developed. Part of what is needed is a new understanding of ecosystem dynamics and the role of disturbance regimes, and another part involves the actual management of dynamic systems and disturbance regimes (e.g., Peters et al., this volume). Ideally the approach to be followed would be for the ecologists involved to work with ecosystem managers and other conservation managers to develop a body of theory appropriate to the new dynamic view of ecosystem behavior, and relevant to the needs of conservation management.

Communicating the Message

If new scientific knowledge is to influence conservation policy and management, that knowledge must be communicated effectively to the decision makers. Accomplishing this requires a set of attitudes and abilities that are rarely part of scientists' training and education. The following paragraphs indicate some of the areas of knowledge that are useful in this communication endeavor.

Knowledge of how policy and law are made is a useful first step. The process is different at local, state, national, and international levels, and an understanding of how the process works and where it is necessary to insert ecological information into the process is a necessary foundation. An intellectual model of policy making is an important tool for scientists, because unless they know the process, they cannot know where and how they can make the most effective contribution. The processes by which policy and law are made should be part of the education of ecologists and other scientists concerned with the environment.

An understanding of the specific needs of the intended recipients of the information is another important fundamental. A scientific dissertation on the theory of ecosystem disturbances may be totally ineffective, or even counterproductive, when a hurried decision maker is looking for the answer to a specific narrow question.

It helps if the scientist can develop an appreciation of the level of scientific sophistication of the decision makers involved, so he or she can then make the presentation accordingly. President Ford had essentially no scientific background, whereas President Carter was a qualified physicist with an active interest in scientific matters. Consequently, presentation of scientific issues to these two presidents required wholly different approaches.

Decision makers are usually in search of quite specific information, not general background knowledge. If they are working on a law or policy, they need to know why it will work or will not work, in very specific terms. If it needs revision, they want to know how, specifically, to revise it. They are often concerned about what will be the effects or impacts of the proposed action, and again, they need very specific information. Consequently, to be most effective the scientists should seek to provide the specific type of information that is needed.

In general, decision makers must make decisions quickly and the decision that is needed often is in terms of absolutes. They need to be as correct as possible, but often there is a greater incentive for them to make any decision within a specific time limit than to make a better decision that requires a longer time. On the other hand, the training and incentive systems for scientists are nearly 180 degrees different. Most scientists think in terms of shades of gray rather than absolute black and white, and their incentive systems reward correctness in the long term rather than quick and possibly wrong decisions in the short term. However, if the scientist is to be effective in influencing or assisting the decision maker, the scientist must learn to present the message in a scientifically responsible form, but in the form and time frame that is required by the decision maker.

If scientists are unable or unwilling to present their information in the terms and form needed by the decision makers, they lose the opportunity to convey their information to the policy process and to influence it. In effect, they are handing over the opportunity to influence policy to the lawyers and politicians.

The Central Issue of Uncertainty

In the new ecological paradigm, ecosystems are described as probabilistic rather than deterministic and they are characterized by uncertainty rather than predictability. A central problem, then, is how to communicate uncertainty effectively among scientists, decision makers, and managers. One approach to the problem has been developed in conservation biology, where ecological uncertainty has been broken into three categories which affect population dynamics in different ways: *demographic stochasticity*, which involves uncorrelated (i.e., uncorrelated among individuals) variability in chance events such as birth and death; *environmental stochasticity*, involving correlated variability in chance events caused by individuals experiencing a similar environment that includes both biotic and abiotic events; and *catastrophes*, a correlated variability of large magnitude that occurs at a low frequency. When the single concept of *uncertainty* is broken down into these three categories it enables a substantially more precise approach to the problem of understanding population dynamics (and of ecosystem dynamics as well) because each has its own set of effects.

There is very general lack of appreciation of how to deal with scientific uncertainty. Consequently there is need to identify the bounds or conditions under which decisions can be made in the face of uncertainty, identify the means to deal with uncertainty (such as risk analysis), and get agreement between those involved (scientists, managers, users and decision makers) to use these means.

Management in the presence of uncertainty requires first defining the objectives of management and the time scale of the objectives; then identifying what are the managers' options, that is, what are the decisions they can take that can be varied (i.e., catch limits, restrictions on various activities, monitoring); then analyzing the available data to determine what is known and what is not.

It is essential to recognize the importance of distributions rather than single points, that is, that in the face of uncertainty it is not possible to identify a single point (e.g., the MSY), but, instead, a range or distribution of points and a consequent range of effects of subsequent management efforts. Rigorous modeling is required to define the safety or precautionary approach, which involves modeling to simulate the whole management system, applying sensitivity analysis to determine the sensitive elements in the system, and finally defining through the models the likely outcomes of various management options. In all cases, it is important to make the uncertainties explicit (Ralls and Taylor, this volume).

An integral part of this approach is to incorporate an adaptive management strategy and, in effect, to treat management as an experiment. That is, establish an initial management strategy, execute it, monitor the results along with other parameters as necessary, and adjust the management as needed to accomplish the original objective (Christensen, this volume).

In addition, an effective approach to decision making in the face of uncertainty must also involve alternatives analysis, which takes into account human factors as well as nonhuman considerations (O'Brien, this volume). There is growing recognition that conservation policy must take social and economic factors into account. Socioeconomic factors normally determine whether or not a conservation regime will be implemented, regardless of how sound it is scientifically. Consequently, any approach to conservation that does not take socioeconomic factors into account probably will not succeed.

In some cases in the past, scientists or managers hesitated to try to convey uncertainty to decision makers in the fear that they would not understand or accept it. In truth, however, many decision makers are well accustomed to dealing with uncertainty and risk assessment. These are integral elements, for example, of the economic and political judgments and decisions that they are constantly required to make. Consequently, many decision makers are probably more comfortable in the face of uncertainty than are most scientists and managers.

Getting the Message to Managers

Whereas decision makers determine policy, managers have the responsibility to implement it. Consequently, conveying the new paradigm to managers is a key element in influencing and effecting conservation action. Scientists' resistance to change was discussed in the section on the issue of consensus. Resistance to change and virtually all the other problems described there also apply to managers. As a consequence, effectively conveying the implications of the new paradigm to managers, and more, getting them to act on them, represents a very major challenge.

The challenge often is increased by the resistance of industries or agencies for whom the managers work. In most cases recognizing and incorporating uncertainty into conservation management regimes will require greater safeguards and consequently may involve lower levels of yield or immediate economic profit.

In addition to conveying the new ecological knowledge and information to present managers, it is important also to reach the educational systems that produce future managers. Here too, there is immense resistance to change, again for the same reasons of the institutions' and academics' professional and psychic investments in the status quo. For example, even though there is now widespread agreement that MSY is not an appropriate basis for fisheries management, it is still taught by almost all fisheries colleges in North America and Europe.

Conclusion

The new paradigm in ecology essentially makes obsolete most previous conservation theory. This paradigm provides the best available foundation for the practice of conservation, and scientists now have the opportunity to develop new conservation theory and to contribute to its incorporation in conservation policy and practice. In view of the pressing global challenges to our environment, however, this situation presents scientists with both a unique opportunity and a heavy responsibility to create an effective link between ecology and conservation policy.

32

The Central Scientific Challenge for Conservation Biology

John Harte

Our conference goal has been to improve the scientific basis of conservation—to forge a new research and policy agenda that reflects accurately both new ecological insights and the urgency posed by accelerating human threats to the biosphere. Worthy and exciting as is this goal, however, progress toward it is readily threatened by preoccupation with the false choices posed by risks versus benefits, species-level versus ecosystem-level analyses, practical versus intrinsic reasons for conservation, and time-dependent populations versus a balance of nature.

I wish to offer here a perspective that provides both a resolution of these putative dichotomies and, more importantly, a clear picture of how we could most effectively direct our research efforts. This perspective is illustrated in the Figure 32.1, which shows the linkages between human activity and natural ecosystems. The bidirectional nature of these linkages suggests that the critical task of designing a sustainable future is the task of identifying human activities that self-consistently preserve their own life-support system.

At the center of the picture (Fig. 32.1) is the circle labeled ecosystem goods and services. Hardly a conference on environmental concerns goes by in which the importance of healthy ecosystems to our well-being is not mentioned. A list of ecosystem services, like the one presented by Judy Meyer (this volume), is usually displayed, often accompanied with dire warnings of the end of civilization as we know it if ecosystems and the services they provide are allowed to degrade. But that is always the end of the matter. Participants agree with the generalities expressed, and then continue along on their research in population biology or biogeochemistry or whatever their specialty is. The value of protecting ecosystems remains unexpressed because it is remains unanalyzed.

I can speak with some experience about the extraordinarily effective role that a clear depiction of ecosystem services can play in steering public opinion and policy. In 1969, in my first foray from puzzle solving into problem solving, I

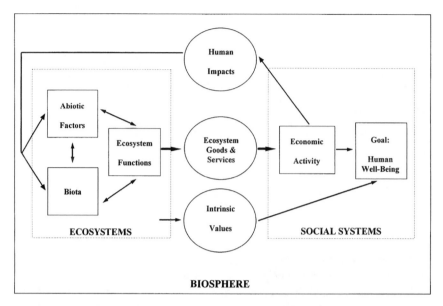

Figure 32.1. The linkages between human activity and natural ecosystems.

participated in a National Academy-sponsored study of impacts of a proposed jetport in the Everglades (Commins et al., 1970). I got interested in the possibility of salt intrusion into the aquifers that provide freshwater to residents of southwestern Florida. It appeared likely that the Big Cypress Swamp would be drained if the jetport were approved, thereby removing the head of freshwater that kept salt water from intruding into the groundwater supplies. With the aid of hydrologic maps and some robust physical calculations, we were able to show that large, undisturbed patches of swamp water were providing an invaluable service to numerous Floridians—the protection of their water supply. People could survive in a heterogeneous landscape, with ecosystem functions intact, but not in a drained, homogenized South Florida. We of course publicized the result (Harte and Socolow, 1971), and thereby helped sway the Department of Transportation to deny approval of the jetport. Today the Big Cypress Swamp is protected, but it narrowly escaped development. At the time, we called the role of the swamp water an example of a "confluence of interest between man and nature" because the phrase "ecosystem service" had not been coined (and the term "man" was still used). But the concept was the same and its political potency was firmly impressed on me.

The lesson is clear. If we do not understand for ourselves, and then educate the public about, the essential role of ecosystems in sustaining the human economy, we have not a ghost of a chance of stanching the worldwide biodiversity hemorrhage. Habitat will disappear and, with it, will go populations and species.

The full task of convincingly characterizing the value of ecosystem services will call for the collaborations of natural and social scientists of all academic backgrounds. Insights from the business community, from local governments, and informed citizens will be important as well.

The science we need to properly value ecosystem services is undoubtedly not all in hand yet, so basic as well as applied work is needed. Making the task of characterizing and valuing ecosystem services more difficult (and more interesting) are the new insights into the importance of spatial and temporal variability in ecosystems and the absence of all-purpose system boundaries (Wiens, this volume). The boundary issue, especially coupled with a growing awareness of global biogeochemical interconnectedness, raises tough questions concerning the allocation of "credit" for an ecosystem service conferred at any particular site. No ecosystem is an island, and all ecosystems everywhere influence any particular ecosystem.

If we choose to value an ecosystem service by estimating the cost of a technological replacement (for example, application of industrial pesticides to replace natural pest control), how do we incorporate temporal variability in the populations of the insect predators?

Why aren't we taking up the challenge of understanding, and informing the public of, the value of ecosystem services? One reason may be that nobody's specialty is the actual evaluation of ecosystem services—the delineation of their contribution to human well- being—and academic rewards come from working in the service of academic specialties. Collaborative research is essential, but it's much easier to put down economists as being untutored in the insights of ecology, and to insist that collaboration must await their retraining, than it is to learn the economics we would need to even begin talking their language and understanding their perspectives.

Another reason is that the shadow of intrinsic value falls over the shoulder of those who would take up the challenge. Intrinsic values are not, of course, belittled merely because practical values are illuminated, but a pervasive dichotomy has made it appear so to many. In fact, a possible amalgamation of the ethical (intrinsic values) perspective and the practical (ecosystem services) perspective emerges from the insight that we have no right to commit the act of depriving humanity, including future generations, of the practical benefits of ecological wealth. Moreover, I surely am not the only person who finds it intrinsically satisfying to know that my material sustenance is dependent on nature. And, as the everglades experience showed, the results of an analysis of ecosystem services need not be expressed in monetary units to be useful.

Returning to the Figure 32.1 and the systems perspective on sustainability it provides, we can put to rest the other unnecessary dichotomies that erupted at this conference. Take the issue of risks versus benefits. To the extent we understand ecosystem services—their dependence on healthy ecosystems and the benefits they provide—we can interconvert risk and benefit. The benefit of not draining a swamp is the risk of doing so. Absent an understanding of ecosystem benefits, of the practical or intrinsic type, the dichotomy will continue to haunt us.

Take species versus ecosystems. A narrow species focus easily leads to the argument that ecosystem function is a concept so fuzzy and disconnected from information about individual species that if our only goal is to protect functions, then we will lose species. An exaggerated ecosystem approach can lead to the naive argument that the only way to protect species is to protect stocks and flows of energy and matter. But this divergence represents an attitude problem, not a serious and scientifically defensible dichotomy. Although ecologists are generally not able to predict in detail how loss of particular species will reduce the capacity of an ecosystem to provide human society with essential ecosystem services, there is unanimity among ecologists that as biodiversity is lost, that capacity is degraded. More to the point, as our scientific knowledge of ecosystem functions accumulates, the evidence increasingly suggests that each species does play a unique role in carrying out ecological functions. For example, work on soil arthropods indicates that even nearly indistinguishable congeners carry out distinct biochemical roles in soil nutrient cycling and the maintenance of soil fertility (Faber and Verhoef, 1991). An individual species may not be observed to play a detectable role in maintaining ecosystem functions during a brief study period. But over many years, as interannual climate and other kinds of variablility are experienced, the odds that individual species play unique roles in maintaining such functions will steadily increase. The concept of a redundant species lacks scientific merit, and there is no defensible scientific basis for believing that such a concept will gain credence in the future.

The practical reality is that we cannot save ecosystem functions if species disappear and we cannot save species if ecosystem functions are degraded. We cannot even understand how to save either without understanding both. Ecology is the study of both species and functions, of both abiotic and biotic constituents and interactions. If any belief remains in the usefulness of the dichotomy, consider the fact that an overarching ecological function is the maintenance of the shifting stage on which play out the evolutionary and successional processes that sustain biodiversity.

Next, let us look at the putative dichotomy between "balance of nature" and ecological reality. I share the view of some participants here that the public's street sense of balance is a savvy one, and we ought not to kick the legs out from beneath it with current ecological dogma. This is particularly true because the putative contradiction between balance or equilibrium, on the one hand, and dynamical time-dependence, on the other hand, is not even well founded. Terms like stability, stasis, balance, and equilibrium are readily conflated. Yet no pair of those terms are synonyms, and flux is not the same as disequilibrium or instability or imbalance.

Just as cognizance of ecological heterogeneity across time and space illuminates the importance of species in maintaining ecological functions and the ecological role of ecosystem functions in sustaining human life, so it clarifies the concept of ecological balance. No population can grow without bounds, or produce effluent without bounds; ecological checks and balances, in the guise of negative feedbacks that arise as growth undermines ecological functions, inevitably operate. Similar feedbacks constrain human activities because of their effects on the ecosystems that sustain those activities, as shown in the Figure 32.1. Within this comprehensive picture of the biosphere, the balance of nature means a sustainable future, with human activity matched properly to (in balance with) the biotic and abiotic elements (Babbitt, this volume).

In much the same way that wild populations are never found in steady state, despite the operation of natural checks and balances (negative feedbacks) that prevent their unlimited growth, so a sustainable economic future for humanity does not mean a time-invariant one. Fluxes across the heterogeneous human landscape and ebbs and flows of activity over time will ensure a lively future— one quite distinct from the state of boredom often predicted by economists upset at ecological visions of no growth. Perhaps ecologists can help provide society with compelling visions of the excitement that is possible in a balanced world.

But let us return to the key issue. In only a few generations, under current trends of growth in population, in energy and other resource consumption, and in our needs for effluent sinks, we will have destroyed or seriously degraded most of the planet's life support system—its once natural habitat—and imperiled our very existence. The highest priority for conservation is the protection of habitat from anthropogenic disturbance. Achieving this, particularly in the current political climate, will require public understanding of the linkages between the left and right sides of the figure. Support for habitat protection, and for all our other priorities, will be lacking unless that understanding exists. The figure, a kind of global road map of the human dilemma, reminds us that what is critical for conservation is also critical to humanity. It seems an understatement then to assert that we should get on with the task of finding ways to value nature. Is it just possible that this will be the last conference, ever, to end with a plea that we begin the task?

33

Toward a Comprehensive Conservation Theory

Steward T. A. Pickett, Moshe Shachak,
Richard S. Ostfeld, and Gene E. Likens

Summary

This chapter summarizes a framework to relate the theoretical, conceptual, and empirical base of ecological science to the practice of conservation. The abstractions and idealizations embodied in paradigms and theories have a great deal to do with how ecology is used in conservation, restoration, and management. However, these abstractions must be translated in various ways to the concrete systems that society seeks to conserve. We show how the various contributors to this book fill in different parts of the framework, and how diverse approaches to science come together in this framework. Although the chapters in this book go a long way toward fleshing out the framework, we draw attention to theoretical, empirical, and practical areas in need of further development and unification.

Introduction

The scope of concern in this book has been with the scientific basis of conservation practice. We view both the science and the practice in their largest senses. Ecology ranges from genetics to landscapes, and from organisms to biogeochemistry. Likewise, conservation is a broad topic, comprising protection, management, and restoration. Both these topics invite consideration of communication. Hence a powerful subtext in the book, sometimes rising to the surface, has been the concern with communication between scientists with different concerns and approaches, among scientists, managers, and policy makers, and between professionals and the public. The breadth of concerns encompassed in this book is thus large, and requires a framework for effective analysis. It is the goal of this chapter to provide a single framework in which all the contributions to this book can be viewed. We hope that the framework also points the way toward unification in a science and a practice each of which is diverse and multifaceted. The principal focus is on the science itself, however, since we believe that getting the scientific house in order is required for most effective interaction with managers. We have tried to include sufficient attention to management and input from people engaged in and familiar with management to ensure that the tidy ecological house is hospitable to managers' jobs and concerns.

The title of the book lists biodiversity, ecosystem function, and heterogeneity as the specific subjects within the science that are of concern. This choice has been borne out by the analyses and discussions presented by various authors. They turn out to be universal aspects of ecological systems. The elements, whether they are organisms, genes, communities, or habitats, are addressed in the concern for biological diversity. All systems contain such elements. The enumeration and understanding of the richness of those elements is a fundamental task of ecology and evolutionary biology. Because such elements are the building blocks of natural and managed systems, they also become a fundamental concern for conservation.

Ecosystem function is another of the universals of ecology. The elements of biological diversity are sustained and limited by the webs of trophic and behavioral interactions that take place within ecosystems. The fluxes of energy and materials are intimately interconnected with the kinds and performance of species and populations present in an area.

The third universal is the spatial arena in which the elements of biodiversity exist and interact, and the function of ecosystems takes place. Rather than simply labeling this component as "space," we chose a richer term—heterogeneity. By doing so, we hoped to emphasize that almost never is the spatial arena for organisms and fluxes uniform. Ecology has learned to look for gradients, patchworks, and graded patchworks rather than to expect homogeneous domains.

These three subjects are also the phenomena that conservation practitioners are concerned with. Heterogeneity turns up in the concern with protecting and restoring land. Ecosystem function turns up in the concern for maintaining resource fluxes on which humans and other creatures depend. Biological diversity is the target of concern for maintaining resources, providing important services, and for moral reasons. But all three of these individual targets are linked in pairs and as a whole. Throughout this book, examples have appeared showing how these three phenomena are crucially linked. Biodiversity depends on the degree and temporal pattern of spatial heterogeneity in an area. Biodiversity also governs ecosystem function, and generates or obliterates certain kinds of spatial heterogeneity. Ecosystem function is the flow of resources that change through time and clump or diffuse through space to determine how many and which kinds of organisms can coexist in an area. Ecologists, via training or penchant, may focus on only one, or perhaps two of the three facets of ecological reality. However, it is the linked influences that ultimately are the subject of the entire discipline, and are the target of its most effective application.

A Framework for Integration

We invited the contributors to this volume to explore how their biggest assumptions or their theoretical approaches affected the way their science spoke to the practice of conservation and management. They met this challenge in many different ways. However, when the entire collection is viewed together, a framework does emerge. We first present the entire framework, and then summarize some of the key points made by each chapter in light of the framework.

The framework has a hierarchical structure, moving from the most abstract to the most concrete. Managers and scientists (who are sometimes the same people) are both affected by paradigms. Basically, a paradigm is a worldview of a subject. More specifically, it is the collection of background assumptions, rarely stated or examined, that underwrite a discipline or an approach. Paradigms also include the methodological approach of a discipline, such as modes of argumentation and analysis, the modes of testing theories and hypotheses, and the ways that the knowledge is presented for application, if at all. Hence, habits of mind also contribute to paradigms (Margolis, 1993).

Paradigms have different degrees, of course. Modern science has a very large paradigm, which includes the paradigm of modern evolutionary biology and ecology. Within ecology there are more specific paradigms that govern the different subject areas, such as biogeochemistry, population ecology, and so on. So the concept of paradigm is complex, hierarchical, and multidimensional. Unfortunately, paradigms by their very nature are often invisible in the day-to-day practice and application of a science. Several contributors have noted that these shadowy but important collections of background assumptions and exemplars for problem selection and solution have a subtle but significant impact on the use of ecology in conservation (Wiens[1]; Fiedler, et al.). A scientific paradigm can be constrained in its influence by still larger cultural assumptions and practices (Rogers). The only way to see a paradigm is to stand outside of it.

The framework next exposes the narrower and more explicit foundations for application. These are the theories of the discipline, which include both conceptual and abstract components, and empirical facts and generalizations. Theories are thus "conceptual constructs" that address a specified scale, scope, and domain, and that absorb accepted facts, generate explanatory and predictive models, and spin off specific testable hypotheses (Pickett et al., 1994). The most complete theories will have an explicit hierarchical structure that connects the abstractions to the concrete systems they address. Such translation from the abstract to the concrete is via models with increasingly operational parameters, and via experiments that are themselves a sort of model requiring operationalization of the concepts contained in the hypotheses to be tested.

---

[1]Because we extensively cite the contributors to this volume throughout this chapter, we avoid repetition and refer to those contributions by name of the author(s) only.

So far, we are still embedded in the realm of science. Paradigms, theories, and abstract models lead to testable hypotheses, which generate data, which feed back to the models and theory, leading to their rejection or improvement (Fig. 33.1). A cycle of test and reflection is embedded within a paradigm. But to apply a science, a different pathway needs to be invoked. This is the pathway or cycle of management. The first requirement for effective management is the generation of tools to translate the abstract models and theories generated within science to the practical problems that concern managers. This effort may require a different operationalization than that used within science, and use different variables. Rogers (this volume) notes that managers have a limited number of specific tools, and so the operationalization has to take that very real constraint into account.

In management, just as in science, there is a cycle connecting conceptual tools and their translation through manipulations and definition of management

PARADIGM

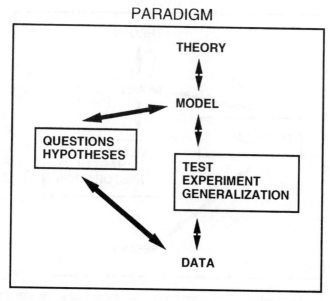

Figure 33.1. A generalized relationship between the conceptual and empirical aspects of science, showing key feedbacks among them. The background assumptions (problem identification, methodological approach, and explanatory strategy, for example) are embraced by the paradigm of the science. Although the paradigm may be vague and poorly articulated, science also employs theories that are relatively general conceptual structures referring to a certain empirical content, and including abstract models, among other components. Models contributing to theory also generate questions and specific hypotheses. Hypotheses and questions lead to tests through experiment, observation, and generalization. Data results from the dialogue between theory and nature that constitute testing. Data are used to refine or generate new questions, models, and in some cases, entire theories. Note that paradigms are only exposed by comparison to other alternative paradigms.

problems, and joined by management results (Fig. 33.2). Thus both science and management employ the same basic strategy of a cycle of action and reflection. There is an expected or desired condition, the theory or general model, which is translated into practice (manipulation for managers and experiment or test for scientists), and evaluated through data or results, which can lead to refinement of the theory or management regime or continuation of a successful management strategy in a relatively unchanging social and environmental setting. This basic similarity was recognized by Holling (1978) in his invention of the concept of adaptive management. That strategy suggests that managers employ an array of management techniques as experimental treatments. With careful description and measurement of the treatment conditions, and monitoring of the results, the success of the different management techniques can be assessed. Management

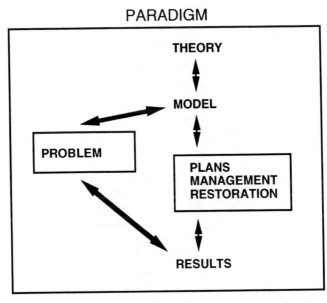

Figure 33.2. A generalized relationship between the conceptual and practical aspects of management, showing key feedbacks among them. The background assumptions (culture, economics, social structure) behind the management problem constitute the paradigm that encompasses the activity. There may be an explicit theory, based on generalizations from prior experience or from natural or social science, that is used to generate models relevant to the particular case. The models are informed by practical problems, and guided by those problems to generate specific plans, management tactics, or restoration methods. The application of plans, methods, and restoration produces results that can be judged against the model and the initial problem. Thus, as in science, there is a cycle of problem generation, action, and evaluation, although the names of the processes are different from the equivalent activities in the realm of science. There is a multifaceted dialogue among goal, conception, action, and result.

can thus evolve to accommodate unexpected interactions or changing conditions. These recommendations for management can be promoted by the partnership between managers and ecologists that was unanimously called for in the Second Cary Conference (Likens, 1989). Joint definition of problems and questions and mutual support are key aspects of the "new partnership" called for by that 1987 Cary Conference.

The consideration of the two cycles, one for science and one for management, and the embedding of both in a paradigmatic and theoretical context suggests the shape of the framework for synthesizing the contributions to this book (Fig 33.3). A paradigm is reflected in an explicit body of general theory. Such a generalized conceptual construct contains general models and embodies empirical information. Translating the theory to specific situations is driven in science by questions, and in management by problems. The translation of these questions or problems requires focused, operational models and hypotheses, and suggests the use of experiments, simulations, and comparisons in science, and of burning, cutting, substrate modification, fertilization, hunting, and enrichment in management. It is this joint scheme that enables us to seek the commonality in the diverse contributions in this book. In the summaries that follow, we integrate the various contributions in this book by mapping their major points and perspectives onto the cycles depicted in Figure 33.3.

Filling the Framework

We began the book with statements by three people intimately involved in the application of scientific knowledge to conservation. Glickman emphasized the need for scientists to communicate effectively to policy makers and managers. She was especially concerned that scientists counter misinformation and avoid withholding their information until the knowledge base was complete. Such withholding will often result in problems being dealt with using other, less scientifically informed input.

Pulliam echoed these concerns, but explicitly recognized the expanding scope of conservation science and the relevance of the modern ecological paradigm for conservation practice. In communicating with managers, ecologists need to recognize concrete goals that managers have and focus on the manipulations that managers can actually accomplish. He further emphasized the cyclic nature of the management process, and hence the importance of the adaptive management strategy.

Continuing the emphasis on the need for communication between managers and scientists, Bean noted that clear scientific conceptualization of ecosystem management was needed, and that there was a missing link between models and methods of manipulation in the case of the Endangered Species Act. Bean pointed out that legal incentives impinge on the management tools available and used

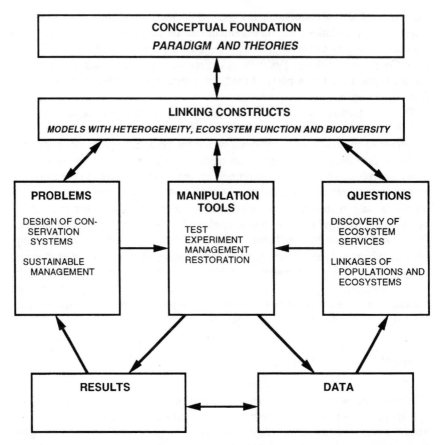

Figure 33.3. Combination of the parallel cycles of science and management. Because both science and management involve cycles comparing desire (in management) or expectation (in science) with the world as it exists or is manipulated, the opportunity to combine the two cycles exists when the topic of action or enquiry is the same. Such is the case in conservation. The joint scheme suggests that the two realms can be joined by sharing a paradigm (or substantial noncontradictory portions thereof), a theory that embraces such key concepts as ecosystem function, biological richness, and the spatially explicit relations within and between the two, and models that operationalize the theoretical generalizations and abstractions. When the manipulations of science and the manipulations of management address the same object, and the cycles between goal and expectation, on the one hand, and result and data, on the other, are closed, adaptive ecosystem management results.

by managers, so that whether and how well the feedback of adaptive management is used is often a legal matter. These three linked statements by Glickman, Pulliam, and Bean set the policy stage for the evaluation and articulation of the ecological basis for conservation.

Another social context impinges on conservation practice. Cohen (Chapter 3) pointed out a powerful and pervasive set of interactions among human population growth, economics, and culture, which impinge on the paradigm and theoretical constructs within which ecologists operate. Perhaps his most powerful message was that the conceptual constructs about human population are woefully and dangerously incomplete. Demographers operate in a paradigmatic world without constraint, and economists likewise operate in a conceptual world that does not deal effectively with the ways that humans are integrated into ecological systems. He proposed new ways of looking at pattern in the spatial distribution of human population on contrasting scales as a way to explore that integration. Economics and demography have no theory of scale, and they do not connect with the environment on which human population depends. He therefore recognized a master problem for the application of ecology to conservation: what is the shape of the relationships among human population, economics, and culture? Ecologists cannot afford to ignore these open questions and the need to interact with demographers and economists to answer them.

Noon, McKelvey, and Murphy illustrated a theoretical shift from the focus on one species at a time to the consideration of several species. They used population viability analysis and metapopulation models and the concept of umbrella species as ways to translate between theory and practical problems of fragmentation. The manipulations they addressed are corridor function and habitat manipulation, given the focus on umbrella species.

The relationship of paradigms and theories to the interaction of scientists and managers was explored by Rogers. These two groups often operate rather independently, and the cultural constraints on the paradigms of each group can be powerful. Largely missing is a conceptual and professional integration that applies the abstractions of science to the practicality of management. Rogers called for greater attention to the technology transfer and provision of professional rewards for building bridges between scientists and managers. He also emphasized the basically similar cyclic ways that both scientists and managers work and promoted adaptive management. Recognizing spatial heterogeneity at various nested scales proved a powerful tool to integrate the views and work of managers and scientists in management of the rivers in the Kruger National Park, South Africa. Discovering what scales and pragmatic handles managers could actually manipulate to maintain ecosystem function and biodiversity enabled better integration of scientific research with management.

The previous set of chapters outlined the needs for scientific growth and improved application of science, and the constraints on such application. These chapters point to the opportunity for enhancing the ecological foundation of conservation practice and point the way to the better communication that will be required. We turn now to the shape a comprehensive conservation theory might take.

Fiedler et al. outlined the paradigm that has emerged in ecology over the last several decades. This paradigm shift represents an evolutionary rather than a revolutionary change, and still exists within the master ecological paradigm that accepts the physiological, evolutionary, and historical limits inherent in ecological systems (Pickett and Ostfeld, 1995; Pickett, Parker, and Fiedler, 1992). Importantly, the new paradigm does not excuse human excesses that exceed those limits in ecological systems. Fiedler et al. indicated that the new paradigm affects the units of conservation and explained some of the split between ecology and management. There is a missing link between the knowledge of the new paradigm and actual practice of conservation and management in many cases.

Wiens recognized that the new paradigm can be considered to be an expansion of the classical ecological worldview and that it does indeed influence conservation practice. The new paradigm views equilibrium dynamics, for example, as a special case of the expanded suite of dynamics. Following the paradigm shift, he noted that static patch models are being replaced and complemented by dynamic patch models in which the matrix is a dynamic element as well. He cautioned that heterogeneity, although always important, will not always be best cast as patchiness. There is a need to develop tools that relate patchiness and population processes and dynamics, as well as tools to visualize these dynamics, so that this information can be translated more effectively to managers. Wiens's chapter pointed out an important research agenda in ecology to supply better information on patch-matrix-species interactions. It is these same tools that would assist the efforts of Noon et al. as well.

Shachak and Pickett presented an example of the role of heterogeneity in managing and restoring a desertified area. It is this example, in which patchiness plays a critical role in communicating with managers and in supplying manipulation tools, that emerged as the outline of the framework used in this chapter.

The next cluster of chapters examined the linkages between biodiversity and other ecological phenomena and attempted to refine the way that conservation biologists look at biodiversity. There is some tension in this group of chapters because sometimes the view is inward, toward biodiversity, and sometimes it is more focused on the connections of biodiversity to its functional context. However, a clear call for a unified scientific understanding of biological diversity in its spatial and functional context emerges.

Nott and Pimm noted the two dominant paradigms within ecology—the ecosystem and the organismal. They cautioned that it is crucial to avoid taking biodiversity as a simple number in erecting a goal for conservation. The bulk of their chapter is an insightful analysis of how an improved assessment of biodiversity can better frame the problems conservation biologists attack. By focusing on areas of endemism, the scarce resources of conservation can be better employed. They suggested that ecosystem function alone cannot index such areas.

The modern paradigm was filled out by Meyer to explicitly include ecosystem principles. The ecosystem and species foci can be unified in a single paradigm. She pointed out that the conceptual tools (e.g., models and research designs) that translate the general theory and the overarching paradigm to practice require a variety of metrics rather than a single one. Functional models of systems are required to assist evaluation of conservation needs and strategies. Such functional models will, like those of Shachak and Pickett, include fluxes and organismal components. Ensuring that important ecological functions are the result of a suite of species can be important insurance against the loss or decline of certain species. Literal equivalence of species is not implied by such a suggestion. Spatial heterogeneity is an important ingredient in the functioning of ecological systems.

Hansson examined the theoretical relationship between patchiness and community structure. He translated this to practical application by evaluating the functional versus structural patchiness and focused on the role of generalists versus specialist species in structuring different systems. This focus led to recommendations for conserving large units of landscape in which the appropriate species strategies can be maintained and suggested important ecological research questions about the role of generalists and specialists in different systems with known or measurable attributes. Hansson thus presented a refinement to biodiversity similar in intent to that of Nott and Pimm.

Kareiva et al. suggested that there already is sufficient theory for supporting conservation. We interpret this to apply to certain aspects of the population paradigm in ecology. They pointed out the importance of the measurement and assessment tools for conservation matching the quality and nature of the data. By accepting the problems as given by conservation, they suggested that presence-absence data will be sufficient and in fact more trustworthy for many analyses. They contended that spatially explicit models require more data of high quality than are usually available, and therefore, such models should be avoided for conservation application. The problems Karieva et al. responded to are fragmentation, data quality, and complexity of analyses. They implied that the results that managers should be concerned with are presence versus absence. Likewise scientists who are concerned with conservation should more widely use presence-absence data and look for lagged habitat responses. A change in the kinds of conceptual tools (such as models, measurement protocols, and data syntheses) employed by scientists is also suggested by this analysis.

Ecosystem management responds critically to models that embrace diversity and function according to Christensen. Concrete models are needed to address the problem of maintaining diversity and function. The principal tool used will be a manipulation of disturbance and disturbance regimes, but the goal is not the mere existence of disturbance (or a particular agent of disturbance) per se.

Scientific questions guiding conservation research should take more effectively into account what managers need to know about system function. Christensen advocated the adaptive management loop as guided by fundamental science. Thus, the parallelism between science and management was emphasized in concept and in practice, and should be exploited by communities consisting of both scientists and managers. This parallelism should ease the consummation of the partnership (Likens, 1989) required of ecologists and managers.

The divergent paradigms within ecology sometimes create conflicts that can be counterproductive. Tartowski et al. made an urgent call to suppress the combative tone if not motive, and to unify the paradigms that focus on energy and material flux, and on organisms, respectively, to construct new conceptual tools for conservation. It is impossible to distinguish between population and ecosystem problems in the real world because they share space, resources, and interactions. This need suggests a unified research and management agenda.

Carl Leopold, in presenting a personal view of his father's land ethic (Leopold, 1949), pointed out a way to conceive the problems of conservation that differs from the other ways presented in this book. Although it shares some concerns with Cohen's (Chapter 3) expanded cultural, economic, and ecological view of human population and hence conservation, Leopold framed the issue in ethical terms. Overlapping the ethical issues with the cultural and economic ones, and with the human demographic issues, suggests a compelling need to unify the science and carry it into the public sphere.

The authors in the next group of chapters were charged with analyzing how knowledge, approaches, and limitations of an area of science might shape the scientific basis of conservation in the future.

Holsinger and Vitt pointed out that integrating genetic and ecological criteria represents an advance in conservation biology. Such a general theoretical linkage translates into better models for relating rarity, genetic diversity, and adaptability. Loss of genetic diversity most often emerges as a symptom of endangerment in such an expanded view. The problems then become those of rarity and impending extinction, and fragmentation of widespread species as a threat to their genetic health. Manipulation tools, such as captive breeding and maintenance of genetic stock, are meant to complement field efforts only. Genetic parameters can be used to index other threats to biodiversity.

Hanski indicated how the new paradigm has informed a new way to look at metapopulation theory. Choosing the right level of complexity in metapopulation models is a key task that will help application to such problems as assessing trajectories toward extinction and the trajectory of a metapopulation of concern. The basic ecological research question suggested by Hanski's analysis is "When is a metapopulation's trajectory toward extinction?"

Ralls and Taylor presented an even-handed analysis of the uses and limits of population viability analysis (PVA). Catastrophic factors have been omitted, and the new paradigm suggests that incorporating stochasticity and uncertainty into

PVA is a key need. PVA, with such improvements, can serve as a useful conceptual tool translating between theory and problems. PVA needs to become more mechanistic to determine whether flux or intrinsic terms impinge on survival and fecundity. Organismal and biogeochemical mechanisms should both be accounted for. This analysis was driven by practical problems and by fundamental ecological questions about the nature of population regulation, the testing of hypotheses suggested by specific PVAs, and the use of PVA to integrate large amounts of data. The testing of PVA closes the scientific loop on a practical problem, and the data should be useful for application.

Barrett and Barrett addressed the structure and function of ecosystems and noted in detail the ways in which the paradigm influences how conservation problems are approached. They sought models to integrate ecosystem function, heterogeneity, and biodiversity in the design of a distributed reserve centered on the Connecticut River. They emphasized processes and community structure and erected a compelling structure in which to consider how scientists, policy makers, and managers reciprocally influence one another. Their analysis was driven by a practical problem of reserve design and ecosystem integrity and was colored by the failure of the classical "fence-building" approach to conservation.

Ewel discussed the theoretical translation of the ecosystem concept to conservation problems. He refined the problem in terms of protecting the structure of life forms in ecosystems and maintaining heterogeneity by preserving species. Given this way of viewing the problem, he noted that heterogeneity influences ecosystem processes, and biologically induced heterogeneity can be a key determinant of community structure. He cautioned that creating heterogeneity can have negative results in certain environmental contexts. Ecologists and managers should communicate to assess how ecosystem processes impinge on life form structure and to ask how a management manipulation tool impinges on the life form structure and the ecosystem function of a system. The tools must be evaluated for untoward effects and be assessed according to their impact on ecological goods and services. Closing the loop in management by adding monitoring and measurement will help with such evaluations. Making communities and ecosystems less susceptible to disturbance is a growing issue of conservation under global environmental change and global transport of organisms.

Gordon et al. connected disturbance scale with scale of system organization and species responses. Both temporal and spatial scales are critical to successful matching of management with conservation goals. Tools must be developed to normalize the responses since not all scales can be manipulated at once or independently. They noted that discovering what to monitor to assess the success of management involves choosing scale properly. They closed the management loop based on results affecting species and system trajectories.

The biogeography of conservation is perhaps the oldest concern of modern conservation biology. In the theoretical realm, Simberloff noted that island biogeography theory has spawned other theories, such as metapopulation and source-sink theory. Nonequilibrium ideas have been important in the growth of conservation biology, and there has been a shift in paradigm. There needs to be a new paradigm on the community level as well. There is a difficulty in translating island biogeography theory and its derivatives into usable tools for conservation. Thus, specific models must be created, and metapopulation theory has greater potential in conservation than island biogeography theory. The rules for reserve design are desired translations to manipulations of the real world.

A key evolutionary perspective was supplied by Thompson. He emphasized the relationship between spatial structure and evolutionary processes. The distribution of interactions in space is an element of biodiversity and should be translated to a target in conservation. He identified new problems for conservationists to be aware of: (1) the loss of interaction biodiversity; (2) the loss of evolutionary potential, which is as much a product of contemporary ecological time as some long time run; and (3) the loss of physical structure and arrangements through which evolutionary interactions play out. These problems inform fundamental questions for ecological and evolutionary research and suggest a needed integration between the two realms.

Having an outline of the emerging conservation theory supplied by the foregoing, the book turned explicitly to application. Of course, conservation ecology is preeminently a problem-oriented discipline, and many chapters have already pointed toward or even deeply analyzed opportunities and limitations in application. This section focuses explicitly on these concerns and raises new issues.

Possingham discussed a decision tool to link conservation theory and practice. It explicitly assumes the new paradigm and employs Markov models as the quantitative decision tool. Management options are dependent on the current state of the systems. The problem was cast in terms of metapopulations, and the manipulations are the choices between the option of creating new patches versus managing within patches. This echoed the contrast in Noon et al. and therefore suggests a general management dilemma.

Pringle calls attention to the alarming degradation of freshwater systems in Central and Eastern Europe, as well as the former Soviet Union, resulting from a traditional lack of legislative protection. She discusses the political context of U.S. freshwater environmental protection, such as the Clean Water Act of 1972, and revisions approved in the U.S. Congress in 1995 that threaten the act's effectiveness. She advocates close cooperation between the scientific community and organizations involved in freshwater conservation, providing case studies from the U.S., Costa Rica, and Puerto Rico. The case studies presented demonstrate the effectiveness of action at the interface between scientists, managers, and educators.

Peters et al. were motivated by the constraints our judicial system places on the application of new scientific insights by federal agencies. This constraint, coupled with the inertia common to all organizations, suggests that explicit legal standards may be the best way to ensure that modern scientific insights come to bear on planning and management of federal lands. They embraced the new paradigm and used the conceptual tool of minimum dynamic area to extract management standards. The adaptive management loop and the inclusion of sound and rigorous scientific methods were embedded in the standards. Although limitations of time and money may be a problem, these ideal steps should be followed closely. Deviation should only be based on open discussion and involvement of the broader scientific community.

The analysis of environmental problems has a paradigm all its own. O'Brien suggested that a new paradigm for framing environmental questions should be employed so that alternative courses of action are considered, not just risks of one or a few such courses. She suggested that environmental analysis is too confined by the traditional method of risk analysis, and hence the problems are often cast poorly.

Zedler addressed the personal costs and responsibilities involved in assisting managers and the public in making informed environmental decisions. She admitted that there are both costs and benefits to effective communication in the conservation sphere, but suggested that being scientific and an advocate are compatible. The advocacy (from the Latin, "to speak to") must be confined to a scientific basis, and the speed of public discourse requires that scientists speak on the basis of the best knowledge available at the time.

O'Neill and Attiwill exposed some of the problems in communicating in a discourse where interest groups and rival camps may seek to discredit the valid input of science. Their experience should be useful in alerting other scientists to the nature of problems that can be encountered in speaking out.

Lovejoy added his voice to those calling for scientists to be more forthcoming in the public discourse and decision-making process that affects conservation and the use of scientific information.

Talbot focused on the need for ecologists to communicate their understanding of nature effectively, and to influence policy. Recognition of the nature of the audience is a key to success. The message should embody the new paradigm, and should better explain and use disturbance regimes in management of ecosystem processes and biological diversity. Although the fundamental scientific knowledge of disturbance regimes in particular areas should be increased, what knowledge we do have should be applied now, and the uncertainty of the natural world clearly explained to the public and policy makers. Major efforts must be made to fight disinformation on such topics as excusing excessive human disturbance, the error of assuming interchangability of species, and confusing scientific ecologists with environmentalists. Ecologists must develop skills in communication, adaptive management, and alternatives assessment.

Harte cautioned ecologists about the problem of imprecise and metaphorical use of terms that may to us appear technical, but to people in other specialties may appear loaded, and to the public may be accidentally or intentionally misinterpreted. He called forcefully for ecology to make a priority of research that specifies what ecosystem goods and services are. It is critical for human well-being to be better articulated as the guiding problem to be solved by conservation and management. This is one clear agenda for ecology and conservation biology.

Because the focus of this book is the scientific basis of conservation, the arena of public values becomes a concern. We note here some important findings about public perceptions of biodiversity and other targets of conservation that bear on the need and nature of communication so often mentioned in this book. The linkage of ecology to the every day concerns of people is highlighted by a survey done by the Consultative Group on Biological Diversity. That group engaged a professional survey firm to assess the views of voters who were neither committed to nor hostile toward the environment. The results are sobering. This politically engaged but moderate public had little deep concern with environmental issues. Introducing other concerns, say crime or jobs, easily dislodged environmental concerns from their pride of place. The term "biodiversity," which we suspect scientists view as a triumph of public relations savvy, seems to the public as a divisive and somehow negative idea. The public is much more engaged by and positive toward the ecosystem or habitat concepts, which they read to indicate connectedness. The dynamics of the natural world are essentially all seen as equivalent. Contemporary human-caused extinctions are confused with the natural extinctions of the past. Extinctions are seen as something that have always happened, and the crisis resulting from the difference between the unusually high rate of contemporary anthropogenic extinction and the concommitant much reduced chance that evolution can operate to fill the voids seems to be missed by the public. And finally, reinforcing Harte's perspective, people do not see what species richness or degradation of ecosystem function have to do with them. This reality should help inform how we cast our research, how we as a discipline decide on at least some of our research effort, and how we communicate our science with the public. Conservation ecology is driven by a practical issue, and that issue has a public face and value. Making that link is a crucial job of the science underwriting conservation.

The Future

What is the future of conservation science? The framework has been hung with many needs for conceptual clarification, model construction, translation from abstract to concrete, and incorporation of paradigms well outside those of ecology. There is a rich agenda that emerges, within our science; among our science, managers, and policy makers; and among vastly different disciplines. There is much to do. But all of this activity addresses what sort of future we want our science to be prepared to help shape. If we are committed to an ethical, or to an integrated social-economic-ecological perspective, then the research will take a certain shape, and integration is a key strategy. If we want the public to be well informed as it makes both incremental decisions in local communities, and grand choices affecting environmental change on planet Earth, then the linkage between heterogeneity, biological richness, and ecosystem function needs to be translated into human terms. We must increase our patience in conversing with the public and increase our effort in seeking scientific perspectives that mesh with their concerns.

We give the last word to Joel Cohen (Epilogue), to invite us all to consider the nature of the future that humankind is preparing. Science can help evaluate various scenarios, and build, maintain, or restore the machinery of nature. Will science rise to the challenge? And what will that future be?

Epilogue

A Vision of the Future

Joel E. Cohen[1]

If conservationists, together with demographers, economists, earth scientists, anthropologists and politicians, could put forward a positive and persuasive vision of the future, they could lead billions of nonscientists to look to conservationists as helpful allies in their search for better lives. They could also give direction and meaning to the daily research that occupies many scientists. Without a positive vision of the future, conservationists are doomed to fight, and probably lose, a series of rearguard actions. So I shall propose an outline of a vision of the coming century—a century that is sure to be very exciting.

A century from now, humankind will live in a managed—or mismanaged—global garden. Most people will live in cities, surrounded by large, thinly populated zones for nature, agriculture, and silviculture. Worldwide, between 100 and 1,000 cities of 5 million to 25 million people each will serve their inhabitants' wants for food, water, energy, waste removal, political autonomy, and cultural and natural amenities. Some cities will serve people who want to live only with people ethnically and culturally like themselves; others will serve people who are attracted by ethnic and cultural diversity. Different cities will gain shifting reputations as being favorable for young people, childrearing, working, or retirement. The efficiency and quality of services that cities provide will depend on the quality of their managements and on the behavioral skills (including manners) of their populations.

As feudal rights and obligations to labor were replaced by individual and collective markets in labor, other present rights and obligations will increasingly be replaced by markets. For example, there will be a worldwide market in permits for permanent residence in cities. The prices of these permits may be tacked on real estate or rental prices, as city managements compete to command market rewards for the public goods that they are able to provide. Countries that insist on a person's right to leave his or her country of birth will have to decide if that implies a person's right to go somewhere else.

Women around the world will demand and receive education and jobs comparable to men's education and jobs. Better education and jobs will accompany increased autonomy for women in the family, economy, and society. This autonomy will lead to increased power in all these domains, from the sexual and psychological to the political. Partly as a consequence of women having attractive alternatives to childbearing and childrearing, the number of children that women bear in a lifetime will decline globally to the replacement level or below. As childbearing will occupy a falling fraction of most women's lengthening lives, women will intensify their demands for other meaningful roles.

Although global human population growth will have ended, some regions will be net exporters of people, whereas others will be net importers. Rising pressures for migration from poorer to richer countries will strain traditionally xenophobic countries like Germany and Japan, as well as traditionally receptive countries like the United States, Australia, and Argentina. Migrations will bring culturally diverse populations into increasing contact. The result will be many frictions as humans learn manners and tolerance. Intermarriages will make a kaleidoscope of human skin colors.

The elderly fraction of the population will increase greatly, and the absolute numbers of elderly still more dramatically. Among the elderly, women will outnumber men by as much as two to one. New social arrangements among the elderly will arise.

The continental shelf, especially off Asia, will be developed to provide food, energy, and perhaps living space. Oceanic food sources will be largely domesticated. The capture of any remaining wild marine animals will be managed like deer hunting now.

The tropical forests that will have survived the onslaught of rapid population growth and economic exploitation between 1950 and 2050 will be preserved as educational and touristic curiosities, like the immensely popular John Muir Woods north of San Francisco. Many forests will be meticulously managed for fiber, food, pharmaceuticals, and fun (that is, recreational exploration). Today's simplified agricultural ecosystems will be replaced by synthetic ecosystems of high complexity. Biological controls and farmer intelligence will maximize yields while nearly eliminating biocidal inputs like today's pesticides and herbicides. Required agricultural inputs of energy and nutrients will be derived from human, animal, and industrial wastes rather than from today's fertilizers and fossil fuels. Unwanted effluents like eroded soil or pesticides and fertilizers in runoff will be eliminated or converted to productive inputs for industrial and urban use.

The atmosphere will also be managed. Rights to add carbon dioxide, methane, and other climatically significant trace gases and particles to the atmosphere will be traded in global markets for ecosystem services. Governments will by then have recognized the potential of atmospheric and many other ecosystem services to generate taxes in support of other common goods. The production of gases will be manipulated as part of food production and wildlife management. For example, lake managers will control the ratio of plankton-eating fish to piscivorous fish to regulate the level of primary productivity in lakes. Genetic engineering of bacteria and farming practices will be used to manipulate agricultural methane production.

People will revalue living nature because they will realize that they do not know how to make old forests, coral reefs, and the diversity of living forms grow as fast as they can multiply machines. Conservation movements will gain renewed force. People will increasingly value nature for its genetic resources and for its aesthetic amenities.

To make intensive management of continents, oceans, and the atmosphere possible and effective will require massive improvements in data collection and analysis and especially in our concepts.

A century hence, we will live on a wired earth. Like the weather stations on land and the observational satellites that now monitor the atmosphere, the oceans of the next century will have a three-dimensional lattice of sensing stations at all depths. The crust of the Earth will receive the same comprehensive monitoring now devoted to weather. Earth, air, and sea will be continuously sensed and their interactions modeled to predict major events such as El Niños, hurricanes, earthquakes, volcanos, megaplumes of hot water from oceanic vents, shifts in major ocean currents, and climatic fluctuations.

Millionfold improvements or more in computing power over the next century will improve the resolution and scope of global models. These models will integrate the atmosphere, crust, and oceans; geochemical and geothermal flows; human and other biological populations, including domestic animals, trees, cereal crops, and infectious diseases; economic stocks and flows, accounting intelligently for the welfare-generating potential of natural resources; informational stocks and flows, including scientific, literary, artistic, and folk traditions; and familial, social, institutional, and political resources and constraints. These comprehensive models will have external forcing factors that remain beyond human control, such as solar flares, and will explicitly represent, though not predict, human decisions.

In spite of improvements in information, concepts, and management, the Earth will still bring surprises. Geophysical surprises will arise from the improved awareness made possible by better planetary monitoring, from inherent instabilities in geophysical systems that are described by the mathematics of chaos and from rising human impacts. Surprising infectious diseases will continue to emerge from the infinite well of genetic variability. Historically, each factor-of-10 increase in the density of human settlements has made possible the survival of new human infections. As more humans contact the viruses and other pathogens of previously remote forests and grasslands, dense urban populations and global travel will increase opportunities for infections to spread.

Economies will be increasingly integrated. Megalopolises will concentrate the talent and resources required for international business. Hardly any complex product will be conceived, financed, engineered, manufactured, sold, used, and retired within the boundaries of a single political unit. Governments will find that a growing fraction of the power to control the economic well-being of their citizens lies outside their borders. Economic integration will give profit to those who can recognize the comparative advantage of other societies and can negotiate mutually beneficial exchanges. Information will become an increasingly valuable commodity. Those who can create it, analyze it, and manage effectively on the basis of it will be at a premium. Information technology and global economic integration will grow hand in hand.

An international common law—not a world government but rather international standards of behavior—will grow stronger and more comprehensive in a progression from technical to commercial to political law. International agreements on vaccination and on metric measures work because they benefit all who abide by them, and many who do not. Growing investments by multinational corporations will force the development of international contract law. Once the regional and global economic customs, institutions, and laws are firm, it will become too costly for nation-states or their successors to ignore them. Legal and economic resolutions of political conflicts will become more efficient than violent ones.

As the peoples of Asia, Latin American, and Africa approach the levels of wealth of Europe and North America, their environmental fatalism and modest demands for food will be replaced by impatience with the accidents of nature, intolerance of environmental mismanagement, and refusal to eat less well than their neighbors. The need for careful global management—particularly management of the living resources that are the focus of our material and altruistic concerns—will become irresistible.

Perhaps I am dreaming when I speculate that geophysical and biological surprises, the revaluation of living nature, our greater dependence on people all over the world, our growing determination to act lawfully and our own aging (individually and as a population) will increasingly inspire in many of us a greater awe for the world, for others, and for ourselves. The problem for scientists is how we can provide the factual and theoretical foundations needed to achieve a managed, wired, and beautiful global garden a century from now—a prospect, as Matthew Arnold (1867:708) put it, "which seems to lie before us like a land of dreams, so various, so beautiful, so new."

Bibliography

Abbott, I. 1980. Theories dealing with land birds on islands. Advances in Ecological Research *11*:329–371.

Adler, F. R., and B. Nuernberger. 1994. Persistence in patchy irregular landscapes. Theoretical Population Biology *45*:41–75.

Adler, R. W., J. C. Landman, and D. M. Cameron. 1993. The Clean Water Act twenty years later. Island Press, Washington DC.

Aguirre, A. A., and E. E. Starkey. 1994. Wildlife disease in U.S. National Parks: historical and coevolutionary perspectives. Conservation Biology *8*:654–661.

Akçakaya, H. R. 1992. Population viability analysis and risk assessment. Pages 148–157 in D. R. McCollough and R. H. Varrett, editors. Wildlife 2001: populations. Elsevier Applied Science, New York.

Allan, J. D. 1995. Stream ecology. Chapman and Hall, New York

Allen, T. F. H., and T. W. Hoekstra. 1992. Toward a unified ecology. New York: Columbia University Press.

Allen, W. H. 1993. The great flood of 1993. BioScience *43*:732–737.

Allendorf, F. W., and R. F. Leary. 1986. Heterozygosity and fitness in natural populations of animals. Pages 57–76 in Conservation biology: the science of scarcity and diversity, M. E. Soulé, editor. Sinauer Associates, Sunderland, MA.

Alpert, P., and J. L. Maron. 1994. The M.S./Ph.D. in natural resources management and biology: a graduate program to link basic science and practical management. Bulletin of the Ecological Society of America *75*:42–44.

Alvarado, A., C. W. Berish, and F. Peralta. 1981. Leaf cutter ant (*Atta cephalotes*) influence on the morphology of Andepts in Costa Rica. Soil Science Society of America Journal *45*:790–794.

Alvarez-Buylla, E. R., and R. Garcia-Barrios. 1993. Models of patch dynamics in tropical forests. Trends in Ecology and Evolution *8*:201–204.

Alverson, W. S., W. Kuhlmann, and D. M. Waller. 1994. Wild forests: conservation biology and public policy. Island Press, Washington, DC.

Anderson, H. M. 1993. Conserving America's freshwater ecosystems: the Wilderness Society's approach. Journal of the North American Benthological Society *12*:194–196.

Anderson, H. M. 1994. Reforming National-Forest policy. Issues in Science and Technology *10*:40–47.

Anderson, R. M., and R. M. May. 1982. Coevolution of hosts and parasites. Parasitology *85*:411–426.

Anderson, R. M., and R. M. May. 1991. Infectious diseases of humans: dynamics and control. Oxford University Press, Oxford.

Anderson, S. H. 1995. Traditional approaches and tools in natural resources management., Pages 61–74 *in* R. L. Knight and S. F. Bates, editors. A new century for natural resources management. Island Press, Washington, DC.

Andrén, H. 1992. Corvid density and nest predation in relation to forest fragmentation: a landscape perspective. Ecology *73*:794–804.

Andrén, H. 1994. Effects of habitat fragmentation on birds and mammals in landscapes with different proportions of suitable habitat: a review. Oikos *71*:355–366.

Andrén, H. 1995. Effects of landscape composition on predation rates at habitat edges. Pages 225–255 *in* L. Hansson, L. Fahrig, and G. Merriam, editors. Mosaic landscapes and ecological processes. Chapman and Hall, London.

Andrew, M. H., and R. T. Lange. 1986a. Development of a new piosphere in arid chenopod shrubland grazed by sheep. I. Changes to the soil surface. Australian Journal of Ecology *11*:395–409.

Andrew, M. H., and R. T. Lange. 1986b. Development of a new piosphere in arid chenopod shrubland grazed by sheep. II. Changes in the vegetation. Australian Journal of Ecology *11*:411–424.

Andrewartha, H. G., and L. C. Birch. 1954. The distribution and abundance of animals. University of Chicago Press, Chicago.

Andrewartha, H. G., and L. C. Birch. 1984. The ecological web. More on the distribution and abundance of animals. University of Chicago Press, Chicago.

Angermeier, P. L., and J. R. Karr. 1994. Biological integrity versus biological diversity as policy directives. BioScience *44*:690–697.

Antonovics, J. 1968. Evolution in closely adjacent plant populations. Heredity *23*:219–238.

Anunsen, C. S., and R. Anunsen. 1993. Response to Scheffer. Conservation Biology *7*:954–957.

Aplet, G. H., R. D. Laven, and P. L. Fiedler. 1992. The relevance of conservation biology to natural resource management. Conservation Biology *6*:298–300.

Arnold, M. 1867. "Dover Beach." Lines 30–32, pages 707–709 *in* M. L. Rosenthal, editor. Poetry in English: an anthology. Oxford University Press, 1987.

Aronson, J., C. Floret, E. Le Floc'h, C. Ovalle, and R. Pontanier. 1993. Restoration and rehabilitation of degraded ecosystems in arid and semi arid lands. II. Case studies in southern Tunisia, central Chile and northern Cameroon. Restoration Ecology *1*:168–187

Ås, S., J. Bengtsson, and T. Eberhard. 1992. Archipelagoes and theories of insularity. Pages 201–251 *in* L. Hansson, editor. Ecological principles of nature conservation: applications in temperate and boreal environments. Elsevier Applied Science, New York.

Attiwill, P. M. 1994a. The disturbance of forest ecosystems: the ecological basis for conservative management. Forest Ecology and Management *63*:247–300.

Attiwill, P. M. 1994b. Ecological disturbance and the conservative management of eucalypt forests in Australia. Forest Ecology and Management *63*:301–346.

Attiwill, P. M., and M. A. Adams. 1993. Nutrient cycling in forests. Tansley Review No. 50. New Phytologist *124*:561–582.

Avery, Dennis. 1993. Quoted in Ronald Bailey, Saving the planet with pesticides, CEI Update, Competitive Enterprise Institute, November, p. 3.

Avise, J. C. 1994a. A rose is a rose is a rose. Page 174 *in* G. K. Meffe and C. R. Carroll, Principles of conservation biology. Sinauer, Sunderland, MA.

Avise, J. C. 1994b. Molecular markers, natural history, and evolution. Chapman and Hall, New York.

Avise, J. C. 1995. Mitochondrial DNA polymorphism and a connection between genetics and demography of relevance in conservation. Conservation Biology *9*:686–690.

Avise, J. C., and R. M. Ball, Jr. 1990. Principles of genealogical concordance in species concepts and biological taxonomy. Pages 45–67 *in* D. Futuyma and J. Antonovics, editors. Oxford surveys in evolutionary biology. Oxford University Press, New York.

Avise, J. C., and W. S. Nelson. 1989. Molecular genetic relationships of the extinct dusky seaside sparrow. Science *243*:646–648.

Barlow, J., and P. Boveng. 1991. Modeling age–specific mortality for marine mammal populations. Marine Mammal Science *7*:50–65.

Begon, M., J. L. Harper, and C. R. Townsend. 1990. Ecology: individuals, populations and communities. 2nd edition. Blackwell, Cambridge, MA.

Bell, M. 1985. The face of Connecticut: people, geology, and the land. Connecticut Geological and Natural History Survey. Bulletin No. 99.

Bell, S. S., E. D. McCoy, and H. R. Mushinsky editors. 1991. Habitat structure: the physical arrangements of objects in space. Chapman and Hall, New York.

Benke, A. C. 1990. A perspective on America's vanishing streams. Journal of the North American Benthological Society *9*:77–88.

Bennett, G., editor. 1991. Towards a European ecological network. Institute for European Environmental Policy, Arnhem.

Benstead, J. P., J. G. March, and C. M. Pringle. 1996. Effects of water withdrawals on the migration of freshwater shrimps in a tropical river, Puerto Rico. Bulletin of the North American Benthological Society *13*:202.

Berger, J. O., and D. A. Berry. 1988. Statistical analysis and the illusion of objectivity. American Scientist *76*:159–165.

Blakesley, J. A., A. B. Franklin, and R. J. Gutierrez. 1992. Spotted owl roost and nest site selection in Northwestern California. Journal of Wildlife Management *56*:388–392.

Boeken, B., and M. Shachak. 1994. Desert plant communities in human made patches—implications for management. Ecological Applications 4:702–716.

Bolling, D. M. 1994. How to save a river: a handbook for citizen action. Island Press, Washington, DC.

Bond, W. J. 1992. Keystone species. Pages 327–354 *in* E. D. Schulze and H. A. Mooney, editors. Biodiversity and ecosystem function. Springer-Verlag, New York.

Bonnicksen, T. M., and E. C. Stone. 1985. Restoring naturalness to national parks. Environmental Management 9:479–486.

Boorman, S. A., and P. R. Levitt. 1973. Group selection on the boundary of a stable population. Theoretical Population Biology 4:85–128.

Bormann, F. H., and G. E. Likens. 1979a. Catastrophic disturbance and the steady state in the northern hardwood forest. American Scientist 67:660–669.

Bormann, F. H., and G. E. Likens. 1979b. Pattern and process in a forested ecosystem. Springer-Verlag, New York.

Boserup, E. 1981. Population and technological change: a study of long-term trends. University of Chicago Press, Chicago.

Botkin, D. B. 1990. Discordant harmonies: a new ecology for the twenty-first century. Oxford University Press, New York.

Bourgeron, P. S., and M. E. Jensen. 1994. An overview of ecological principles for ecosystem management. Pages 45–57 *in* M. E. Jensen and P. S. Bourgeron, editors. Volume II: Ecosystem management: principles and applications. General Technical Report PNW-GTR–318, Portland, OR.

Bowen, B. W., A. B. Meylan, J. P. Ross, C. J. Limpus, G. H. Balzas, and J. C. Avise. 1992. Global population structure and natural history of the green turtle (*Chelonia mydas*) in terms of matriarchal phylogeny. Evolution 46:865–881.

Boyce, M. S. 1992. Population viability analysis. Annual Review of Ecology ans Systematics 23:481–506.

Bradshaw, A. D. 1987a. Restoration: an acid test for ecology. Pages 23–29 *in* W. R. Jordan III, M. E. Gilpin, and J. D. Aber, editors. Restoration ecology: a synthetic approach to ecological research. Cambridge University Press, New York.

Bradshaw, A.D. 1987b. The reclamation of derelict land and the ecology of ecosystems. Pages Pages 53–74 *in* W. R. Jordan III, M. E. Gilpin, and J. D. Aber, editors. Restoration ecology: a synthetic approach to ecological research. Cambridge University Press, New York.

Breen, C., N. Quinn, and A. Deacon. 1994. A description of the Kruger National Park rivers research programme (second phase). Foundation for Research Development. Pretoria.

Breen, C. M., H. Biggs, M. C. Dent, A. Gorgens, J. O'Keeffe, and K. H. Rogers. 1995. Designing a research programme to promote river basin management. Proceedings of the IAWQ Conference on River Basin Management for Sustainable Management. IAWQ Specialist Group on River Basin Management.

Brooks, T. M., S. L. Pimm, and N. J. Collar. In press. The extent of deforestation predicts the number of threatened birds in insular South-east Asia. Conservation Biology.

Brouha, P. 1993. The emerging science-based advocacy role of the American Fisheries Society. Journal North American Benthological Society *12*:215–218.

Brown, A. H. D., and J. D. Briggs. 1991. Sampling strategies for genetic variation in *ex situ* collections of endangered plant species. Pages 99–119 *in* D. A. Falk and K. E. Holsinger, editors. Genetics and conservation of rare plants. Oxford University Press, New York.

Brown, G. E., Jr. 1992. The objectivity crisis. American Journal of Physics *60*:779–781.

Brown, J. H. 1994. Macroecology. University of Chicago Press, Chicago.

Brown, J. H., and A. Kodric-Brown. 1977. Turnover rates in insular biogeography: effect of immigration on extinction. Ecology *58*:445–449.

Bruijnzeel, L. A. 1989. Nutrient cycling in moist tropical forests: the hydrological framework. Pages 383–415 *in* J. Proctor, editor. Mineral nutrients in tropical forest and savanna ecosystems. Special Publications Series of the British Ecological Society No. 9. Blackwell Scientific Publications, Oxford, U.K.

Bull, J. J. 1994. Virulence. Evolution *48*:1423–1437.

Bulmer, M. G. 1985. The mathematical theory of quantitative genetics. Clarendon Press, Oxford.

Bunn, S. E. 1995. Biological monitoring of water quality in Australia: workshop summary and future directions. Australian Journal of Ecology *20*:220–227.

Burbidge, A. A., and N. L. McKenzie. 1989. Patterns in the modern decline of Western Australia's vertebrate fauna: causes and conservation implications. Biological Conservation *50*:143–198.

Burdon, J. J. 1994. The distribution and origin of genes for race-specific resistance to *Melampsora lini* in *Linum marginale*. Evolution *48*:1564–1575.

Burdon, J. J., and A. M. Jarosz. 1992. Temporal variation in the racial structure of flax rust (*Melampsora lini*) populations growing on natural stands of wild flax (*Linum marginale*): local versus metapopulation dynamics. Plant Pathology 41:165–179.

Burdon, J. J., D. R. Marshall, and R. H. Groves. 1980. Isozyme variation in *Chondrilla juncea* L. in Australia. Australian Journal of Botany *38*:193–198.

Burdon, J. J., and J. N. Thompson. 1995. Changed patterns of resistance in a population of *Linum marginale* attacked by the rust pathogen *Melampsora lini*. Journal of Ecology *83*:199–206.

Burgman, M. A., S. Ferson, and H. R. Akçakaya. 1993. Risk assessment in conservation biology. Chapman and Hall, New York.

Burnham, K. R., D. R. Anderson, and G. C. White. 1996. Meta-analysis of vital rates of the northern spotted owl. Studies in Avian Biology *17*:92–101.

Bush, G. L. 1979. Ecological genetics and quality control. Pages 145–152 *in* M. A. Hoy and J. J. McKelvey, Jr., editors. Rockefeller Foundation, New York.

Cade, T. J. 1983. Hybridization and gene exchange among birds in relation to conservation. Pages 288–310 *in* C. M. Schoenwald-Cox, S. M. Chambers, B. MacBryde, and W. L. Thomas, editors. Genetics and conservation: a reference for managing wild animal and plant populations. Benjamin-Cummings, Menlo Park, CA.

Caicco, S. L., J. M. Scott, B. Butterfield, and B. Csuti. 1995. A gap analysis of the management status of the vegetation of Idaho (U.S.A.). Conservation Biology 9:498–511.

Callicott, J. B. 1989. In defense of the land ethic. State University of New York Press, Albany.

Campbell, D. G. 1994. Scale and patterns of community structure in Amazonian forests., Pages 179–197 in P. J. Edwards, R. M. May, and N. R. Webb, editors. Large-scale ecology and conservation biology. Blackwell Scientific Publications, Boston.

Campbell, J. M., and R. B. Kenyon. 1994. Factors influencing fish species diversity in the Great Lakes and the effect of recent introductions on Lake Erie. Pages 368–379 in S. K. Majumdar, F. J. Brenner, J. E. Lovich, J. F. Schalles, and E. W. Miller, editors. Biological diversity: problems and challenges. Pennsylvania Academy of Sciences, Philadelphia.

Canham, C. D., and S. W. Pacala. 1995. Linking tree population dynamics and forest ecosystem processes. Pages 84–93 in C. G. Jones and J. H. Lawton, editors. Linking species and ecosystems. Chapman and Hall, New York.

Caraco, N., J. J. Cole, and G. E. Likens. 1991. A cross-system study of phosphorus release from lake sediments. Pages 241–258 in J. Cole, G. Lovett, and S. Findlay, editors. Comparative analyses of ecosystems. Springer-Verlag, New York.

Careless, R., and L. E. Bernese. 1993. The Tatshenshini Wilderness: under threat of mining. Journal of the North American Benthological Society 12:211–214.

Carpenter, S. R., and J. F. Kitchell, editors. 1993. The trophic cascade in lakes. Cambridge University Press, London.

Caswell, H. 1988. Theory and models in ecology: a different perspective. Bulletin of the Ecological Society of America 69:102–109.

Caughley, G. 1994. Directions in conservation biology. Journal of Animal Ecology 63:215–244.

Chambers, R. 1844. Vestiges of the natural history of creation. Reprinted 1969 Humanities Press, Atlantic Highlands, NJ.

Chase, A. 1995. In a dark wood: the fight over forests and the rising tyranny in ecology. Houghton Mifflin Company, Boston.

Christensen, N. L. 1991. Wilderness and high intensity fire: how much is enough. Proceedings of the Tall Timbers Ecology Conference 17:9–24.

Christensen, N. L. 1995a. Plants in dynamic environments: is "wilderness management" an oxymoron? Pages 63–72 in P. Schullery and J. Varley, editors. Plants and the environment. U.S. Department of the Interior, National Park Service, Washington, DC.

Christensen, N. L. 1995b. Fire and wilderness. International Journal of Wilderness 1:30–34.

Christensen, N. L. 1988. Succession and natural disturbance: paradigms, problems, and preservation of natural ecosystems. Pages 62–86 in J. K. Agee and D. R. Johnson, editors. Ecosystem management for parks and wilderness. University of Washington Press, Seattle.

Christensen, N. L., J. K. Agee, P. F. Brussard, J. Huges, D. H. Knight, G. W. Minshall,

J. M. Peek, S. J. Pyne, F. J. Swanson, J. W. Thomas, S. Wells, S. E. Williams, and H. A. Wright. 1989. Interpreting the Yellowstone fires. BioScience *39*:677–685.

Christensen, N. L., A. Bartuska, J. H. Brown, S. Carpenter, C. D'Antonio, R. Francis, J. F. Franklin, J. A. Mac Mahon, R. F. Noss, D. J. Parsons, C. H. Peterson, M. G. Turner, and R. G. Woodmansee. 1996. The report of the Ecological Society of America Committee on the scientific basis for ecosystem management. Ecological Applications.

Clark, L. R., P. W. Geier, R. D. Hughes, and R. F. Morris. 1967. The ecology of insect populations in theory and practice. Methuen, London.

Clements, F. E. 1916. Plant succession: an analysis of the development of vegetation. Publication Number 242. Carnegie Institute, Washington, DC.

Clements, F.E. 1935. Experimental ecology in the public service. Ecology *16*:342–363.

Cockburn, Alexander. 1993. Ulterior secretary: Babbitt makes me miss Jim Watt. The Washington Post, Aug. 29, 1993, p. C1.

Cohen, J. E. 1995. How many people can the Earth support? Norton, New York.

Cohen, J. E., K. S. Schoenly, K. L. Heong, H. Justo, G. Arida, A. T. Barrion, and J. A. Litsinger. 1994. A food web approach to evaluating the effect of insecticide spraying on insect pest population dynamics in a Philippine irrigated rice ecosystem. Journal of Applied Ecology *31*:747–763.

Cole, J. J., B. L. Peierls, N. F. Caraco, and M. L. Pace. 1993. Nitrogen loading of rivers as a human-driven process. Pages 141–157 *in* M. J. McDonnell, and S. T. A. Pickett, editors. Humans as components of ecosystems: the ecology of subtle human effects and populated areas. Springer-Verlag, New York.

Collins, J. P., and H. M. Wilbur. 1979. Breeding habits and habitats of the E. S. George Reserve, Michigan, with notes on the local distribution of fishes. Occasional Papers of the Museum of Zoology, University of Michigan *686*:1–34.

Colwell, R. K. 1984. What's new? Community ecology discovers biology. Pages 387–396 *in* P. W. Price, C. N. Slobodchikoff, and W. S. Gaud, editors. A new ecology: novel approaches to interactive systems. John Wiley & Sons, New York.

Commins, E., J. Fay, J. Feiveson, M. Goldberger, W. Hall, J. Harte, L. Hinkle, W. Liinville, G. MacDonald, W. Matthews, H. Menard, and R. Socolow. 1970. Environmental problems in South Florida, pages 200–259. National Academy of Sciences Environmental Study Group, National Technical Information Service, Washington DC.

Commoner, B. 1972. The closing circle. Confronting the environmental crisis. Jonathan Cape, London.

Conant, R., and J. T. Collins. 1991. A field guide to reptiles and amphibians of eastern and central North America. 3rd edition. Houghton-Mifflin, Boston.

Congressional Research Service (CRS). 1994. Ecosystem management: federal agency activities. CRS Report for Congress 94–339 ENR. Congressional Research Service, Library of Congress, Washington, DC.

Connell, J. H. 1978. Diversity in tropical rain forests and coral reefs. Science *199*: 1302–1310.

Connell, J. H., and Keough, M. J. 1985. Disturbance and patch dynamics of subtidal marine animals on hard substrata. Pages 125–151 *in* S. T. A. Pickett and P. S. White, editors. The ecology of natural disturbance and patch dynamics. Academic Press, Orlando.

Connell, J. H., and R. O. Slatyer. 1977. Mechanisms of succession in natural communities and their role in community stability and organization. American Naturalist *111*:1119–1144.

Connor, E. F., and E. D. McCoy. 1979. The statistics and biology of the species-area relationship. American Naturalist *113*:791–833.

Conroy, M. J., Y. Cohen, F. C. James, Y.G. Matsinos, and B. A. Maurer. 1995. Parameter estimation, reliability and model improvement for spatially explicit models of animal populations. Ecological Applications *5*:17–19.

Cooper, W. E. 1990. Aquatic research and water quality trends in the United States and Poland. Pages 297–314 *in* W. Grodzinski, E. B. Cowling, and A. I. Breymeyer, editors. Ecological risks: perspectives from Poland and the United States. National Academy Press, Washington, DC.

Cortwright, S. C. 1987. Impacts of species interactions and geographical-historical factors on larval amphibian community structure. Dissertation, Indiana University, Bloomington, Indiana.

Costanza, R., J. Audley, R. Borden, P. Elkins, C. Folke, S. O. Funtowicz, and J. Harris. 1995. Sustainable trade: a new paradigm for world trade. Environment *37*:16–44.

Council on Environmental Quality. 1986. Code of Federal Regulations *40*:Parts 1500–1508.

Cowles, H. C. 1899. The ecological relations of the vegetation of the sand dunes of Lake Michigan. Botanical Gazette *27*:214–216.

Cox, J., R. Kautz, M. MacLaughlin, and T. Gilbert. 1994. Closing the gaps in Florida's wildlife habitat conservation system. Florida Game and Freshwater Fish Commission, Tallahassee.

Coyle, K. J. 1993. The new advocacy for aquatic species conservation. Journal of the North American Benthological Society *12*:185–188.

Crawley, M. J. 1987. What makes a community invasible? Pages 429–453 *in* A. J. Gray, M. J. Crawley, and P. J. Edwards, editors. Colonization, succession and stability. Oxford University Press, Oxford.

Crispin, S. 1994. A report on the Great Lakes Biodiversity Data System. The Nature Conservancy, Chicago, and The Nature Conservancy of Canada, Toronto.

Croker, Thomas C., Jr. 1990. Longleaf pine—myths and facts. Pages 2–10 *in* R. M. Farrar, Jr., editor. Proceedings of the symposium on the management of longleaf pine; 1989 April 4–6; Long Beach, MS. Gen. Tech. Rep. SO–75. U.S. Dept. of Agriculture, Forest Service, Southern Forest Experiment Station, New Orleans, LA. (hereinafter, "1989 Proceedings").

Cronon, W. 1995. Uncommon ground: toward reinventing nature. Norton, New York.

Crow, J., and M. Kimura. 1970. An introduction to population genetics theory. Burgess Publishing Company, Minneapolis, MN.

Crow, T. R., A. Haney, and D. M. Waller. 1994. Report on the scientific roundtable on biological diversity convened by the Chequamegon and Nicolet National Forests. General Technical Report NC-166. North Central Forest Experiment Station, U.S. Forest Service, St. Paul, MN.

Cullen, J. M. 1978. Evaluating the success of the program for the biological control of *Chondrilla juncea* L. Pages 117-121 *in* T. E. Freeman, editor. Proceedings of the IV international symposium on the biological control of weeds. University of Florida, Gainesville (cited in Heap 1993).

Cullen, J. M. 1991. The current status of research on the biological control of skeleton weed. Proceedings of the Australian workshop on the management of skeleton weed. Walkup, Victoria, Australia (cited in Heap 1993).

Cullen, P. 1990. The turbulent boundary between water science and water management. Freshwater Biology *24*:201–209.

Curnutt, J., J. Lockwood, H.-K. Luh, P. Nott, and G. Russell. 1994. Hotspots and species diversity. Nature *367*:326–327.

Cushman, J. H., C. L. Boggs, S. B. Weiss, D. D. Murphy, A. W. Harvey, and P. R. Ehrlich. 1994. Estimating female reproductive success of a threatened butterfly: influence of emergence time and hostplant phenology. Oecologia *99*:194–200.

Daehler, C. C., and D. R. Strong. 1994. Native plant biodiversity vs. the introduced invaders: status of the conflict and future management options. Pages 92–113 *in* S. K. Majumdar, F. J. Brenner, J. E. Lovich, J. F. Schalles, and E. W. Miller, editors. Biological diversity: problems and challenges. Pennsylvania Academy of Sciences, Philadelphia.

Daily, G. C. 1995. Restoring value to the world's degaded lands. Science *269*:350–354.

D'Angelo, D. J., J. R. Webster, S. V. Gregory, and J. L. Meyer. 1993. Transient storage in Appalachian and Cascade mountain streams as related to hydraulic characteristics. Journal of the North American Benthological Society *12*:223–235.

Dasmann, R. F. 1984. Environmental conservation. 5th edition. John Wiley & Sons, New York.

Davis, F. W, P. A. Stine, and D. M. Stoms. 1994. Distribution and conservation status of coastal sage scrub in southwestern California. Journal of Vegetation Science *5*:743–756.

Davis, M. B. 1986. Climatic instability, time lags, and community disequilibrium. Pages 269–284 *in* J. Diamond and T. J. Case., editors. Community ecology. Harper and Row, New York.

Davis, W. 1992. Shadows in the sun. Essays on the spirit of place. Lone Pine Publishing, Edmonton, Canada.

Day, J. R., and Possingham. H. P. 1996. A stochastic metapopulation model with variability in patch size and position. Theoretical Population Biology *48*:333–360.

DeAngelis, D. L., and J. C. Waterhouse. 1987. Equilibrium and nonequilibrium concepts in ecological models. Ecological Monographs *57*:1–21.

DeAngelis, D. L., and L. J. Gross, editors. 1992. Individual-based models and approaches in ecology. Chapman and Hall, New York.

Delcourt, H. R., and P. A. Delcourt. 1991. Quaternary ecology. Chapman and Hall, New York.

DeMauro, M. M. 1993. Relationship of breeding system to rarity in the lakeside daisy (*Hymenoxys acaulis* var. *glabra*). Conservation Biology 7:542–550.

Demeny, P. 1994. Population and development. International Conference on Population and Development 1994. International Union for the Scientific Study of Population, Liège, Belgium.

den Boer, P. J. 1981. On the survival of populations in a heterogeneous and variable environment. Oecologia 50:39–53.

Dennis, B., P. L. Munholland, and J. M. Scott. 1991. Estimation of growth and extinction parameters for endangered species. Ecological Monographs 61:115–143.

Dewberry, T.C. 1995. The Knowles Creek report: 1992–94. Pacific Rivers Council, Eugene, OR.

Dewberry, T. C. 1996. Can we diagnose the health of ecosystems? Northwest Science and Photography, Florence, OR.

Dewberry, T. C., and C. Pringle. 1994. Lotic conservation and science: moving towards common ground to protect our stream resources. Journal of the North American Benthological Society 13:399–404.

Dobkin, D. S., I. Olivieri, and P. R. Ehrlich. 1987. Rainfall and the interaction of microclimate with larval resources in the population dynamics of checkerspot butterflies *Euphydryas editha* inhabiting serpentine grassland. Oecologia 71:161–166.

Dobson, A., and P. Hudson. 1995. The interaction between the parasites and predators of red grouse *Lagopus lagopus scoticus*. Ibis 137:S87–96.

Doppelt, B., M. Scurlock, C. Frissell, and J. Karr. 1993. Entering the watershed. Island Press, Washington, DC.

Doppelt, R. 1993. The vital role of the scientific community in new river conservation strategies. Journal of the North American Benthological Society 12:189–193.

Drake, J. A. 1990. Communites as assembled structures: do rules govern patterns? Trends in Ecology and Evolution 5:159–164.

Drake, J. A., H. A. Mooney, F. di Castri, R. H. Groves, F. J. Kruger, M. Rejmánek, and M. Williamson, editors. 1989. Biological invasions. A global perspective. Wiley, New York.

Duff, D. A. 1993. Conservation partnerships for coldwater fisheries habitat. Journal of the North American Benthological Society 12:206–210.

Dunning, J. B., Jr., D. J. Stewart, B. J. Danielson, B. R. Noon, R. L. Root, R. H. Lamberson, and E. E. Stevens. 1995. Spatially explicit population models: current forms and future uses. Ecological Applications 5:3–11.

Durrett, R., and S. A. Levin. 1994a. Stochastic spatial models: a user's guide to ecological applications. Philosophical Transactions of the Royal Society London B 343:329–350.

Durrett, R., and S. Levin. 1994b. The importance of being discrete (and spatial). Theoretical Population Biology 46:363–394.

Dwyer, G., S. A. Levin, and L. Buttel. 1990. A simulation model of the population dynamics and evolution of myxomatosis. Ecological Monographs 60:423–447.

Dynesius, M., and C. Nilsson. 1994. Fragmentation and flow regulation of river systems in the northern third of the world. Science *266*:753–762.

Easterbrook, G. 1995. A moment on the Earth. Viking, New York.

Ebenard, T. 1995. Conservation breeding as a tool for saving animal species from extinction. Trends in Ecology and Evolution *10*:438–443.

Eberhard, S. A., E. E. Roos, and L. E. Towill. 1991. Strategies for long-term management of germplasm collections. Pages 135–145 *in* D. A. Falk and K. E. Holsinger, editors. Genetics and conservation of rare plants. Oxford University Press, New York.

Ebert, D. 1994. Virulence and local adaptation of a horizontally transmitted parasite. Science *264*:1084–1086.

Echeverria, J. D., P. Barrow, and R. Roos-Collins. 1989. Rivers at risk: the concerned citizen's guide to hydropower. Island Press, Washington, DC.

Edmondson, W. T. 1991. The uses of ecology: Lake Washington and beyond. University of Washington Press. Seattle.

Edwards, P. J., R. M. May, and N. R. Webb, editors. 1994. Large-scale ecology and conservation biology. Blackwell Scientific Publications, Boston.

Egerton, F. N. 1993. The history and present entanglements of some general ecological perspectives. Pages 9–23, *in* M. J. McDonnell and S. T. A. Pickett, editors. Humans as components of ecosystems: the ecology of subtle human effects and populated areas. Springer-Verlag, New York.

Eidsvik, H. K. 1992. Strengthening protected areas through philosophy, science and management: a global perspective. Pages 9–18 *in* J. H. M. Willison, S. Bondrup-Nielsen, C. Drysdale, T. B. Herman, N. W. P. Munro, and T. L. Pollock, editors. Science and the management of protected areas. Elsevier Science Publishers, Amsterdam.

Ellstrand, N. C. 1992. Gene flow by pollen: implications for plant conservation genetics. Oikos *63*:77–86.

Ellstrand, N. C., and D. R. Elam. 1993. Population genetic consequences of small population size: implications for plant conservation. Annual Review of Ecology and Systematics *24*:217–242.

Elton, C. 1927. Animal ecology. MacMillan, New York.

Esseen P.-A., B. Ehnström, L. Ericson, and K. Sjöberg. 1992. Boreal forests - the focal habitats of Fennoscandia. Pages 252–325 *in* L. Hansson, editor. Ecological principles of nature conservation. Elsevier, Barking.

Etnier, D. A., and W. C. Starnes. 1993. The fishes of Tennessee. University of Tennessee Press, Knoxville.

Evans, E. P. 1989. Evolutionary ethics and animal psychology.

Evenari, M., L. Shanan, and N. Tadmor. 1983. The Negev: the challenge of a desert. Oxford University Press, London.

Ewald, P. W. 1994. Evolution of infectious diseases. Oxford University Press, Oxford.

Ewel, J. J. 1986a. Designing agricultural ecosystems for the humid tropics. Annual Review of Ecology and Systematics *17*:245–271.

Ewel, J.J. 1986b. Invasibility: lessons from South Florida. Pages 214–230 *in* H. A. Mooney and J. A. Drake, editors. Ecology of biological invasions of North America and Hawaii. Springer-Verlag, New York.

Ewel, J. J. 1987. Restoration as the ultimate test of ecological theory. Pages 31–33 *in* W. R. Jordan III, M. E. Gilpin, and J. D. Aber, editors. Restoration ecology: a synthetic approach to ecological research. Cambridge University Press, New York.

Ewel, J. J., and S. W. Bigelow. 1996. Plant life forms and tropical ecosystem functioning. Pages 101–126 *in* G. H. Orians, R. Dirzo, and J.H. Cushman, editors. Biodiversity and ecosystem processes in tropical forests. Springer-Verlag, New York.

Ewel, J. J., M. J. Mazzarino, and C. W. Berish. 1991. Tropical soil fertility changes under monocultures and successional communities of different structure. Ecological Applications *1*:289–302.

Ewens, W. J. 1979. Mathematical population genetics. Springer-Verlag, Berlin.

Ewens, W. J., P. J. Brockwell, J. M. Gani, and S. I. Resnick. 1987. Minimum viable population sizes in the presence of catastrophes. Pages 59–69 *in* M. E. Soulé, editor. Viable Populations for conservation. Cambridge University Press, Cambridge.

Faber, J. H., and H. A. Verhoef. 1991. Functional differences between closely-related soil arthropods with respect to decomposition processes in the presence or absence of pine tree roots. Soil Biology and Biochemistry *23*:15–23.

Fagerstrom, T. 1987. On theory, data and mathematics in ecology. Oikos *50*:258–261.

Fahrig, L. 1988. A general model of populations in patchy habitats. Applied Mathematics and Computation *27*:53–66.

Fahrig, L. 1990. Interacting effects of disturbance and dispersal on individual selection and population stability. Comments Theoretical Biology *1*:275–297.

Fahrig, L. 1991. Simulation methods for developing general landscape-level hypotheses of single species dynamics. Pages 417–442 *in* M. G. Turner, and R. H. Gardner, editors. Quantitative methods in landscape ecology. Springer-Verlag, New York.

Fahrig, L. 1992. Relative importance of spatial and temporal scales in a patchy environment. Theoretical Population Biology *41*:300–314.

Fahrig, L., and J. Paloheimo. 1988. Determinants of local population size in patchy habitats. Theoretical Population Biology *34*:194–212.

Falk, D. A. 1987. Integrated conservation strategies for endangered plants. Natural Areas Journal *7*:118–123.

Falk, D. A. 1990. The theory of integrated conservation strategies for biological diversity. Pages 5–10 *in* R. S. Mitchell, C. J. Sheviak, and D. J. Leopold, editors. Ecosystem management: rare species and significant habitats. New York State Museum, Albany.

Falk, D. A. 1992. From conservation biology to conservation practice: strategies for protecting plant diversity. Pages 397–431 *in* P. L. Fiedler and S. K. Jain, editors. Conservation biology. Chapman and Hall, New York.

Fenner, F., and K. Myers. 1978. Myxoma virus and myxomatosis in retrospect: the first quarter century of a new disease, Pages 539–570 *in* E. Kurstak and K. Maromorosch, editors. Viruses and environment. Academic Press, New York.

Fiedler, P. L., R. A. Leidy, R. D. Laven, N. Gershenz, and L. Saul. 1993. The contemporary paradigm in ecology and its implications for endangered species conservation. Endangered Species Update *10*:7–12.

Fire Management Policy Review Team. 1988. Report on fire management policy. U.S. Department of Agriculture and U.S. Department of the Interior, Washington, DC.

Fleischner, T. L. 1994. Ecological costs of livestock grazing in western North America. Conservation Biology *8*:629–644.

Fletcher, J. E., and W. P. Martin. 1948. Some effects of algae and molds in the rain-crust of desert soils. Ecology *29*:95–100.

Forman, R. T. T. 1995. Land mosaics: the ecology of landscapes and regions. Cambridge University Press, Cambridge.

Forman, R. T. T., and M. Godron. 1986. Landscape ecology. John Wiley & Sons, New York.

Forman, R. T. T., and P. N. Moore. 1992. Theoretical foundations for understanding boundaries in landscape mosaics. Pages 236–248 *in* A. J. Hansen and F. di Castri, editors. Landscape boundaries, consequences for biotic diversity and ecological flows. Springer Verlag, New York.

Foster, B. 1993. The importance of British Columbia to global biodiversity. Pages 65–81 *in* M. A. Fenger, E. H. Miller, J. A. Johnson, and E. J. R. Williams, editors. Our living legacy: proceedings of a symposium on biological diversity. Royal British Columbia Museum, Victoria.

Foster, D. R. 1988. Species and stand response to catastrophic wind in central New England, U.S.A. Journal of Ecology *76*:135–151.

Frampton, G. 1993. Quoted in Forum profile. Environmental Forum, Nov./Dec. 1993, p. 28.

Frankham, R. 1995. Conservation genetics. Annual Review of Genetics *29*:305–327.

Franklin, J. F., and R. T. T. Forman. 1987. Creating landscape patterns by forest cutting: ecological consequences and principles. Landscape Ecology *1*:5–18.

Franklin, J. F. 1993a. Lessons from old-growth. Journal of Forestry *91*:10–13.

Franklin, J. F. 1993b. Preserving biodiversity: species, ecosystems, or landscapes? Ecological Applications *3*:202–205.

Franklin, I. R. 1980. Evolutionary change in small populations. Pages 135–150 *in* M. E. Soulé and B. A. Wilcox, editors. Conservation biology: an evolutionary-ecological perspective. Sinauer Associates, Sunderland, MA.

Frissell, C. A., W. L. Liss, C. E. Warren, and M. D. Hurley. 1986. A hierarchical framework for stream habitat classification: viewing streams in a watershed context. Environmental Management *10*:199–214.

Frost, T. M., S. R. Carpenter, A. R. Ives, and T. K. Kratz. 1995. Species compensation and complementarity in ecosystem function. Pages 224–229 *in* C. G. Jones and J. H. Lawton, editors. Linking species and ecosystems. Chapman and Hall, New York.

Fuggle, R. F., and Rabie, M. A. 1992. Environmental management in South Africa. Juta and Co., Cape Town.

Gabriel, W., and R. Bürger. 1992. Survival of small populations under demographic stochasticity. Theoretical Population Biology *41*:44–71.

Gadd, B. 1995. Handbook of the Canadian Rockies. Corax Press, Jasper, Alberta.

Gardner, R. H., B. T. Milne, M. G. Turner, and R. V. O'Neill. 1987. Neutral models for the analysis of broad-scale landscape pattern. Landscape Ecology *1*:19–28.

Gardner, R. H., R. V. O'Neill, M. G. Turner, and V. H. Dale. 1989. Quantifying scale-dependent effects of animal movement with simple percolation models. Landscape Ecology *3*:217–227.

Gardner, R. H., M. G. Turner, V. H. Dale, and R. V. O'Neill. 1992. A percolation model of ecological flows. Pages 259–269 *in* A. J. Hansen, and F. di Castri, editors. Landscape boundaries: consequences for biodiversity and ecological flows. Springer-Verlag, New York.

Gaston, K. J. 1994. Rarity. Chapman and Hall, London.

Gayes, P. T. and H. J. Bokuniewicz. 1991. Estuarine paleoshorelines in Long Island Sound, New York. Journal of Coastal Research, Special Issue *11*:39–54.

Giblin, A. E., K. H. Foreman, and G. T. Banta. 1995. Biogeochemical processes and marine benthic community structure: Which follows which? Pages 36–44 *in* C. G. Jones and J. H. Lawton, editors. Linking species and ecosystems. Chapman and Hall, New York.

Gilbert, F.S. 1980. The equilibrium theory of island biogeography: fact or fiction? Journal of Biogeography *7*:209–235.

Gilchrest, Wayne. 1995. Quoted in Greenwire: The Environmental News Daily. *5*(2). no. 2, Gill, D. E. 1978. The metapopulation ecology of the red-spotted newt, *Notophthalmus viredescens*. Ecological Monographs *48*:145–166.

Gilpin, M. E., and I. Hanski, editors. 1991. Metapopulation dynamics: empirical and theoretical investigations. Academic Press, London.

Gilpin, M. E., and M. E. Soulé. 1986. Minimum viable populations: the process of species extinctions. Pages 13–34 *in* M. E. Soulé, editor. Conservation biology: the science of scarcity and diversity. Sinauer Associates, Sunderland, MA.

Ginzburg, L. R., S. Ferson, and H. R. Akçakaya. 1990. Reconstructibility of density dependence and the conservative assessment of extinction risks. Conservation Biology *4*:63–70.

Givnish, T. J., E. S. Menges, and D. F. Schweitzer. 1988. Minimum area requirements for long-term conservation of the Albany pine bush and Karner blue butterfly: an assessment. Report to the City of Albany and the New York State Department of Environmental Conservation.

Gleason, H. A. 1917. The structure and development of the plant association. Bulletin of the Torrey Botanical Club *43*:463–481.

Gleason, H. A. 1926. The individualistic concept of the plant association. Bulletin of the Torrey Botanical Club *53*:7–26.

Gleick, P. H. 1993. Water in crisis: A guide to the world's freshwater resources. Oxford University Press, New York.

Goldsmith, B., editor. 1991. Monitoring for conservation and ecology. Chapman and Hall, New York.

Goldstein, D. B., and K. E. Holsinger. 1992. Maintenance of polygenic variation in spatially structured populations: a role for local mating and genetic redundancy. Evolution 46:412–429.

Goodman, D. 1987. The demography of chance extinction. Pages 11–34 *in* M. E. Soulé, editor. Viable populations for conservation. Cambridge University Press, Cambridge.

Gosz, J. R. 1993. Ecotone hierarchies. Ecological Applications 3:369–376.

Gould, S. J. 1986. Evolution and the triumph of homology, or why history matters. American Scientist 74:60–69.

Gould, S. J. 1989. Wonderful life. The Burgess shale and the nature of history. Norton and Company, New York.

Graham, R. W. 1986. Response of mammalian communities to environmental changes during the late Quaternary. Pages 300–313 *in* J. Diamond and T. J. Case, editors. Community ecology. Harper and Row, New York.

Grant, P.R. 1977. Review of Lack's Island biology. Bird-Banding 48:296–300.

Gregory, S. V., F. J. Swanson, W. A. McKee, and K. W. Cummins. 1991. An ecosystem perspective on riparian zones. BioScience 41:540–551.

Griffin, C., B. Harvey, F. Heliotis, B. Howes, K. Lajtha, B. McKee, M. Palmer, C. Pringle, G. Vellidis, and A. Zale. 1991. Scientific report of the U.S.-Romanian summer program for young investigators in ecology/environmental sciences. Report to the National Research Council's Office of Eastern European Affairs.

Griffiths, R., and B. Tiwari. 1995. Sex of the last wild Spix's macaw. Nature 375:454.

Grimm, N. B. 1995. Why link species and ecosystems? A perspective from ecosystem ecology. Pages 5–15 *in* C. G. Jones and J. H. Lawton, editors. Linking species and ecosystems. Chapman and Hall, New York.

Grodzinski, W., E. B. Cowling, and A. I. Breymeyer, editors. 1990. Ecological risks: perspectives from Poland and the United States. National Academy Press, Washington, DC.

Groffman, P. M., and G. E. Likens, editors. 1994. Integrated regional models: interactions between humans and their environment. Chapman and Hall, New York.

Gromiec, M. J. 1990. River water quality assessment and management in Poland. Pages 315–332 *in* W. Grodzinski, E. B. Cowling, and A. I. Breymeyer, editors. Ecological risks: perspectives from Poland and the United States. National Academy Press, Washington, DC.

Groombridge, B., editor. 1992. Global biodiversity. World Conservation Monitoring Centre. Chapman and Hall, London.

Grossman, D. H., K. L. Goodin, C. L. Reuss. 1994. Rare plant communites of the conterminous United States: an initial survey. The Nature Conservancy, Arlington, VA.

Grumbine, E. 1990. Protecting biological diversity through the greater ecosystem concept. Natural Areas Journal 10:114–120.

Grumbine, R. E. 1994. What is ecosystem management? Conservation Biology 8:27–38.

Gumbel, E. J. 1958. Statistics of extremes. Columbia University Press, New York.

Gurney, W. S. C., and R. M. Nisbet. 1978. Single species population fluctuations in patchy environments. American Naturalist *112*:1075–1090.

Gurtz, M. E., and T. A. Muir, editors. 1994. Report of the interagency biological methods workshop. U.S. Geological Survey, Raleigh, NC.

Gutting, G., editor. 1980. Paradigms and revolutions. University of Notre Dame Press, Notre Dame, IN.

Gyllenberg, M., and I. Hanski. 1992. Single-species metapopulation dynamics: a structured model. Theoretical Population Biology *42*:35–61.

Hacker, R. B. 1987. Species responses to grazing and environmental factors in an arid halophytic shrubland community. Australian Journal of Botany *35*:135–150.

Haila, Y. 1986. On the semiotic dimension of ecological theory: the case of island biogeography. Biological Philosophy *1*:377–387.

Haila, Y. 1991. Implications of landscape heterogeneity for bird conservation. Pages 2286–2291 in Acta XX Congressus Internationalis Ornithologici. Ornithological congress Trust Board, Wellington, New Zealand.

Haila, Y., and O. Järvinen. 1982. The role of theoretical concepts in understanding the ecological theatre: a case study on island biogeography. Pages 261–278 in E. Saarinen, editor. Conceptual issues in ecology. D. Reidel, Dordrecht.

Haining, R. 1990. Spatial data analysis in the social and environmental sciences. Cambridge University Press, Cambridge, England.

Hall, A. V., B. de Winter, S. P. Fourie, and T. H. Arnold. 1984. Threatened plants in Southern Africa. Biological Conservation *28*:5–20.

Halley, J. M. 1996. Ecology, evolution, and 1/f noise. Trends in Ecology and Evolution *11*:33–37.

Hamel, G., and C. K. Prahalad. 1989. Strategic intent. Harvard Business Review *89*:63–76.

Hansen, A. J., and F. di Castri, editors. 1992. Landscape boundaries: consequences for biotic diversity and ecological flows. Springer-Verlag, New York.

Hansen, A. J., F. di Castri, and R. J. Naiman. 1988. Ecotones: what and why? Biology International 17: 9–46.

Hanski, I. 1985. Single-species dynamics may contribute to long-term rarity and commonness. Ecology *66*:335–343.

Hanski, I. 1991a. Single-species metapopulation dynamics. Pages 17–38 in M. Gilpin and I. Hanski, editors. Metapopulation dynamics. Academic Press, London.

Hanski, I. 1991b. Single-species metapopulation dynamics: concepts models and observations. Biological Journal of the Linnean Society *42*:17–38.

Hanski, I. 1994a. Spatial scale, patchiness and population dynamics on land. Philosophical Transactions of the Royal Society London B *343*:19–25.

Hanski, I. 1994b. A practical model of metapopulation dynamics. Journal of Animal Ecology *63*:151–162.

Hanski, I. 1994c. Patch-occupancy dynamics in fragmented landscapes. Trends in Ecology and Evolution *9*:131–135.

Hanski, I., and M. E. Gilpin. 1991. Metapopulation dynamics: brief history and conceptual domain. Pages 3–16 *in* M. E. Gilpin and I. Hanski, editors. Metapopulation dynamics. Academic Press, London.

Hanski, I., and M. Gilpin, editors. 1996. Metapopulation biology: ecology, genetics and evolution. Academic Press, London.

Hanski, I., and M. Gyllenberg. 1993. Two general metapopulation models and the core-satellite species hypothesis. American Naturalist 142:17–41.

Hanski, I., and M. Kuussaari. 1995. Butterfly metapopulation dynamics. Pages 149–171 *in* N. Cappuccino and P. W. Price, editors. Population dynamics. New approaches and synthesis. Academic Press, San Diego.

Hanski, I., and D. Simberloff. 1996. The metapopulation approach, its history, conceptual domain and application to conservation. *In* I. Hanski and M. Gilpin, editors. Metapopulation biology: ecology, genetics and evolution. Academic Press, London.

Hanski, I., A. Moilanen, and M. Gyllenberg. 1996a. Minimum viable metapopulation size. American Naturalist *147*:527–541.

Hanski, I., A. Moilanen, T. Pakkala, and M. Kuussaari. 1996b. The quantitative incidence function model and persistence of an endangered butterfly metapopulation. Conservation Biology *10*:578–590.

Hanski, I., T. Pakkala, M. Kuussaari, and G. Lei. 1995a. Metapopulation persistence of an endangered butterfly in a fragmented landscape. Oikos *72*:21–28.

Hanski, I., J. Pöyry, T. Pakkala, and M. Kuussaari. 1995b. Multiple equilibria in metapopulation dynamics. Nature *377*:618–621.

Hanski, I., and C.D. Thomas. 1994. Metapopulation dynamics and conservation: a spatially explicit model applied to butterflies. Biological Conservation *68*:167–180.

Hansson, L., editor. 1992a. Ecological principles of nature conservation: applications in temperate and boreal environments. Elsevier Applied Science, New York.

Hansson, L. (1992b. Landscape ecology of boreal forests. Trends in Ecology and Evolution *7*:229–302.

Hansson, L. 1994. Vertebrate distributions relative to clearcut edges in a boreal forest landscape. Landscape Ecology *9*:105–115.

Hansson, L., L. Fahrig, and G. Merriam, editors. 1995. Mosaic landscapes and ecological processes. Chapman and Hall, London.

Hardin, G. 1968. The tragedy of the commons. Science *162*:1243–1248.

Harper, J. 1992. Foreword. Pages xi–xviii *in* P. L. Fiedler and S. K. Jain, editors. Conservation biology. Chapman and Hall, New York.

Harper, J. L. 1967. A Darwinian approach to plant ecology. Journal of Ecology *55*:247–270.

Harper, J. L. 1987. The heuristic value of ecological restoration. Pages 35–45 *in* W. R. Jordan III, M. E. Gilpin, and J. D. Aber, editors. Ecological restoration: a synthetic approach to ecological research. Cambridge University Press, New York.

Harris, L. D. 1984. The fragmented forest. University of Chicago Press, Chicago.

Harrison, P. 1992. The third revolution: environment, population and a sustainable world. I. B. Tauris, London.

Harrison, P. 1993. Wildlife and people: scrambling for space. People and the Planet 2(3):6–9.

Harrison, S. 1989. Long distance dispersal and colonization in the Bay checkerspot butterfly *Euphydryas editha bayensis*. Ecology *70*:1236–1243.

Harrison, S. 1991a. Local extinction in a metapopulation context: an empirical evaluation. Biological Journal of the Linnean Society *42*:73–88.

Harrison, S. 1991b. Local extinction in a metapopulation context: an empirical evaluation. Pages 73–88 *in* M. Gilpin, and I. Hanski, editors. Metapopulation dynamics: empirical and theoretical investigations. Academic Press, London.

Harrison, S. 1994. Metapopulations and conservation. Pages 111–128 *in* P. J. Edwards, R. M. May, and N. R. Webb, editors. Large-scale ecology and conservation biology. Eds. Blackwell Scientific Publications, Oxford.

Harrison, S., and L. Fahrig. 1995. Landscape pattern and population conservation. Pages 293–308 *in* L. Hansson, L. Fahrig, and G. Merriam, editors. Mosaic landscapes and ecological processes. Chapman and Hall, New York.

Harrison, S., D.D. Murphy, and P.R. Ehrlich. 1988. Distribution of the Bay checkerspot butterfly, *Euphydryas editha bayensis*: evidence for a metapopulation model. American Naturalist *132*:360–382.

Harrison, S., and A. D. Taylor. 1996. Empirical evidence for metapopulation dynamics: A critical review. *In* I. Hanski and M. E. Gilpin, editors. Metapopulation biology: ecology, genetics and evolution. Academic Press, New York.

Hart, T. B. 1990. Monospecific dominance in tropical rain forests. Trends in Ecology and Evolution *5*:6–11.

Harte, J., and R. Socolow, 1971. Patient Earth. Holt Rinehart, and Winston, New York.

Harvey, H. T., H. S. Shellhammer, and R. E. Stecker. 1980. Giant Sequoia ecology: fire and reproduction. U.S. Department of the Interior National Park Service Scientific Monograph no. 12.

Hassell, M. P., H. N. Comins, and R. M. May. 1991. Spatial structure and chaos in insect population dynamics. Nature *353*:255–258.

Hassell, M. P., H. C. J. Godfray, and H. N. Comins. 1993. Effects of global change on the dynamics of insect host- parasitoid interactions. Pages 402–423 *in* P. M. Kareiva, J. G. Kingsolver, and R. B. Huey, editors. Biotic interactions and global change. Sinauer Associates, Sunderland, MA.

Hastings, A., and S. Harrison. 1994. Metapopulation dynamics and genetics. Annual Review of Ecology and Systematics *25*:167–188.

Hawkins, C. P., J. L. Kershner, P. A. Bisson, M. D. Bryant, L. M. Decker, S. V. Gregory, D. A. Mccullough, C. K. Overton, G. H. Reeves, R. J. Steedman, and M. K. Young. 1993. A hierarchical approach to classifying stream habitat features. Fisheries *18*:3–11.

Heap, J. W. 1993. Control of rush skeletonweed (*Chondrilla juncea*) with herbicides. Weed Technology *4*:954–959.

Hedrick, P. W. 1994. Purging inbreeding depression and the probability of extinction: full-sib mating. Heredity *73*:363–372.

Hein, D. 1995. Traditional education in natural resources. Pages 75–87 *in* R. L. Knight and S. F. Bates, editors. A new century for natural resources management. Island Press, Washington, DC.

Heliotis, F., G. Vellidis, D. Bandacu, and C. Pringle. 1994. The Danube Delta: historical wetland drainage and potential for restoration. Pages 759–768 *in* W. J. Mitsch, editor. Global wetlands: old world and new. Elsevier Press, New York.

Henderson, W., and C. W. Wilkins. 1975. The interaction of bushfires and vegetation. Search 6:130–133.

Hengeveld, R. 1990. Dynamic biogeography. Cambridge University Press, Cambridge.

Henningham, J. 1988. Looking at television news. Longman Cheshire, Melbourne.

Henningham, J. 1995. Who are Australia's science journalists? Search 26:89–94.

Herre, E. A. 1993. Population structure and the evolution of virulence in nematode parasites of fig wasps. Science 259:1442–1445.

Hilborn, R. 1992. Hatcheries and the future of salmon in the Northwest. Fisheries 17:5–8.

Hill, G. E. 1995. Ornamental traits as indicators of environmental health. Bioscience 45:25–31.

Hillbricht-Ilkowska, A. 1990. Assessment of trophic impact on the lake environment in Poland: A proposal and case study. Pages 283–296 *in* W. Grodzinski, E. B. Cowling, and A. I. Breymeyer, editors. Ecological risks: perspectives from Poland and the United States. National Academy Press, Washington, DC.

Hobbs, R. J. 1992. The role of corridors in conservation: solution or bandwagon? Trends in Ecology and Evolution 7:389–392.

Hobbs, R. J., and L. F. Huenekke 1992. Disturbance, diversity, and invasion: implications for conservation. Conservation Biology 6:324–337.

Hobbs, R. J., D. A. Saunders, and A. R. Main. 1993. Conservation management in fragmented systems. Pages 279–296 in R. J. Hobbs and D. A. Saunders, editors. Reintegrating fragmented landscapes. Springer-Verlag, New York.

Holden, J., J. Peacock, and T. Williams. 1993. Genes, crops, and the environment. Cambridge University Press, Cambridge.

Holland, M. M., P. G. Risser, and R. J. Naiman, editors. 1991. Ecotones. The role of landscape boundaries in the management and restoration of changing environments. Chapman and Hall, New York.

Holling, C. S. 1978. Adaptive environmental assessment and management. John Wiley & Sons, New York.

Holling, C. S. 1986. The resilience of terrestrial ecosystems, local surprise and global change. Pages 292–317 *in* W. C. Clark and R. E. Munn, editors. Sustainable development of the biosphere. Cambridge University Press, Cambridge, England.

Holling, C. S. 1992. Cross-scale morphology, geometry and dynamics of ecosystems. Ecological Monographs 62:447–502.

Holling, C. S. 1993. Investing in research for sustainability. Ecological Applications 3:552–555.

Holling, C. S., D. W. Schindler, B. W. Walker, and J. Roughgarden. 1995. Biodiversity in the functioning of ecosystems: and ecological synthesis. Pages 44–83 *in* C. Perrings, K. G. Maler, C. Folke, B. O. Jansson, and C. S. Holling, editors. Biodiversity loss: economic and ecological issues. Cambridge University Press, New York.

Holsinger, K. E. 1991. Conservation of genetic diversity in rare and endangered plants. Pages 626–633 *in* E. C. Dudley, editor. The unity of evolutionary biology: the proceedings of the Fourth International Congress of Systematic and Evolutionary Biology, editor. pp. 626–633. Dioscorides Press, Portland, OR.

Holsinger, K. E. 1993. The evolutionary dynamics of fragmented plant populations. Pages 198–216 *in* P. Kareiva, J. Kingsolver, and R. Huey, editors. Biotic interactions and global change. Sinauer Associates, Sunderland, MA.

Holsinger, K. E. 1995a. Population biology for policy makers: promises and paradoxes. BioScience *45*(Supplement):S10–S20.

Holsinger, K. E. 1995b. Conservation programs for endangered plant species. Pages 385–400 *in* W. A. Nierenberg, editor. Encyclopedia of environmental biology, vol. 1. Academic Press, San Diego.

Holsinger, K. E., and L. D. Gottlieb. 1991. Conservation of rare and endangered plants: principles and prospects. Pages 195–208 *in* D. A. Falk and K. E. Holsinger, editors. Genetics and conservation of rare plants. Oxford University Press, New York.

Holt, S. J., and L. M. Talbot. 1978. New Principles for the conservation of wild living resources. Wildlife Monographs *59*:1–33.

Houston, D. B., E. G. Schreiner, and B. B. Moorhead, editors. 1994. Mountain goats in Olympic National Park: biology and management of an introduced species. National Park Service, Denver.

Hudson, R. R. 1990. Gene genealogies and the coalescent process. Oxford Surveys in Evolutionary Biology Evol. Biol. *7*:1–44.

Hull, V. J., and R. H. Groves. 1973. Variation in *Chondrilla juncea* in south-eastern Australia. Australian Journal of Botany *12*:112–135.

Hunter, M. L. 1990. Wildlife, forests, and forestry: principles of managing forests for biological diversity. Regents/Prentice Hall, Engelwood Cliffs, NJ.

Huryn, A. D., and J. B. Wallace. 1987. Local geomorphology as a determinant of macrofaunal production in a mountain stream. Ecology *68*:1932–1942.

Huston, M. A. 1994. Biological diversity: the coexistence of species in changing landscapes. Cambridge University Press, New York.

International Union for the Conservation of Nature and Natural Resources (I.U.C.N.). 1980. World Conservation Strategy. I.U.C.N., Gland, Switzerland.

Intriligator, M. D. 1971. Mathematical optimisation and economic theory. Prentice-Hall, Englewood Cliffs, NJ.

IUDZG/CBSG (IUCN/SSG). 1993. The world zoo conservation strategy: the role of the zoos and aquaria of the world in global conservation. Chicago Zoological Society, Chicago.

IUCN. 1980. World Conservation Strategy. IUCN/UNEP/WWF, Gland, Switzerland.

Jehl, J. R., Jr., and K. C. Parkes. 1983. "Replacements" of landbird species on Socorro Island, Mexico. Auk *100*:551–559.

Jensen, D. B. 1994. Testimony, Harmful non-indigenous species in the United States, Senate Governmental Affairs Committee, 15 March.

Johnson, K. 1993. Reconciling rural communities and resource conservation. *Environment 35*:16–33.

Johnstone, C. A. 1995. Effects of animals on landscape pattern. Pages 57–80 *in* R. Hanson, L. Farhig and G. Merriam, editors. Mosaic landscapes and ecological processes. Chapman and Hall, London.

Jolly, C. L., and B. B. Torrey, editors. 1993. Population and land use in developing countries. National Research Council Workshop. National Academy Press, Washington, DC.

Jones, C. G., and J. H. Lawton, editors. 1995. Linking species and ecosystems. Chapman and Hall, New York.

Jones, C. G., J. H. Lawton, and M. Shachak. 1993. Organisms as ecosystem engineers. Oikos *69*:373–386.

Jones, G. E. 1987. The conservation of ecosystems and species. Croom Helm, New York.

Jones, J. R., L. Rittenhouse, R. Martin, and J. Vieira, editors. 1993. Ecosystem management: beyond the rhetoric. Colorado State University, Fort Collins, CO.

Jordan, C. F., editor. 1987. Amazonian rain forests ecosystem disturbance and recovery. Springer-Verlag, New York.

Joshi, A., and J. N. Thompson. 1995. Alternative routes to the evolution of competitive ability in two competing species of *Drosophila*. Evolution *49*:616–625.

Joshi, A., and J. N. Thompson. Evolution of broad and specific competitive ability in novel versus familiar environments in *Drosophila* species. Evolution *50*:188–194.

Kajak, Z. 1992. The river Vistula and its floodplain valley (Poland): its ecology and importance for conservation. Pages 35—50 *in* P. Boon, P. Calow, and G. E. Petts, editors. River conservation and management. John Wiley & Sons, New York.

Kalisz, P. J. 1986. Soil properties of steep Appalachian old fields. Ecology *67*:1011–1023.

Kalisz, P. J., and E. L. Stone. 1984. Soil mixing by scarab beetles and pocket gophers in north-central Florida. Soil Science Society of America Journal *48*:169–172.

Kareiva, P. 1990. Population dynamics in spatially complex environments: theory and data. Philosophical Transactions of the Royal Society London B *330*:175–190.

Kareiva, P., and U. Wennergren. 1995. Connecting landscape patterns to ecosystem and population processes. Nature *373*:299–302.

Kareiva, P. M., and M. Anderson. 1988. Spatial effects and species interactions: the wedding of models and experiments. Pages 33–52 *in* A. Hastings, editor. Community ecology. Springer-Verlag, New York.

Karr, J. 1993. Advocacy and responsibility. Conservation Biology *7*:8–9.

Kautz, R. S., D. T. Gilbert, and G. M. Mauldin. 1993. Vegetative cover in Florida based on 1985–1989 Landsat Thematic Mapper imagery. Florida Scientist *56*:135–154.

Keddy, P. A. 1991. Working with heterogeneity: an operator's guide to environmental gradients. Pages 181–201 *in* J. Kolasa and S. T. A. Pickett, editors. Ecological heterogeneity. Springer-Verlag, New York.

Keddy, P. A., H. T. Lee, and I. C. Wisheu. 1993. Choosing indicators of ecosystem integrity: wetlands as a model system. Pages 61–79 *in* S. Woodley, J. Kay, and G. Francis, editors. Ecological integrity and the management of ecosystems. St. Lucie Press, Ottawa.

Keeley, J. E. 1981. Reproductive cycles and fire regimes. Pages 231–277 *in* H. A. Mooney, T. M. Bonnickson, N. L. Christensen, J. E. Lotan, and W. A. Reiners, editors. 1981. Fire regimes and ecosystem properties. U.S. Department of Agriculture Forest Service, Washington, DC.

Keiter, R. B., and M. S. Boyce, editors. 1991. The greater Yellowstone ecosystem: redefining America's wilderness. Yale University Press, New Haven.

Kelly, J. R., and M. A. Harwell. 1990. Indicators of ecosystem recovery. Environmental Management *14*:527–545.

Kempf, M. 1993. A new way to oversee the public's forests? American Forests *99*:28–31.

Keystone Center. 1993. National ecosystem management forum: meeting summary. The Keystone Center, Keystone, CO.

Klijn, F., and H. A. Udo de Haes. 1994. A hierarchical approach to ecosystems and its implications for ecological land classification. Landscape Ecology *9*:89–104.

Kolasa, J., and S. T. A. Pickett, editors. 1991. Ecological heterogeneity. Springer-Verlag, New York.

Kotliar, N. B., and J. A. Wiens. 1990. Multiple scales of patchiness and patch structure: a hierarchical framework for the study of heterogeneity. Oikos *59*:253–260.

Kuhn, T. 1970. The structure of scientific revolutions. 2nd edition. University of Chicago Press, Chicago.

Kuussaari, M., M. Nieminen, and I. Hanski. In press. An experimental study of migration in the butterfly *Melitaea cinxia*. Journal of Animal Ecology.

Lack, D. 1976. Island biology illustrated by the land birds of Jamaica. Blackwell, Oxford.

Lacy, R. C. 1992. The effects of inbreeding on isolated populations: are minimum viable population sizes predictable? Pages 277–296 *in* P. L. Fiedler and S. K. Jain, editors. Conservation biology: the theory and practice of nature conservation and management. Chapman and Hall, New York.

Lacy, R. C. 1993. Vortex: A computer simulation model for population viability analysis. Wildlife Research *20*:45–65.

Lacy, R. C. In press. Putting population viability analysis to work in endangered species recovery and small population management. *In* S. Fleming, editor. Conserving species dependent on older forests: a population viability workshop. Parks Canada, Fundy National Park, Alma, New Brunswick.

Laidlaw, K. 1996. The implementation of a volunteer stream monitoring program in Costa Rica. Masters Thesis, Institute of Ecology, University of Georgia, Athens.

Lamberson, R. H., R. McKelvey, B. R. Noon, and C. Voss. 1992. A dynamic analysis of northern spotted owl viability in a fragmented forest landscape. Conservation Biology 6:505–512.

Lamberson, R. H., B. R. Noon, C. Voss, and K. S. McKelvey. 1994. Reserve design for territorial species: the effects of patch size and spacing on the viability of the Northern Spotted Owl. Conservation Biology 8:185–195.

Lande, R. 1975. The maintenance of genetic variability by mutation in a polygenic character with linked loci. Genetic Research 26:221–234.

Lande, R. 1987. Extinction thresholds in demographic models of territorial populations. American Naturalist 130:624–635.

Lande, R. 1988a. Demographic models of the northern spotted owl (*Strix occidentalis caurina*). Oecologia 75:601–607.

Lande, R. 1988b. Genetics and demography in biological conservation. Science 241:1455–1460.

Lande, R. 1993. Risks of population extinction from demographic and environmental stochasticity and random catastrophes. American Naturalist 142:911–927.

Lande, R. 1994. Risk of population extinction from fixation of new deleterious mutations. Evolution 48:1460–1469.

Landers, J. L., N. A. Byrd, and R. Komarek. 1990. A holistic approach to managing longleaf pine communities, *in* 1989 Proceedings, p. 137.

Landres, P. B., J. Verner, and J. W. Thomas. 1988. Ecological uses of vertebrate indicator species: a critique. Conservation Biology 2:316–328.

Launer, A. E., and D. D. Murphy. 1994. Umbrella species and the conservation of habitat fragments: a case of a threatened butterfly and a vanishing grassland ecosystem. Biological Conservation 69:145–153.

Lawton, J. H. 1988. Population dynamics: more time means more variation. Nature 334:563.

Lawton, J. H. 1995. Ecological experiments with model systems. Science 269:328–331.

Lawton, J. H., and C. G. Jones. 1995. Linking species and ecosystems: organisms as ecosystem engineers. Pages 141–150 *in* C. G. Jones and J. H. Lawton, editors. Linking species and ecosystems. Chapman and Hall, New York. 150.

Lawton, J. H., S. Naeem, R. M. Woodfin, V. K. Brown, A. Gange, H. J. C. Godfray, P. A. Heads, S. Lawler, D. Magda, C. D. Thomas, L. J. Thompson, and S. Young. 1993. The Ecotron: a controlled environmental facility for the investigation of population and ecosystem processes. Philosophical Transactions of the Royal Society London B 341:181–194.

Lawton, J. H., S. Nee, A. J. Letcher, and P. H. Harvey. 1994. Animal distributions: patterns and process. Pages 41–58 *in* P. J. Edwards, R. M. May, and N. R. Webb, editors. Large-scale ecology and conservation biology. Blackwell Scientific Press, Oxford.

Ledig, F. T. 1986. Heterozygosity, heterosis, and fitness in outbreeding plants. Pages 77–104 *in* M. E. Soulé, editor. Conservation biology: the science of scarcity and diversity. Sinauer Associates, Sunderland, MA.

Lee, K.N. 1993. Compass and gyroscope: integrating science and politics for the environment. Island Press, Washington, DC.

Lefkovitch, L. P., and L. Fahrig. 1985. Spatial characteristics of habitat patches and population survival. Ecological Modeling *30*:297–308.

Lehmkuhl, J. F., and M. G. Raphael. 1993. Habitat pattern around northern spotted owl locations on the Olympic Peninsula, Washington. Journal of Wildlife Management *57*:302–315.

Leopold, A. 1933. The conservation ethic. Journal of Forestry *31*:634–643.

Leopold, A. 1949. A Sand County almanac. Oxford University Press, New York.

Leopold, A. S., S. A. Cain, C. M. Cottam, J. N. Gabrielson, and T. L. Kimball. 1963. Wildlife management in the national parks. American Forests *69*:32–35, 61–63.

Levin, S. A. 1989. Challenges in the development of a theory of community and ecosystem structure and function. Pages 242–255 *in* J. Roughgarden, R. M. May, and S. A. Levin, editors. Perspectives in ecological theory. Princeton University Press, Princeton.

Levin, S. A. 1992. The problem of pattern and scale in ecology. Ecology *73*:1943–1967.

Levin, S. A., and R. T. Paine. 1974. Disturbance, patch formation, and community structure. Proceedings of the National Academy of Science, USA *71*(7):2744–2747.

Levins, R. 1969. Some demographic and genetic consequences of environmental heterogeneity for biological control. Bulletin of the Entomological Society of America *15*:237–240.

Levins, R. 1970. Extinction. Pages 77–107 *in* M. Gerstenhaber, editor. Some mathematical questions in biology, second symposium on mathematical biology. American Mathematical Society, Providence, RI.

Likens, G. E., editor. 1989. Long-term studies in ecology: approaches and alternatives. Springer-Verlag, New York.

Likens, G. E. 1991. Human-accelerated environmental change. BioScience *41*:130.

Likens, G. E. 1992. The ecosystem approach: its use and abuse. Ecology Institute, Oldendorf/Luhe, Germany.

Likens, G. E., and F. H. Bormann. 1995. Biogeochemistry of a forested ecosystem. 2nd edition. Springer-Verlag, New York.

Likens, G. E., F. H. Bormann, R. S. Pierce, J. S. Eaton, and N. M. Johnson. 1977. Biogeochemistry of a forested ecosystem. Springer-Verlag, New York.

Lindenmayer, D. B., and H. P. Possingham. 1994. The risk of extinction: ranking management options for Leadbeater's possum using PVA. Centre for Resource and Environmental Studies, ANU, Canberra, Australia.

Liu, J. 1993. Discounting initial population sizes for prediction of extinction probabilities in patchy environments. Ecological Modeling *70*:51–61.

Liu, J., J. Dunning, and H. Pulliam. 1995. Potential effects of a forest management plan on Bachman's sparrows: linking a spatially explicit model with GIS. Conservation Biology *9*:62–75.

Lockwood, J.L., and S. L. Pimm. 1994. Species: would any of them be missed? Current Biology *4*:455–457.

London, B., M. O'Brien, J. Noell, and J. Bonine. 1985. An NCAP position document: risk analysis. NCAP News *4*(4):33.

Lowe, I. 1995. Are we really that smart? New Scientist *147*:47.

Lubchenco, J., A. M. Olson, L. B. Brubaker, S. R. Carpenter, M. M. Holland, S. P. Hubbell, S. A. Levin, J. A. Macmahon, P. A. Matson, J. M. Melillo, H. A. Mooney, C. H. Peterson, H. R. Pulliam, L. A. Real, P. J. Regal, and P. G. Risser. 1991. The sustainable biosphere inititative: an ecological research agenda. Ecology *72*:371–412.

Ludanskiy, M. L., D. McDonald, and D. MacNeill. 1993. Impact of the zebra mussel, a bivalve invader. BioScience *43*:533–544.

Ludwig, D., R. Hilburn, and C. Walters. 1993. Uncertainty, resource exploitation and conservation: lessons from history. Science *260*:17–26.

Ludwig, D., D. D. Jones, and C. S. Holling. 1978. Qualitative analysis of insect outbreak systems: the spruce budworm and the forest. Journal of Animal Ecology *44*:315–332.

Lyman, R. L. 1988. Significance for wildlife management of the late Quaternary biogeography of mountain goats (*Oreamnos americanus*) in the Pacific Northwest, U.S.A. Arctic and Alpine Research *20*(1):13–23.

Lyman, R. L. 1994. The Olympic mountain goat controversy: a different perspective. Conservation Biology *8*:898–901.

MacArthur, R. H., and J. W. MacArthur. 1961. On bird species diversity. Ecology *42*: 594–598.

MacArthur, R. H., H. Recher, and M. Cody. 1966. On the relation between habitat selection and species diversity. American Naturalist *100*:319–332.

MacArthur, R. H., and E. O. Wilson. 1967. The theory of island biogeography. Princeton University Press, Princeton.

Mace, G. M., and R. Lande. 1991. Assessing extinction threats: toward a reevaluation of IUCN threatened species categories. Conservation Biology *5*:148–157.

MacKay, H. 1994. A Prototype decision support system for the Kruger National Park Rivers Research Programme. Report No. 74/95, Water Research Commission, Pretoria.

Magnuson, J. J. 1990. Long-term ecological research and the invisible present. BioScience *40*:495–501.

Maguire, L. A. 1986. Using decision analysis to manage endangered species populations. Journal of Environmental Management *22*:345–360.

Maguire, L. A. 1991. Risk analysis. Conservation Biology *5*:123–125.

Maguire, L. A., U. S. Seal, and P. F. Brussard. 1987. Pages 141–158 *in* M. E. Soulé, editor. Viable populations for conservation. Cambridge University Press, Cambridge, UK.

Maguire, L. A., and C. Servheen. 1992. Integrating biological and sociological concerns in endangered species management: augmentation of grizzly bear populations. Conservation Biology *6*:426–434.

Malcolm, S. 1995. Earth issues. Forest regrowth does not guarantee full recovery. The Age, 14 February, p. 12.

Malecki, R. A., B. Blossey, S. D. Hight, D. Schroeder, L. T. Kok, and J. R. Coulson. 1993. Biological control of purple loosestrife. BioScience *43*:680–686.

Mangel, M. 1995. Social interactions, nonlinear dynamics and task allocation in groups. Trends in Ecology and Evolution *10*:347.

Mangel, M., and C. Tier. 1993a. A simple direct method for finding persistence times of populations and application to conservation problems. Proceedings of the National Academy of Sciences, USA *90*:1083–1086.

Mangel, M., and C. Tier. 1993b. Dynamics of metapopulations with demographic stochasticity and environmental catastrophes. Theoretical Population Biology *44*:1–31.

Mangel, M., and C. Tier. 1994. Four facts every conservation biologist should know about persistence. Ecology *75*:607–614.

Mangel, M., L. M. Talbot, G. K. Meffe, M. T. Agardy, D. L. Alverson, J. Barlow, D. B. Botkin, G. Budowski, T. Clark, J. Cookie, R. H. Crozier, P. K. Dayton, D. L. Elder, C. W. Fowler, S. Funtowicz, J. Giske, R. J. Hofman, S. J. Holt, S. R. Kellert, L. A. Kimball, D. Ludwig, K. Magnusson, B. S. Malayang III, C. Mann, E. A. Norse, S. P. Northridge, W. F. Perrin, C. Perrings, R. M. Peterman, G. B. Rabb, H. A. Regier, J. E. Reynolds III, K. Sherman, M. Sissenwine, T. D. Smith, A. Starfield, R. J. Taylor, M. F. Tillman, C. Toft, J. R. Twiss, Jr., J. Wilen, and T. P. Young. 1996. Principles for the conservation of wild living resources. Ecological Applications *6*:338–362.

Mann, C. C., and M. L. Plummer. 1995. Noah's choice: the future of endangered species. Knopf, New York.

March, J. G., J. P. Benstead, and C. M. Pringle. 1996. Migration of freshwater shrimp larvae: elevational and diel patterns in two tropical river drainages, Puerto Rico. Bulletin of the North American Benthological Society *13*:161.

Marek, M. J., and A. T. Kassenberg 1990. The relationship between strategies of social development and environmental protection, Pages 41–59 *in* W. Grodzinski, E. B. Cowling, and A. I. Breymeyer, editors. Ecological risks: perspectives from Poland and the United States. National Academy Press, Washington, DC.

Margalef, R. 1968. Perspectives in ecological theory. University of Chicago Press, Chicago.

Margolis, H. 1993. Paradigms and barriers: how habits of mind govern scientific beliefs. University of Chicago Press, Chicago.

Marks, P., J. D. Allen, C. Canham, K. Van Cleve, A. Covich, and F. James. 1993. Scientific peer review of the ecological aspects of forest ecosystem management: an ecological, economic and social assessment. A report of the Ecological Society of America, Washington, DC.

Marquette, C. M., and Bilsborrow, R. 1994. Population and the environment in developing countries: literature survey and research bibliography. ESA/P/WP.123. Population Division, United Nations, New York.

Marsh, G. P. 1864. Man and nature; or, physical geography as modified by human action. Reprinted 1965, Belknap Press of Harvard University Press, Cambridge.

Master, L. 1990. The imperiled status of North American aquatic animals. The Nature Conservancy Biodiversity Network News *3*:1–8.

May, R. M. 1988. Conservation and disease. Conservation Biology *2*:28–30.

May, R. M. 1994a. Ecological science and the management of protected areas. Biodiversity and Conservation *3*:437–448.

May, R.M. 1994b. The effects of spatial scale on ecological questions and answers. Pages 1–17 *in* P. J. Edwards, R. M. May, and N. R. Webb, editors. Large-scale ecology and conservation biology. Blackwell Scientific Publications, Boston.

May, R. M., and R. M. Anderson. 1988. The transmission dynamics of human immunodeficiency virus (HIV). Philosophical Transactions of the Royal Society London B *321*:565–607.

May, R. M., and R. M. Anderson. 1990. Parasite-host coevolution. Parasitology *100*:S89–S101.

Mazumder, A., W. D. Taylor, D. J. Mcqueen, and D. R. S. Lean. 1990. Effects of fish and plankton on lake temperature and mixing depth. Science *247*:312–315.

McCleese, W. 1994. Testimony, Harmful non-indigenous species in the United States, Senate Governmental Affairs Committee, 15 March.

McDonald, K. A., and J. H. Brown. 1992. Using montane mammals to model extinctions due to global change. Conververvation Biology *6*:409–415.

Mcdonald, M. 1995. A combination in behalf of restoration: the Coalition to Restore Urban Waters. Restoration and Management Notes *13*:98–103.

McDonnell, M. J., and S. T. A. Pickett, editors. 1993. Humans as components of ecosystems: the ecology of subtle human effects and populated areas. Springer-Verlag, New York.

McIlvanie, S. K. 1942. Grass seedling establishment and productivity—overgrazed and protected range soils. Ecology *2*:228–231.

McIntosh, R. P. 1987. Pluralism in ecology. Annual Review of Ecology and Systematics *18*: 321–341.

McKelvey, K., B. E. Noon, and R. H. Lambertson. 1992. Conservation planning for species occupying fragmented landscapes: the case of the northern spotted owl. Pages 424–450 *in* P. M. Kareiva, J. G. Kingsolver, and R. B. Huey, editors. Biotic interactions and global change. Sinauer, Sunderland, MA.

McLaughlin, J. F., and J. Roughgarden. 1993. Species interactions in space. Pages 59–98 *in* R. E. Ricklefs and D. Schluter, editors. Species diversity in ecological communities: historical and geographic perspectives. University of Chicago Press, Chicago.

McNaughton, S. J. 1989. Ecosystems and conservation in the twenty-first century. Pages 109–120 *in* D. Western and M. C. Pearl, editors. Conservation for the twenty-first century. Oxford University Press, New York.

McNeill, S. E., and P. G. Fairweather. 1993. Single large or several small marine reserves? An experimental approach with seagrass fauna. Journal of Biogeography *20*:429–440.

Meffe, G. K., and C. R. Carroll. 1994. Principles of conservation biology. Sinauer, Sunderland, MA.

Meffe, G. K., A. H. Ehrlich, and D. Ehrenfeld. 1993. Editorial: human population control: the missing agenda. Conservation Biology *7*(1):1–3.

Meine, C. 1995. The oldest task in human history. Pages 7–35 *in* R. L. Knight and S. F. Bates, editors. A new century for natural resources management. Island Press, Washington, DC.

Menges, E. S. 1988. Conservation biology of Furbish's lousewort. Holcomb Research Institute Report No. 126. Indianapolis.

Menges, E. S. 1990. Population viability analysis for an endangered plant. Conservation Biology 4:52–62.

Menges, E. S. 1991. The application of minimum viable population theory to plants. Pages 45–61 *in* D. A. Falk and K. E. Holsinger, editors. Genetics and conservation of rare plants. Oxford University Press, New York.

Menges, E. S., S. C. Gawler, and D. M. Waller. 1985. Population biology of the endemic plant, Furbish's lousewort (*Pedicularis furbishiae*), 1984 reseach. Report to the U.S. Fish and Wildlife Service. Holcomb Research Institute Report No. 40. Indianapolis.

Menges, E. S., D. M. Waller, and S. C. Gawler. 1986. Seed set and seed predation in *Pedicularis furbishiae*, a rare endemic of the St. John River, Maine. American Journal of Botany 73:1168–1177.

Merriam, G. 1991. Corridors and connectivity: animal populations in heterogeneous environments. Pages 133–142 *in* D. A. Saunders and R. J. Hobbs, editors. Nature conservation 2: the role of corridors. Surrey Beatty, Chipping Norton, Australia.

Metting, B., and W. R. Rayburn. 1983. The influence of microalgal conditions on selected Washington soils: an empirical study. Soil Science Society of America Journal 47:682–685.

Metzler, K. J., and Tiner, R. W. 1992. Wetlands of Connecticut. State Geological and Natural History Survey of Connecticut. Report of investigations No. 13.

Meyer, J. L. 1994. The dance of nature: new concepts in ecology. Chicago Kent Law Review 69:875–886.

Meyer, J. L., and R. T. Edwards. 1990. Ecosystem metabolism and turnover of organic carbon along a blackwater continuum. Ecology 71:668–677.

Meyers, N. 1991. Biologists as policymakers? Environmental Conservation 18:6.

Meylan, A. B., B. W. Bowen, and J. C. Avise. 1990. A genetic test of the natal homing versus social facilitation models for green turtle migration. Science 248:724–727.

Miller, R. R., J. D. Williams, and J. E. Williams. 1989. Extinctions of North American fishes during the past century. Fisheries 14:22–38.

Mills, L. S., C. Baldwin, M. Wisdom, J. Citta, D. J. Mattson, and K. Murphy. 1996. Factors leading to different viability predictions for a grizzly bear data set. Conservation Biology 10:863–873.

Milne, B. T. 1991. Heterogeneity as a multiscale characteristic of landscapes. Pages 69–84 *in* J. Kolasa and S. T. A. Pickett, editors. Ecological heterogeneity. Springer-Verlag, New York.

Minnich, R. A. 1988. Chaparral fire history in San Diego County and adjacent northern Baja California: an evaluation of natural fire regimes and the effects of suppression management. Pages 37–48 *in* S. C. Keeley, editor. The California chaparral: paradigms rexamined. Natural History Museum of Los Angeles County, Los Angeles.

Minshall, G. W., R. C. Petersen, K. W. Cummins, T. L. Bott, J. R. Sedell, C. E. Cushing, and R. L. Vannote. 1983. Interbiome comparison of stream ecosystem dynamics. Ecological Monographs 53:1–25.

Mock, P. J. 1992. Population viability analysis for the California gnatcatcher within the MSCP study area. Ogden Environmental and Energy Services Co., Inc. San Diego, CA. (Revised, February 1993.)

Mock, P. J., and D. Bolger. 1992. Ecology of the California gnatcatcher for Rancho San Diego. Ogden Environmental and Energy Services Co., Inc. San Diego, CA. Project No. 110970000.

Mönkkönen, M., and D. A. Welsh. 1994. A biogeographical hypothesis on the effects of human caused landscape changes on the forest bird communities of Europe and North America. Annales Zoologica Fennici *31*:61–70.

Moon, B. P., A. W. van Niekerk, G. L. Heritage, K. H. Rogers, and C. S. James. In press. A geomorphological approach to the management of rivers in the Kruger National Park: the case of the Sabie River. Transactions of the Institute of British Geographers.

Moore, A. D. 1990. The semi-Markov process: a useful tool in the analysis of vegetation dynamics for management. Journal of Environmental Management *30*:111–130.

Moore, R. M., and J. A. Robertson. 1964. Studies on skeletonweed: competition from pasture plants. CSIRO, Division of Plant Industry, Field Station Record No. 3:69–73.

Moulton, M. P., and S. L. Pimm. 1983. The introduced Hawaiian avifauna: biogeographical evidence for competition. American Naturalist *121*:669–690.

Mukai, T. 1979. Polygenic mutation. Pages 177–196 *in* J. N. Thompson, Jr., and J. M. Thoday, editors. Quantitative genetic variation. Academic Press, New York.

Muller, H. J. 1964. The relation of recombination to mutational advance. Mutation Research *1*:2–9.

Murcia, C. 1995. Edge effects in fragmented forests: implications for conservation. Trends in Ecology and Evolution *10*:58–62.

Murdoch, W. W. 1969. Switching in general predators: experiments on predator specificity and stability of prey populations. Ecological Monographs *39*:335–354.

Murphy, D. D., K. E. Freas, and S. B. Weiss. 1990. An environment-metapopulation approach to population viability analysis for a threatened invertebrate. Conservation Biology *4*: 51–51.

Murphy, D. D., and B. R. Noon. 1992. Integrating scientific methods with habitat conservation planning: reserve design for the northern spotted owl. Ecological Applications *2*:3–17.

Myers, A. A., and P. S. Giller. 1988. Introduction. Pages 3–12 *in* A. A. Myers and P. S. Giller, editors. Analytical biogeography. Chapman and Hall, London.

Myers, N. 1988. Threatened biotas: "hotspots" in tropical forests. The Environmentalist *8*:1–20.

Myers, N. 1990. The biodiversity challenge: expanded hot-spots analysis. The Environmentalist *10*:243–256.

Myers, N. 1993. The question of linkages in environment and development. BioScience *43*:302–310.

Myers, R. L. 1985. Fire and the dynamic relationship between Florida sandhill and sand pine scrub vegetation. Bulletin of the Torrey Botanical Club *112*:241–252.

Naeem, S., and R. K. Colwell. 1991. Ecological consequences of heterogeneity of consumable resources. Pages 114–255 *in* J. Kolasa and S. T. A. Pickett, editors. Ecological heterogeneity. Springer-Verlag, New York.

Naeem, S., L. J. Thompson, S. P. Lawler, J. H. Lawton, and R. M. Woodfin. 1994. Declining biodiversity can alter the performance of ecosystems. Nature *368*:734–737.

Naiman, R. J., C. A. Johnston, and J. C. Kelley. 1988. Alteration of North American streams by beaver. BioScience *38*:752–762.

Naiman, R. J., J. J. Magnuson, D. M. McKnight, and J. A. Stanford. 1995. The freshwater imperative: a research agenda. Island Press, Washington, DC.

Naiman, R. J., J. M. Melillo, M. A. Lock, T. E. Ford, and S. R. Reice. 1987. Longitudinal patterns of ecosystem processes and community structure in a subarctic river continuum. Ecology *68*:1139–1156.

Naiman, R. J., and K. H. Rogers. Submitted. Large animal and the maintenance of system level characteristics in river corridors. BioScience.

Nash, R. F. 1989. The rights of nature: a history of environmental ethics. University of Wisconsin Press, Madison.

National Research Council. 1986. Ecological knowledge and environmental problem-solving: concepts and case studies. National Academy Press, Washington, DC.

National Research Council. 1992. Conserving biodiversity: a research agenda for development agencies. National Academy Press, Washington, DC.

National Research Council. 1993. Setting priorities for land conservation. National Research Council, Washington, DC.

National Research Council. 1995. Science and the endangered species act. National Academy of Sciences Press, Washington, DC.

National Wildlife Federation (NWF). 1994. Grazing to extinction: endangered, threatened and candidate species imperiled by livestock grazing on western public lands: *8*. National Wildlife Federation, Washington, DC.

Naumann, M. 1994. A water-use budget for the Caribbean National Forest of Puerto Rico. Special report to the U.S. Forest Service, International Institute of Tropical Forestry, Rio Piedras, Puerto Rico.

Nee, S. 1994. How populations persist. Nature *367*:123–124.

Nee, S., and R. M. May. 1992. Dynamics of metapopulations: habitat destruction and competitive coextince. Journal of Animal Ecology *61*:37–40.

Nehlsen, W., J. E. Williams, and J. A. Lichatowich. 1991. Pacific salmon at the crossroads: stocks at risk from California, Oregon, Idaho, and Washington. Fisheries *16*:4–21.

Nei, M., T. Maruyama, and R. Chakraborty. 1975. The bottleneck effect and genetic variability in populations. Evolution *29*:1–10.

Neilson, R. P., and L. H. Wullstein. 1983. Biogeography of two southwest American oaks in relation to atmospheric dynamics. Journal of Biogeography *10*:275–297.

NEWFS (New England Wild Flower Society, Inc.). 1992. New England plant conservation program. Wild Flower Notes *7*(1):1–79.

Newman, F. W. 1853. Phases of faith, or, passages from the history of my creed. J. Chapman, London.

Newson, M. D. 1992. River conservation and catchment management: A UK perspective. Pages 385–396 *in* P. J. Boon, P. Calow, and G. E. Petts, editors. River conservation and management. John Wiley & Sons, New York.

Nip, M. I., and H. A. U. de Haes. 1995. Ecosystem approaches to environmental quality assessment. Environmental Management *19*:135–145.

Noon, B., and K. McKelvey. 1992. Stability properties of the spotted owl metatpopulation in southern California. Pages 187–206 *in* J. Verner, K. S. McKelvey, B. R. Noon, technical editors. The California spotted owl: a technical assessment of its current status. U.S. Forest Service Technical Report PSW–133.

Noon, B. R., and K. S. McKelvey. In press. A common framework for conservation planning: Linking individual and metapopulation models. *In* D. R. McCullough, editor. Metapopulations and wildlife conservation management. Island Press, Washington DC. Island Press.

North, R. D. 1995. End of the green crusade. New Scientist *145*:38–41.

Norton, B. G. 1992. A new paradigm for environmental management, Pages 23–41 *in* R. Costanza, B. G. Norton, and B. D. Haskell, editors. Ecosystem health. Island Press, Washington, DC.

Noss, R. F. 1983. A regional landscape approach to maintain diversity. BioScience *33*:700–706.

Noss, R. F. 1990. Indicators for monitoring biodiversity: a hierarchial approach. Conservation Biology *4*:355–364.

Noss, R. F., and A. Y. Cooperrider. 1994. Saving nature's legacy: protecting and restoring biodiversity. Island Press, Washington, DC.

Noss, R. F., and L. D. Harris. 1986. Nodes, networks, and MUMs: preserving diversity at all scales. Environmental Management *10*:299–309.

Noss, R. F., and D. D. Murphy. 1995. Endangered species left homeless in Sweet Home. Conservation Biology *9*:229–231.

Noss, R. F., and R. L. Peters. 1995. Endangered ecosystems: a status report on America's vanishing habitat and wildlife. Defenders of Wildlife, Washington, DC.

Noss, R. F., E. T. LaRoe, and J. M. Scott. 1995. Endangered ecosystems of the United States: a preliminary assessment of loss and degradation. U.S. Department of the Interior, National Biological Service, Washington, DC.

Nott, M. P., E. Rogers, and S. Pimm. 1994. Modern extinctions in the kilo-death range. Current Biology *5*:14–17.

Nunney, L., and K. A. Campbell. 1993. Assessing minimum viable population size: demography meets population genetics. Trends in Ecology and Evolution *8*:234–239.

Nunney, L., and D. R. Elam. 1994. Estimating the effective population size of conserved populations. Conservation Biology *8*:175–184.

O'Brien, M. 1993. Being a scientist means taking sides. BioScience *43*:706–708.

Odum, E. P. 1969. The strategy of ecosystem development. Science *164*:262–270.

Oliver, C. D., and B. C. Larson. 1990. Forest stand dynamics. McGraw-Hill, New York.

O'Neill, G. 1993. From ashes to ashes. Time Australia 8(36):58–59.

O'Neill, R. V. 1989. Perspectives in hierarchy and scale. Pages 140–156 *in* R. M. May and J. Roughgarden, editors. Ecological theory. Princeton University Press, Princeton.

O'Neill, R. V., D. L. DeAngelis, J. B. Waide, and T. F. H. Allen. 1986. A hierarchical concept of ecosystems. Princeton University Press, Princeton.

Opdam, P., R. Van Apeldoorn, A. Schotman, and J. Kalkhoven. 1993. Population responses to landscape fragmentation. Pages 141–171 *in* C. C. Vos and P. Opdam, editors. Landscape ecology of a stressed environment. Chapman and Hall, London.

Orians, G. H. 1993. Endangered at what level? Ecological Applications 3:206–208.

Osborn, F. 1948. Our plundered planet. Little, Brown, Boston.

Pace, M. L., S. R. Carpenter, and P. A. Soranno. 1995. Population variability in experimental ecosystems. Pages 61–71 *in* C. G. Jones and J. H. Lawton, editors. Linking species and ecosystems. Chapman and Hall, New York.

Pagel, M. D., R. M. May, and A. R. Collie. 1991. Ecological aspects of the geographical distribution and distribution of mammalian species. American Naturalist 137:791–815

Palmer, T. 1994. Lifelines: the case for river conservation. Island Press, Washington, DC.

Parker, G. G. 1985. The effect of disturbance on water and solute budgets of hillslope tropical rainforest of northeastern Costa Rica. PhD thesis. University of Georgia, Athens.

Parker, I. C. S. 1983. The Tsavo story: an ecological case history. Pages 37–50 *in* R. N. Owen-Smith, editor. Management of large mammals in African conservation areas. Haum, Pretoria.

Parker, V. T. 1993. Conservation issues in land management. Pages 53–60 *in* J. E. Keeley, editor. Interface between ecology and land development in California. Southern California Academy of Sciences, Los Angeles.

Patton, P. C., and G. S. Horne. 1991. A submergence curve for the Connecticut River estuary. Journal of Coastal Research Special Issue 11:181–196.

Pauly, D., and V. Christenson. 1995. Primary production required to sustain global fisheries. Nature 374:255–257.

Pearce, D. W., and J. J. Warford. 1993. World without end: economics, environment, and sustainable development. Published for the World Bank. Oxford University Press, New York.

Pearson, S. M., M. G. Turner, R. H. Gardner, and R. V. O'Neill. 1995. An organism-based perspective of habitat fragmentation. Pages 77–95 *In* R. C. Szaro and D. W. Johnston, editors. Biodiversity in managed landscapes: theory and practice. Oxford University Press, Oxford.

Pearson, S. M., M. G. Turner, L. L. Wallace, and W. H. Romme. 1995. Winter habitat use by large ungulates following fire in northern Yellowstone National Park. Ecological Applications 5:744–755.

Pechmann, J. H. K., and H. M. Wilbur. 1994. Putting amphibian declines in perspective: natural fluctuations and human impacts. Herpetologica 50:65–84.

Peierls, B. L., N. F. Caraco, M. L. Pace, and J. J. Cole. 1991. Human influence on river nitrogen. Nature *350*:386–387.

Peterjohn, B. J. 1994. The North American breeding bird survey. Birding December:387–398.

Peters, R. H. 1991. A critique for ecology. Cambridge University Press, Cambridge.

Peters, R. L., and J. D. S. Darling. 1985. The greenhouse effect and nature reserves. Bioscience *35*:707–717.

Phipps, M. J. 1992. From local to global: the lesson of cellular automata. Pages 165–187 *in* D. L. DeAngelis and L. J. Gross, editors. Individual-based models and approaches in ecology. Chapman and Hall, New York.

Pickett, S. T. A. 1980. Non-equilibrium coexistence of plants. Bulletin of the Torrey Botanical Club *197*:238–248.

Pickett, S. T. A., and M. L. Cadenasso. 1995. Landscape ecology: spatial heterogeneity in ecological systems. Science *269*:331–334.

Pickett, S. T. A., J. Kolasa, J. J. Armesto, and S. L. Collins. 1989. The ecological concept of disturbance and its expression at various hierarchial levels. Oikos *54*:129–136.

Pickett, S. T. A., J. Kolasa, and C. G. Jones. 1994. Ecological understanding: the nature of theory and the theory of nature. Academic Press, San Diego.

Pickett, S. T. A., and R. S. Ostfeld. 1995. The shifting paradigm in ecology. Pages 261–279 *in* R. L. Knight and S. F. Bates, editors. A new century for natural resources management. Island Press, Washington, DC.

Pickett, S. T. A., and V. T. Parker. 1994. Avoiding the old pitfalls: opportunities in a new discipline. Restoration Ecology *2*:75–79.

Pickett, S. T. A., V. T. Parker, and P. L. Fiedler. 1992. The new paradigm in ecology: Implications for conservation biology above the species level, Pages 65–88 *in* P. L. Fiedler and S. K. Jain, editors. Conservation biology: the theory and practice of nature conservation preservation and management. Chapman and Hall, New York.

Pickett, S. T. A., and K. H. Rogers. In press. Patch dynamics: the transformation of lanscape structure and function. *In* J. A. Bissonette, editor. A primer in landscape ecology. Springer-Verlag, New York.

Pickett, S. T. A., and J. N. Thompson. 1978. Patch dynamics and the design of nature reserves. Biological Conservation *13*:27–37.

Pickett, S. T. A., and P. S. White, editors. 1985. The ecology of natural disturbance and patch dynamics. Academic Press, Orlando, FL.

Pienaar, U. de V. 1983. Management by intervention: the pragmatic/economic option. Pages 23–36 *in* R. N. Owen-Smith, editor. Management of large mammals in African conservation areas. Haum, Pretoria.

Pimentel, D. 1986. Biological invasions into North American forests. Pages 84–98 *in* H. A. Mooney and J. A. Drake, editors. Ecology of biological invasions of North America and Hawaii. Springer-Verlag, New York.

Pimentel, D., and C. W. Hall. 1989. Food and natural resources. Academic Press, San Diego.

Pimm, S. L. 1988. Energy flow and trophic structure. Pages 263–278 *in* L. R. Pomeroy and J. J. Alberts, editors. Concepts of ecosystem ecology: a comparative view. Springer-Verlag, New York.

Pimm, S. L. 1991. The balance of nature? University of Chicago Press, Chicago.

Pimm, S. L., and R. A. Askins. 1995. Forest losses predict bird extinctions in eastern North America. Proceedings National Academy of Sciences U.S.A. *92*:9343–9347.

Pimm, S. L., M. P. Moulton, and L. J. Justice. 1994. Bird extinctions in the central Pacific. Philosophical Transactions of the Royal Society London B *344*:27–33.

Pimm, S. L., G. J. Russell, J. L. Gittleman, and T. M. Brooks. 1995. The future of biodiversity. Science *269*:347–350.

Pingali, P. L., and H. Binswanger. 1987. Population density and agricultural intensification: a study of the evolution of technologies in tropical agriculture. Pages 27–56 *in* D. G. Johnson and R. D. Lee, editors. 1987 Population growth and economic development: issues and evidence. University of Wisconsin Press, Madison. (Originally issued in 1984 as: Report ARU 22. World Bank, Agriculture and Rural Development Department, Washington, DC.)

Poff, N. L., and J. D. Allan. 1995. Functional organization of stream fish assemblages in relation to hydrological variability. Ecology *76*:606–627.

Pojar, J. 1993. Terrestrial diversity of British Columbia, Pages 177–210 *in* M. A. Fenger, E. H. Miller, J. A. Johnson, and E. J. R. Williams, editors. Our living legacy: proceedings of a symposium on biological diversity. Royal British Columbia Museum, Victoria, BC.

Popline. 1992. Central America's shrinking forests. Popline *14*(September–October):4.

Possingham, H. P. (1996). Decision theory and biodiversity management: how to manage a metapopulation. *In* P. Wellings, editor. Frontiers in population ecology. CSIRO Publishing, Canberra.

Possingham, H. P., and I. Davies. 1995. ALEX: A model for the viability analysis of spatially structured populations. Biological Conservation *73*:143–150.

Possingham, H. P., D. B. Lindenmayer, and T. W. Norton. 1993. A framework for the improved management of threatened species based on Population Viability Analysis (PVA). Pacific Conservation Biology *1*:39–45.

Postel, S. 1992. Last oasis: facing water scarcity. W. W. Norton and Company, New York.

Power, M.E. 1992. Hydrologic and trophic controls of seasonal algal blooms in northern California rivers. Archives of Hydrobiology *125*:385–410.

Prendergast, J. R., R. M. Quinn, J. H. Lawton, B. C. Eversham, and D. W. Gibbons. 1993. Rare species, the coincidence of diversity hotspots and conservation strategies. Nature *365*:335–337.

Pressey, R. L., C. J. Humphries, C. R. Margules, R. I. Vane-Wright, and P. H. Williams. 1993. Beyond opportunism: key principles for systematic reserve selection. Trends in Ecology and Evolution *8*:124–128.

Primack, R. B. 1993. Essentials of conservation biology. Sinauer, Sunderland, MA.

Prince, S. D., R. N. Carter, and K. J. Dancy. 1985. The geographical distribution of prickly lettuce (*Lactuca serriola*). II. Characteristics of populations near its distributional limit in Britain. Journal of Ecology *73*:39–48.

Pringle, C. M. 1988. History of conservation efforts and initial exploration of the lower extension of Parque Nacional Braulio Carrillo. Pages 225–241 *in* F. Almeda and C. Pringle, editors. Tropical rainforests: diversity and conservation. Allen Press, Lawrence, KA.

Pringle, C. M. 1991a. U.S.-Romanian environmental reconaissance of the Danube Delta. Conservation Biology *5*:442–447.

Pringle, C. M. 1991b. Geothermal waters surface at La Selva Biological Station, Costa Rica: volcanic processes introduce chemical discontinuities into lowland tropical streams. Biotropica *23*:523–529.

Pringle, C. M. 1996. Atyid shrimp (Decapoda: Atyidae) influence spatial heterogeneity of algal communities over different scales in tropical montane streams, Puerto Rico. Freshwater Biology *35*:125–140.

Pringle, C. M., and N. G. Aumen. 1993. Current issues in freshwater conservation: introduction to a symposium. Journal of the North American Benthological Society *12*:174–176.

Pringle, C. M., and G. A. Blake. 1994. Quantitative effects of atyid shrimp (Decapoda: Atyidae) on the depositional environment in a tropical stream: use of electricity for experimental exclusion. Canadian Journal of Fisheries and Aquatic Sciences *51*:1443–1450.

Pringle, C. M., G. A. Blake, A. P. Covich, K. M. Buzby, and A. Finley. 1993c. Effects of omnivorous shrimp in a montane tropical stream: sediment removal, disturbance of sessile invertebrates and enhancement of understory algal biomass. Oecologia *93*:1–11.

Pringle, C. M., R. J. Naiman, G. Bretschko, J. R. Karr, M. W. Oswood, J. R. Webster, R. L. Welcomme, and M. J. Winterbourn. 1988. Patch dynamics in lotic systems: the stream as a mosaic. Journal of the North American Benthological Society *7*:503–524.

Pringle, C. M., P. Paaby-Hansen, P. D. Vaux, and C. R. Goldman. 1986. In situ nutrient assays of periphyton growth in a lowland Costa Rican stream. Hydrobiologia *134*:207–213.

Pringle, C. M., C. F. Rabeni, A. Benke, and N. G. Aumen. 1993a. The role of aquatic science in freshwater conservation: cooperation between the North American Benthological Society and organizations for conservation and resource management. Journal of the North American Benthological Society *12*:177–184.

Pringle, C. M., G. L. Rowe, F. J. Triska, J. F. Fernandez, and J. West. 1993d. Landscape linkages between geothermal activity, solute composition and ecological response in streams draining Costa Rica's Atlantic slope. Limnology and Oceanography *38*:753–774.

Pringle, C. M., and F. J. Triska. 1991. Effects of geothermal waters on nutrient dynamics of a lowland Costa Rican stream. Ecology *72*:951–965.

Pringle, C. M., G. Vellidis, F. Heliotis, D. Bandacu, and S. Cristofor. 1993b. Environmental problems in the Danube Delta. American Scientist *81*:350—361.

Proctor, J.D. 1995. Whose nature? the contested moral terrain of ancient forests. Pages

269–297 *in* W. Cronon, editor. Uncommon ground: toward reinventing nature. Norton, New York.

Pulliam, H. R. 1988. Sources, sinks, and population regulation. American Naturalist *132*:652–661.

Pulliam, H. R., J. B. Dunning, Jr., and J. Liu. 1992. Population dynamics in complex landscapes: a cast study. Ecological Applications 2:165–177.

Pulliam, H. R., and N. M. Haddad. 1994. Human population growth and the carrying capacity concept. Bulletin of the Ecological Society of America 75:141–157.

Pyne, S. J. 1982. Fire in America. Princeton Unversity Press, Princeton, NJ.

Rahel, F. J. 1990. The hierarchical nature of community persistence: a problem of scale. American Naturalist *136*:328–344.

Ralls, K., and A. M. Starfield. 1995. Choosing a management strategy: two structured decision-making methods for evaluating the predictions of stochastic simulation models. Conservation Biology 9:175–181.

Rapoport, E. H. 1982. Areography. Pergamon Press, Oxford.

Reich, Robert. 1994. Quoted in Reich, Redefining "Competitiveness." The Washington Post, Sept. 24, 1994, p. D1.

Reid, T. S., and D. D. Murphy. 1995. Providing a regional context for local conservation action. BioScience 5:84–92.

Reid, W. V., and K. R. Miller. 1989. Keeping options alive: the scientific basic for conserving biodiversity. World Resources Institute, Washington, DC.

Reiners, W. A. 1981. Nitrogen cycling in relation to ecosystem succession. Ecological Bulletins (Stockholm) *33*:507–528.

Reiners, W.A. 1988. Achievements and challenges in forest energetics. Pages 75–114 *in* L. R. Pomeroy and J. J. Alberts, editors. Concepts of ecosystem ecology: a comparative view. Springer-Verlag, New York.

Remmert, H., editor. 1991. The mosaic-cycle concept of ecosystems. Springer-Verlag, New York.

Reynolds, J. W. 1994. Decision and order: Sierra Club *et al.* v. Floyd J. Marita. Sierra Club V. Marita, 843 F. Supp. 15 (E.D. Wis. 1995), aff'd., 46 F.3d 606 (7th Cir. 1995).

Rhoades, D. F., and R. G. Cates. 1976. A general theory of plant antiherbivore chemistry. Pages 169–213 *in* J. W. Wallace and R. L. Mansell, editors. Biochemical interaction between plants and insects. Plenum Press, New York.

Rhymer, J., and D. Simberloff. 1996. Extinction by hybridization and introgression. Annual Review of Ecology and Systematics *27*:83–109.

Richerson, P. J. 1977. Ecology and human ecology: a comparison of theories in the biological and social sciences. American Ethnologist *4*:1–26.

Richter, B. D. 1993. Ecosystem-level conservation at the Nature Conservancy: growing needs for applied research in conservation biology. Journal of the North American Benthological Society *12*:197–200.

Richter-Dyn, N., and N. S. Goel. 1972. On the extinction of a colonising species. Theoretical Population Biology *3*:406–433.

Ricklefs, R. E. 1977. Review of Lack's Island biology. Auk *94*:794–797.

Ricklefs, R.E., and D. Schluter, editors. 1993. Species diversity in ecological communities. University of Chicago Press, Chicago.

Rieseberg, L. H. 1991. Hybridization in rare plants: insights from case studies in *Cercocarpus* and *Helianthus*. Pages 171–181 *in* D. A. Falk and K. E. Holsinger, editors. Genetics and conservation of rare plants. Oxford University Press, New York.

Rieseberg, L. H., and D. Gerber. 1995. Hybridization in the Catalina Island mountain mahogany (*Cercocarpus traskiae*): RAPD evidence. Conservation Biology *9*:199–203.

Ripple, W. J., D. H. Johnson, K. T. Hershey, and E. C. Meslow. 1991. Old–growth and mature forest near spotted owl nests in western Oregon. Journal of Wildlife Management *55*:316–318.

Risser, P. G. 1995a. Biodiversity and ecosystem function. Conservation Biology *9*:742–746.

Risser, P. G. 1995b. The status of the science examining ecotones. BioScience *45*:318–325.

Rodgers, J. 1985. Bedrock geological map of Connecticut. Connecticut Geological and Natural History Survey, Department of Environmental Protection, and the U.S. Geological Survey, Department of the Interior.

Rogers, C. E., T. E. Thompson, and G. J. Seiler. 1982. Sunflower species of the United States. National Sunflower Association, Bismarck, ND.

Rogers, K. H. 1992. "Applied" interdisciplinary research and the academic: an ecologists perspective. Inaugural lecture. University of the Witwatersrand Archives, Johannesburg.

Rosenberg, D. M., and V. H. Resh, editors. 1993. Freshwater biomonitoring and benthic macroinvertebrates. Chapman and Hall, New York.

Rosenhead, J. 1989. Robustness analysis: keeping your options open. Pages 192–218 *in* J. Rosenhead, editor. Rational analysis for a problematic world. John Wiley & Sons, Chichester.

Rosenzweig, M. L. 1995. Species diversity in space and time. Cambridge University Press, Cambridge.

Ross, J., and M. F. Sanders. 1984. The development of genetic resistance of myxomatosis in wild rabbits in Britain. Journal of Hygiene *92*:255–261.

Rossi, R. E., D. J. Mulla, A. G. Journel, and E. H. Franz. 1992. Geostatistical tools for modeling and interpreting ecological spatial dependence. Ecological Monographs *62*:277–314.

Roszak, T. 1972. Where the wasteland ends: politics and transcendence in post-industrial society. Doubleday, New York.

Roughgarden, J., S. Gaines, and S. Pacala. 1987. Supply-side ecology: the role of physical transport processes. Pages 459–486 *in* P. Giller and J. Gee, editors. Organization of communities: past and present. Blackwell Scientific Publications, Oxford.

Ruckelshaus, M., C. Hartway, and P. Kareiva. In press. Assessing the data requirements of spatially explicit dispersal models. Conservation Biology.

Rudel, T. I. 1991. Relationships between population and environment in rural areas of developing countries. Population Bulletin of the United Nations *31/32*:52–69.

Ruesink, J. L., I. M. Parker, M. J. Groom, and P. M. Kareiva. 1995. Reducing the risks of non-indigenous species introductions. BioScience 45:465–477.

Ruggiero, L. F., G. D. Hayward, and J. R. Squires. 1994. Viability analysis in biological evaluations: concepts of population viability analysis, biological population, and ecological scale. Conservation Biology 8:364–372.

Rummel, J. D., and J. Roughgarden. 1985. A theory of faunal buildup for competition communities. Evolution 39:1009–1033.

Sato, K., and Y. Iwasa. 1993. Modeling of wave regeneration in subalpine Abies forests: population dynamics with spatial structure. Ecology 74:1538–1550.

Savory, A. 1988. Holistic resource management. Island Press, Washington, DC.

Schaffer, M. L. 1981. Minimum population sizes for species conservation. BioScience 31: 131–134.

Schall, J. J., and E. R. Pianka. 1978. Geographical trends in numbers of species. Science 201:679–686.

Scheffer, V.B. 1993. The Olympic goat controversy: a perspective. Conservation Biology 7:916–919.

Schemske, D. W., B. C. Husband, M. H. Ruckleshaus, C. Goodwillie, I. M. Parker, and J. G. Bishop. 1994. Evaluating approaches to the conservation of rare and endangered plants. Ecology 75:584–606.

Schindler, D. W. 1987. Detecting ecosystem responses to anthropogenic stress. Canadian Journal of Fisheries and Aquatic Sciences 44(Suppl. 1):6–25.

Schindler, D. W. 1990. Experimental perturbations of whole lakes as tests of hypotheses concerning ecosystem structure and function. Oikos 57:25–41.

Schindler, D. W. 1995. Linking species and communities to ecosystem management: a perspective from the experimental lakes experience. Pages 326–335 in C. G. Jones and J. H. Lawton, editors. Linking species and ecosystems. Chapman and Hall, New York.

Schoener, T. W. 1991. Extinction and the nature of the metapopulation: a case study. Acta Oecologia 12:53–75.

Schoenly, K. G., J. E. Cohen, K. L. Heong, G. S. Arida, A. T. Barrion, and J. A. Litsinger. 1995. Quantifying the impact of insecticides on food web structure of rice-arthropod populations in a Philippine farmer's irrigated field: a case study. Pages 343–351 in K. Winemiller and G. Polis, editors. Food webs: integration of patterns and dynamics. Chapman and Hall, London.

Schonewald-Cox, C. M., and J. W. Bayless. 1986. The boundary model: a geographic analysis of design and conservation of nature reserves. Biological Conservation 38:305–322.

Schowalter, T. D., and P. Turchin. 1993. Southern pine beetle infestation development: interaction between pine and hardwood basal areas. Forest Science 39:201–210.

Schullery, P. 1989. The fires and fire policy. BioScience 39:686–694.

Schulze, E.-D., and H. A. Mooney, editors. 1994. Biodiversity and ecosystem function. Springer-Verlag, Berlin.

Scott, J. M., B. Csuti, J. D. Jacobi, and F. E. Estes. 1987. Species richness, a geographic approach to protecting future biological diversity. BioScience *37*:782–788.

Scott, J. M., F. Davis, B. Csuti, R. Noss, B. Butterfield, C. Groves, H. Anderson, S. Caicco, F. D'Erchia, T. C. Edwards, Jr., J. Ulliman, and R. G. Wright. 1993. GAP analysis: a geographic approach to protection of biological diversity. Wildlife Monographs *123*:1–41.

Sears, P. B. 1935. Deserts on the march. Reprinted 1988, Island Press, Washington, DC.

Seed, J. 1995. Deep ecology of herbs and forests. International Journal of Ecoforestry *11*: 24–28.

Seligman, N. G., and A. Perevolotsky. 1994. Has intensive grazing by domestic livestock degraded the Old World Mediterranean rangelands? Pages 93–103 *in* M. Arianoutsou and R. H. Groves, editors. Plant animal interactions in Mediterranean type ecosystems. Kluwer, Dordrect, Germany.

Shachak, M., and S. Brand. 1991. Relations among spatiotemporal heterogeneity, population abundance, and variability in a desert. Pages 202–223 *in* J. Kolasa and S. T. A. Pickett, editors. Ecological heterogeneity. Springer-Verlag, New York.

Shafer, C. L. 1990. Nature reserves: island theory and conservation practice. Smithsonian Institution Press, Washington, DC.

Shaffer, M. L. 1987. Minimum viable populations: coping with uncertainty. Pages 69–86 *in* M. E. Soulé, editor. Viable populations for conservation. Cambridge University Press, Cambridge.

Shorrocks, B., and I. R. Swingland, editors. 1990. Living in a patchy environment. Oxford University Press, Oxford.

Short, J., and A. Smith. 1994. Mammal decline and recovery in Australia. Journal of Mammalogy *75*:288–297.

Shrader-Frechette, K. S. 1994a. Applied ecology and the logic of case studies. Philosophy of Science *61*:228–249.

Shrader-Frechette, K. S. 1994b. Science, environmental risk assessment and the frame problem. BioScience *44*:548–551.

Shrader-Frechette, K. S., and E. D. McCoy. 1993. Method in ecology: strategies for conservation. Cambridge University Press, New York.

Simberloff, D. 1974. Equilibrium theory of island biogeography and ecology. Annual Review Ecology and Systematics *5*:161–182.

Simberloff, D. 1982a. Big advantages of small refuges. Natural History *91*(4):6- 14.

Simberloff, D. 1982b. A succession of paradigms in ecology: essentialism to materialism and probablism. Pages 63–99 *in* E. Saarinen, editor. Conceptual issues in ecology. Reidel (Kluwer), Boston.

Simberloff, D. 1983. Biogeography: the unification and maturation of a science. Pages 411–455 *in* A. M. Brush and G. A. Clark, Jr., editors. Perspectives in Ornithology. Cambridge University Press, Cambridge.

Simberloff, D. 1988. The contribution of population and community biology to conservation science. Annual Review of Ecology and Systematics *19*:473–512.

Simberloff, D. 1994a. Habitat fragmentation and population extinction of birds. Ibis *137*:S105–S111.

Simberloff, D. 1994b. The ecology of extinction. Acta Palaeontologica Polonica *38*:159–174.

Simberloff, D. 1995. Introduced species. Pages 323–336 *In* Encyclopedia of environmental biology, Vol. 2. Academic Press, San Diego.

Simberloff, D., and J. Cox. 1987. Consequences and costs of conservation corridors. Conservation Biology *1*:63–71.

Simberloff, D., and J.-L. Martin. 1991. Nestedness of insular avifaunas: simple summary statistics masking complex species patterns. Ornis Fennica *68*:178–192.

Simberloff, D., J. A. Farr, J. Cox, and D. W. Mehlman. 1992. Movement corridors: conservation bargains or poor investments? Conservation Biology *6*:493–504.

Singer, M. C. 1972. Complex components of habitat suitability within butterfly colony. Science *176*:75–77.

Singer, M. C., C. D. Thomas, and C. Parmesan. 1993. Human-induced evolution of insect-host associations. Nature *366*:681–683.

Sjogren-Gulve, P. 1994. Distribution and extinction patterns within a northern metapopulation of the pool frog, *Rana lessonae*. Ecology *75*:1357–1367.

Skaggs, R. W., and W. J. Boecklen. 1996. Extinctions of montane mammals reconsidered: Putting a global-warming scenario on ice. Biodiversity and Conservation *5*:759–778.

Skelly, D. K., E. E. Werner, and S. C. Cortwright. In press. Volatile amphibian distributions: a role for succession. Conservation Biology.

Slocombe, D. S. 1993. Environmental planning, ecosystem science, and ecosystem approaches for integrating environment and development. Environmental Management *17*:289–303.

Smith, A. F. M., and A. E. Gelfand. 1992. Bayesian statistics without tears: a sampling-resembling perspective. The American Statistician *46*:84–88.

Smith, T. B., L. A. Freed, J. Kaimanu Lepson, and J. H. Carothers. 1995. Evolutionary consequences of extinctions in populations of a Hawaiian honeycreeper (*Vestiaria coccinea*). Conservation Biology *9*:107–113.

Snyder, G. 1980. The real work. New Directions Books, New York.

Society of American Foresters. 1993. Task force report on sustaining long-term forest health and productivity. Society of American Foresters, Bethesda, MD.

Soulé, M. E., editor. 1987. Viable populations for conservation. Cambridge University Press, Cambridge.

Soulé, M. E. 1994. A California rescue plan. Defenders Fall:36–39.

Soulé, M. E., and D. Simberloff. 1986. What do genetics and ecology tell us about the design of nature reserves? Biological Conservation *35*:19–40.

Soulé, M. E., and B. A. Wilcox, editors. 1980. Conservation biology: an evolutionary-ecological perspective. Sinauer Associates, Sunderland, MA.

Srodowiska, U. 1992. Environmental protection 1992, Glowny Urzad Statystyczay, Main Statistical Office, Warsaw.

Stacey, P. B., and M. Taper. 1992. Environmental variation and the persistence of small populations. Ecological Applications 2:18–29.

Starfield, A. M., and A. L. Bleloch. 1986. Building models for conservation and wildlife management. Macmillan, New York.

Starfield, A. M., J. D. Roth, and K. Ralls. 1995. "Mobbing" in Hawaiian monk seals (*Monachus schauinslani*): the value of simulation modeling in the absence of apparently crucial data. Conservation Biology 9:166–174.

Steadman, D. W. 1995. Prehistoric extinctions of Pacific island birds: biodiversity meets Zooarchaeology. Science 267:1123–1131.

Steele, L. W. 1989. Managing technology—the strategic view. McGraw-Hill, New York.

Stephenson, N. L., D. J. Parsons, and T. W. Swetnam. 1991. Restoring natural fire to the sequoia-mixed conifer forest: should intense fire play a role? Proceedings of the Tall Timbers Fire Ecology Conference 17:321–338.

Stone, E. L. 1975. Effects of species on nutrient cycles and soil change. Philosophical Transactions of the Royal Society of London, Series B: Biological Sciences 271:149–162.

Sugg, Ike. 1993. Property wrongs, CEI Update, Competitive Enterprise Institute, Nov. 1993, p. 6.

Supkoff, D. M., D. B. Joley, and J. J. Marois. 1988. Effect of introduced biological control organisms on the density of *Chondrilla juncea* in California. Journal of Applied Ecology 25:1089–1095.

Sutherland, W. J., and D. A. Hill. 1995. Managing habitats for conservation. Cambridge University Press, Cambridge.

Swank, W. T., and D. A. Crossley, Jr., editors. 1988. Forest hydrology and ecology at Coweeta. Springer-Verlag, New York.

Swetnam, T. W. 1993. Fire history and climate change in giant sequoia groves. Science 262:885–889.

Taylor, A. D. 1991. Studying metapopulation effects in predator-prey systems. Pages 305–323 *in* M. Gilpin and I. Hanski, editors. Metapopulation dynamics: empirical and theoretical investigations. Academic Press, London.

Taylor, B. L. 1995. The reliability of using population viability analysis for risk classification of species. Conservation Biology 9:551–558.

Taylor, P. D., L. Fahrig, K. Henein, and G. Merriam 1993. Connectivity is a vital element of landscape structure. Oikos 68:571–573.

Templeton, A. R. 1991. Off-site breeding of animals and implications for plant conservation strategies. Pages 182–194 *in* D. A. Falk and K. E. Holsinger, editors. Genetics and conservation of rare plants. Oxford University Press, New York.

Templeton, A. R., and B. Read. 1983. The elimination of inbreeding depression in a captive herd of Speke's gazelle. Pages 241–261 *in* C. M. Schonewald-Cox, S. M. Chambers, B. MacBryde and L. Thomas, editors. Genetics and conservation: a reference for managing wild animal and plant populations. Benjamin-Cummings, Menlo Park, CA.

Thomas, C. D. 1994. Local extinctions, colonizations and distributions: habitat tracking by British butterflies. Pages 319–336 *in* S. Leather, A. Watt, and N. Mills, editors. Individuals, populations and patterns in ecology. Interact Ltd., Andover, U.K.

Thomas, C. D., and I. Hanski. In press. Butterfly metapopulations. *In* I. Hanski and M. E. Gilpin, editors. Metapopulation biology. Academic Press, London.

Thomas, J. W., and the Forest Ecosystem Management Assessment Team. 1993. Forest ecosystem management: an ecological, economic, and social assessment. Pacific Northwest Research Station, La Grande, OR.

Thomas, J. W., E. D. Forsman, J. B. Lint, B. R. Noon, and J. Verner. 1990. A conservation strategy for the northern spotted owl. Report to the Interagency Scientific Committee to address the conservation of the northern spotted owl. Portland, OR.

Thompson, J. N. 1988. Variation in interspecific interactions. Annual Review of Ecology and Systematics *19*:65–87.

Thompson, J. N. 1994a. The coevolutionary process. University of Chicago Press, Chicago.

Thompson, J. N. 1994b. The geographic mosaic of evolving interactions. Pages 419–431 *in* S. R. Leather, A. D. Watt, N. J. Mills, and K. F. A. Walters, editors. Individuals, populations and patterns. Intercept Press, Andover, England.

Thompson, J. N., and J. J. Burdon. 1992. Gene-for-gene coevolution between plants and parasites. Nature *360*:121–125.

Thompson, J. N., and O. Pellmyr. 1992. Mutualism with pollinating seed parasites amid co-pollinators: constraints on specialization. Ecology *73*:1780–1791.

Tidelands Strategic Plan. The Nature Conservancy, Connecticut Chapter, Middletown, CT.

Tilman, D., and S. W. Pacala. 1993. The maintenance of species richness in plant and geographical perspectives. Pages 13–25 *in* R. E. Ricklefs and D. Schluter, editors. Species diversity in ecological communities: historical and geographical perspectives. University of Chicago Press, Chicago.

Tilman, D. 1994. Competition and biodiversity in spatially structured habitats. Ecology *75*: 2–16.

Tilman, D., and Downing, J. A. 1994. Biodiversity and stability in grasslands. Nature *367*:363–365.

Tisdell, C. A. 1991. Economics of environmental conservation. Elsivier, Amsterdam.

Triska, F. J., C. M. Pringle, G. Zellweger, J. H. Duff, and R. J. Avanzino. 1993. Dissolved inorganic nitrogen composition, transformation, retention and transport in naturally phosphate-enriched and unenriched tropical streams. Canadian Journal of Fisheries and Aquatic Sciences *50*:665–675.

Turner, B. L., W. C. Clark, R. W. Kates, J. F. Richards, J. T. Matthews, and W. B. Meyer, editors. 1990. The Earth as transformed by human action: global and regional changes in the biosphere over the past 300 years. Cambridge University Press, New York.

Turner, M. G. 1989. Landscape ecology: the effect of pattern on process. Annual Review of Ecology and Systematics *20*:171–197.

Turner, M. G., G. J. Arthaud, R. T. Engstrom, S. J. Hejl, J. Liu, S. Loeb, and K. McKelvey. 1995. Usefulness of spatially explicit population models in land management. Ecological Applications *5*:12–16.

Turner, M. G. and V. H. Dale. 1991. Modeling landscape disturbance. Pages 323–351 *in* M. G. Turner and R. H. Gardner, editors. Quantitative methods in landscape ecology. Springer-Verlag, New York.

Turner, M. G., R. H. Gardner, V. H. Dale, and R. V. O'Neill. 1989. Predicting the spread of disturbance across heterogeneous landscapes. Oikos *55*:121–129.

Turner, M. G., W. H. Romme, R. H. Gardner, R. V. O'Neill, and T. K. Kratz. 1994. A revised concept of landscape equilibrium: disturbance and stability on scaled landscapes. Landscape Ecology *8*:213–227.

Ulanowicz, R. E. 1992. Ecosystem health and trophic flow networks. Pages 190–206 *in* R. Costanza, B. G. Norton, and B. D. Haskell, editors. Ecosystem health. Island Press, Washington, DC.

UNESCO. 1974. Report of the task force on criteria and guidelines for the choice and establishment of biosphere reserves. UNESCO. MAB Report Series Ç22.

United Nations Development Programme. 1992. Human development report 1992. Oxford University Press, New York.

United Nations Development Programme. 1994. Human development report 1994. Oxford University Press, New York.

United Nations Food and Agriculture Organisation (UN FAO). 1995. The state of world fisheries and aquaculture. Rome, Italy. D/V 5550 E/1/2.95/4000.

United Nations Population Division. 1992. Long-range world population projection: Two centuries of population growth, 1950–2150. UNPD, New York.

Urban, D. L., R. V. O'Neill, and H. H. Shugart, Jr. 1987. Landscape ecology. BioScience *37*: 119–127.

U.S. Army Corps of Engineers. 1993. Water need for Puerto Rico: San Juan metropolitan region. Report Number 2040–12 to the Puerto Rican Water and Sewage Authority.

U.S. Department of the Interior, National Park Service. 1978. Fire management guidelines, NPS 18.

U.S. DOI and U.S. DOA. 1995. Federal wildland fire management: policy and program review. U.S. Department of the Interior and U.S. Department of Agriculture, Washington, DC.

U.S. Fish and Wildlife Service (U.S. FWS). 1994a. Sonoran pronghorn recovery plan revision. U.S. Fish and Wildlife Service, Albuquerque, NM.

U.S. FWS. 1994b. Spectacled Eider recovery plan. Prepared by the Spectacled Eider Recovery Team for the U.S. Fish and Wildlife Service, Anchorage, Alaska.

U.S. FWS. 1994c. Recovery program: endangered and threatened species, 1994. Report to Congress. U.S. Fish and Wildlife Service, Washington, DC

Usher, M. B. 1986. Wildlife conservation evaluation: attributes, criteria and values. Pages 3–43 *in* M. B. Usher, editor. Wildlife conservation evaluation. Chapman and Hall, London.

Valle, C. A. 1995. Effective population size and demography of the rare flightless Galapagos cormorant. Ecological Applications *5*:601–617.

van Coller, A. L. 1992. Riparian vegetation of the Sabie river: relating spatial distribution patterns to the physical environment. MSc. Thesis, University of the Witwatersrand, Johannesburg, South Africa.

van den Bosch, F., R. Hengeveld, and J. A. J. Metz. 1992. Analysing the velocity of animal range expansion. Journal of Biogeography 19:135–150.

van Niekerk, A. W., G. L. Heritage and B. P. Moon. 1995. River classification for management: The geomorphology of the Sabie river in the eastern Transvaal. South African Geographical Journal 77:68–76.

van Schaik, C. P., J. W. Terborgh, and S. J. Wright. 1993. The phenology of tropical forests: adaptive significance and consequences for primary consumers. Annual Review of Ecology and Systematics 24:353–377.

van Vliet, B. M., and A. Gerber. 1992. Strategies for promoting the transfer of technology in the field of water and effluent treatment. Pages 316–325 *in* Proceedings of the Water Week Conference. Publication of the Water Research Commission. Pretoria, South Africa.

Vargas, R. J. 1995. History of municipal water resources in Puerto Viejo, Saraqipui, Costa Rica: a socio-political perspective. Masters Thesis, Institute of Ecology, University of Georgia, Athens.

Verboom, J., K. Lankester, and J. A. J. Metz. 1991. Linking local and regional dynamics in stochastic metapopulation models. Biological Journal of the Linnean Society 42:39–55.

Vickerman, S.. and K. A. Smith. 1995. Recommendations for implementing gap analysis. Draft report for the National Biological Service.

Vitousek, P. M. 1994. Beyond global warming: Ecology and global change. Ecology 75:1861–1876.

Vitousek, P. M., and J. S. Denslow. 1986. Nitrogen and phosphorus availability in treefall gaps of a lowland tropical rainforest. Journal of Ecology 74:1167–1178.

Vitousek, P. M., P. R. Ehrlich, A. H. Ehrlich, and P. H. Matson. 1986. Human appropriation of the products of photosynthesis. Bioscience 36:368–373.

Vitousek, P. M., and D. U. Hooper. 1992. Biological diversity and terrestrial ecosystem biogeochemistry. Pages 3–14 *in* E. D. Schulze and H. A. Mooney, editors. Biodiversity and ecosystem function. Springer-Verlag. New York.

Vitousek, P. M., and L. R. Walker. 1989. Biological invasion by Myrica faya in Hawai'i: plant demography, nitrogen fixation, ecosystem effects. Ecological Monographs 59:247–265.

Vogt, W. 1948. Road to survival. W. Sloan Associates, New York.

Wagner, F. H., and C. E. Kay. 1993. "Natural" or "healthy" ecosystems: are U.S. national parks providing them? Pages 257–270 *in* M. J. McDonnell and S. T. A. Pickett, editors. Humans as components of ecosystems: the ecology of subtle human effects and populated areas. Springer-Verlag, New York.

Walker, B. 1989. Diversity and stability in ecosystem conservation. Pages 121–130 *in* D. Western and M. C. Pearl, editors. Conservation for the twenty-first century. Oxford University Press, New York.

Walker, B. H. 1992. Biological diversity and ecological redundancy. Conservation Biology 6: 18–23.

Walker, B. H. 1995. Conserving biological diversity through ecosystem resilience. Conservation Biology 9:747–752.

Walker, D. 1989. Diversity and stability. Pages 115–146 *in* J. M. Cherrett, editor. Ecological concepts. Blackwell, Oxford.

Walker, J., C. H. Thompson, I. F. Fergus, and B. R. Tunstall. 1981. Plant succession and soil development in coastal sand dunes of subtropical eastern Australia. Pages 107–131 *in* D. C. West, H. H., Shugart, and D. B. Botkin, editors. Forest succession: concepts and application. Springer-Verlag, New York.

Wallace, J. B., J. W. Grubaugh, and M. R. Whiles. In press. Biotic indices and stream ecosystem processes: results from an experimental study. Ecological Applications.

Wallace, J. B., D. S. Vogel, and T. F. Cuffney. 1986. Recovery of a headwater stream from an insecticide-induced community disturbance. Journal of the North American Benthological Society 5:115–126.

Waller, D. M., D. M. O'Malley, and S. C. Gawler. 1987. Genetic variation in the extreme endemic *Pedicularis furbishiae* (Scrophulariaceae). Conservation Biology 1:335–340.

Walters, C. J. 1986. Adaptive management of renewable resources. MacMillan, New York.

Walters, C. J. and C. S. Holling. 1990. Large-scale experiments and learning by doing. Ecology 71:2060–2068.

Walters, J. R., C. K. Copeyon, and J. H. Carter III. 1992. Test of the ecological basis of cooperative breeding in red-cockaded woodpeckers. Auk 109:90–97.

Wapshere, A. J., A. Hasan, W. K. A. Wahba, and L. Caresche. 1974. Ecology of Chondrilla juncea in the western Mediterranean. Journal of Applied Ecology 11:783–800.

Ward, J. V., and J. A. Wiens. In press. Ecotones of riverine ecosystems: rôle and typology, spatio-temporal dynamics and river regulation. *In* M. Zalewski, F. Schiemer, and J. Thorpe, editors. Fish and land/inland water ecotones—the need for integration of fishery science, hydrobiology, and landscape ecology. UNESCO/ Parthenon Publishing.

Ware, S., C. Frost, and P. D. Doerr. 1993. Southern mixed hardwood forest: the former longleaf pine forest. Pages 447–493 *in* W. H. Martin, S. G. Boyce, and A. C. Echternacht, editors. Biodiversity of the southeastern United States. Lowland terrestrial communities. John Wiley & Sons, New York.

Warren, M. S. 1992. The conservation of British butterflies. Pages 246–274 *in* R. L. H. Dennis, editor. The ecology of butterflies in Britain. Oxford University Press, Oxford.

Warshall, P. 1994. The biopolitics of the Mt. Graham red squirrel (*Tamiasciuris hudsonicus grahamensis*). Conservation Biology 8:977–988.

Watkinson, A. R., and W. J. Sutherland. 1995. Sources, sinks, and pseudo-sinks. Journal of Animal Ecology 64:126–130.

Weaver, J. E., and T. J. Fitzpatrick. 1934. The prairie. Ecological Monographs 4:109–295.

Weiner, J. 1995. On the practice of ecology. Journal of Ecology 83:153–158.

Weiss, S. B., D. D. Murphy, P. R. Ehrlich, and C. F. Metzler. 1993. Adult emergence phenology in checkerspot butterflies: the effects of macroclimate, topoclimate, and population history. Oecologia 96:261–270.

Weiss, S. B., D. D. Murphy, and R. R. White. 1988. Sun, slope and butterflies: topographic determinants of habitat quality for Euphydryas editha . Ecology 69:1486–1496.

West, N. E. 1989. Spatial pattern—functional interactions in shrub dominated plant communities. Pages 283–305 in C. M. McKell, editor. The biology and utilization of shrubs. Academic Press, London.

West, N. E. 1990. Structure and function of microphytic soil crusts in wildland ecosystems of arid to semi-arid regions. Advances in Ecological Research 20:179–223.

West, N. E. 1993. Biodiversity of rangelands. Journal of Range Management 46:2–13.

Western, D., M. C. Pearl, S. L. Pimm, B. Walker, I. Atkinson, and D. S. Woodruff. 1989. An agenda for conservation action. Pages 304–323 in D. Western and M. C. Pearl, editors. Conservation for the twenty-first century. Oxford University Press, New York.

White, D. W., W. Worthen, and E. W. Stiles. 1990. Woodlands in a post-agricultural landscape in New Jersey. Bulletin of the Torrey Botanical Club 117:256–265.

White, P. S. 1979. Pattern, process, and natural disturbance in vegetation. The Botanical Review 45(3):229–299.

White, P. S., and S. T. A. Pickett. 1986. Natural disturbance and patch dynamics: an introduction. Pages 3–13 in S. T. A. Pickett and P. S. White, editors. The ecology of natural disturbance and patch dynamics. Academic Press, Orlando.

Whittaker, R. H. 1965. Dominance and diversity in land plant communities. Science 147: 250–260.

Whittaker, R. H., and G. E. Likens. 1975. The biosphere and man. Pages 305–328 in Helmut Lieth and Robert H. Whittaker, editors. Primary productivity of the biosphere. Springer-Verlag, New York.

Wiens, J. A. 1977. On competition and variable environments. American Scientist 65: 590–597.

Wiens, J. A. 1984. On understanding a non-equilibrium world: myth and reality in community patterns and processes. Pages 439–457 in D. R. Strong, Jr., D. Simberloff, L. G. Abele, and A. B. Thistle, editors. Ecological communities: conceptual issues and the evidence. Princeton University Press, Princeton, NJ.

Wiens, J. A. 1989a. The ecology of bird communities. Cambridge University Press, Cambridge.

Wiens, J. A. 1989b. Spatial scaling in ecology. Functional Ecology 3:385–397.

Wiens, J. A. 1992. What is landscape ecology, really? Landscape Ecology 7:149–150.

Wiens, J. A. 1995a. Landscape mosaics and ecological theory. Pages 1–26 in L. Hansson, L. Fahrig, and G. Merriam, editors. Mosaic landscapes and ecological processes. Chapman and Hall, London.

Wiens, J. A. 1995b. Habitat fragmentation: island v landscape perspectives on bird conservation. Ibis 137:S97–S104.

Wiens, J. A. 1996. Metapopulation dynamics and landscape ecology. In I. Hanski and M. Gilpin, editors. Metapopulation biology: ecology, evolution, genetics. Academic Press, London.

Wiens, J. A. In press. Wildlife in patchy environments: metapopulations, mosaics, and management. *In* D. McCullough, editor. Metapopulations and wildlife conservation management. Island Press, Washington, DC.

Wiens, J. A., C. S. Crawford, and J. R. Gosz. 1985. Boundary dynamics: a conceptual framework for studying landscape ecosystems. Oikos *45*:421–427.

Wiens, J. A., N. C. Stenseth, B. Van Horne, and R. A. Ims. 1993. Ecological mechanisms and landscape ecology. Oikos *66*:369–380.

Wilcove, D. S. 1985. Nest predation in forest tracts and the decline of migratory songbirds. Ecology *66*:1211–1214.

Wilcove, D. S., and M. J. Bean. 1994. The big kill. Environmental Defense Fund, Washington, DC.

Wilcove, D. S., M. J. Bean, and P. C. Lee. 1992. Fisheries management and biological diversity: problems and opportunities. Pages 373–383 *in* Biological diversity in aquatic management, reprint of Special Session 6 from Transactions of the 57th North American Wildlife and Natural Resources Conference. Wildlife Management Institute, Washington, DC.

Wilcove, D., C. H. McLellan, and A. P. Dobson. 1986. Habitat fragmentation in the temperate zone. Pages 237–256 *in* M. E. Soulé, editor. Conservation biology: the science of scarcity and diversity. Sinauer Associates, Sunderland, MA.

Williams, J. D., M. L. Warren, Jr., K. S. Cummings, J. L. Harris, and R. J. Neves. 1992. Conservation status of freshwater mussels of the United States and Canada. Fisheries *18*:6–22.

Williams, J. E., J. E. Johnson, D. A. Hendrickson, S. Contreras-Balderas, J. D. Williams, M. Navarro-Mendoza, D. E. McAllister, and J. E. Deacon. 1989. Fishes of North America endangered, threatened, or of special concern. Fisheries *14*:2–20.

Williams, T. 1995. The Blackstone now runs blue. Audubon *Nov-Dec*:26–32.

Williamson, M. 1981. Island populations. Oxford University Press, Oxford.

Williamson, M. 1988. Relationship of species number to area, distance and other variables. Pages 91–115 *in* A. A. Myers and P. S. Giller, editors. Analytical biogeography. Chapman and Hall, London.

Williamson, M. 1989a. The MacArthur and Wilson theory today: true but trivial. Journal of Biogeography *16*:3–4.

Williamson, M. 1989b. Natural extinction on islands. Philosophical Transactions of the Royal Society London B *325*:457–468.

Wilson, E. O. 1992. The diversity of life. Harvard University Press, Cambridge, MA.

With, K. A. In press. The application of neutral landscape models in conservation biology. Conservation Biology.

With, K. A., R. H. Gardner, and M. G. Turner. In press. Landscape connectivity and population distributions in heterogeneous environments. Oikos.

Woodley, S. 1993. Monitoring and measuring ecosystem integrity in Canadian National Parks. Pages 155–176 *in* S. Woodley, J. J. Kay, and G. Francis, editors. Ecological integrity and the management of ecosystems. St. Lucie Press, Ottawa.

Woodley, S., J. J. Kay, and G. Francis, editors. 1993. Ecological integrity and the management of ecosystems. St. Lucie Press, Ottawa.

Woodwell, G. M., and H. H. Smith, editors. 1969. Diversity and stability in ecological systems. Brookhaven Symposium in Biology, no. 22. Brookhaven National Laboratory, Atomic Energy Commission, Brookhaven, New York.

Woody, T. 1993. Grassroots in action: the Sierra Club's role in the campaign to restore the Kissimmee River. Journal of the North American Benthological Society *12*:201–205.

Wootton, J. T. 1992. Indirect effects, prey susceptibility, and habitat selection: impacts of birds on limpets and algae. Ecology *73*:981–991.

World Conservation Monitoring Centre. 1992. Global biodiversity: status of the Earth's living resources. Chapman and Hall, London.

World Resources Institute. 1992. World resources data base diskette 199–293. World Resources Institute, Washington, DC.

World Resources Institute. 1994. World resources 1994–95. Oxford University Press, New York and Oxford.

Worster, D. 1994. Nature and the disorder of history. Environmental History Review *18*:1–15.

Wright, R. G., J. G. MacCracken, and J. Hall. 1994. An ecological evaluation of proposed new conservation areas in Idaho: evaluating proposed Idaho national parks. Conservation Biology *8*:207–216.

Wright, S. 1932. The roles of mutation, inbreeding, crossbreeding and selection in evolution. Proceedings of the Sixth International Congress of Genetics *1*:356–366.

Wright, S. 1943. Isolation by distance. Genetics *28*:114–138.

Wright, S. 1946. Isolation by distance under diverse systems of mating. Genetics *31*:39–59.

Wrobel, S., editor. 1989. Zanieczyszczenia atmosfery a degradacja wod, Materialy sympozjum. Zaklad Ochrony Przyrody i Zasobow nauralnych, Krakow.

Wu, J., and O. L. Loucks. 1996. From balance of nature to hierarchical patch dynamics: A paradigm shift in ecology. The Quarterly Review of Biology *71*:In press.

Young, T. P. 1994. Natural die-offs of large mammals: implications for conservation. Conservation Biology *8*:410–418.

Zaady, E., and M. Shachak. 1994. Microphytic soil crust and ecosystem leakage in the Negev Desert. American Journal of Botany *81*:109.

Index

A Sand County Almanac, 194
Acid rain, 308, 356
Acid rain and deposition, 306
Acidification, 140, 144
Act, 27
Actea spicata (baneberry), 151
Activism, 297, 346, 345, 349, 350
  conservation activism, 346, 345, 348, 349
Adaptations, 289
Adaptive landscapes, 101, 102
Adaptive management, 7, 66, 174, 183, 186,
  191, 235, 240, 247, 263, 324, 336, 340,
  367, 377, 388, 391, 394, 397
Adaptive management strategy, 389
Additive genetic variance, 205
Advocacy, 311, 397
Africa, 32, 33, 40, 257, 403
African, 40
African savannas, 63
Agonostomus monticola, 317
Agricultural, 40, 41, 42, 126, 138, 150, 152,
  170, 309, 318
Agricultural lands, 155
Agricultural pest, 21
Agriculture, 29, 36, 38, 40, 41, 42, 48, 152,
  153, 169, 257
Agroecology, 315
Agroecosystems, 42
Air quality, 183
Åland islands, 220, 225
Alga, 256
Algae, 111, 137, 139
Algal, 140, 317
Alien ecosystems, 127
Alien plants, 258

Alien species, 286, 364
Alien systems, 127
Aliens, 257, 258
Alligators, 189
All-Union Institute of Plant Industry, 203
Alternatives assessment, 342, 343, 344
Amazon, 141
Ambystoma maculatum, 164
Ambystoma tigrinum, 164
American Farm Bureau, 18
American Fisheries Society, 307
American Rivers, 312
American Rockies, 292
Amphibians, 157, 160, 161, 162, 164, 165,
  166, 213, 255, 292
Amphidromous shrimps, 318
Anglican Church, 194
Anguilla rostrata, 317
Annual sunflower (*Helianthus annuus*), 209
Antarctica, 38
Anthropologists, 400
Antibiotic resistance, 191
Ants, 151, 155
Appalachians, 254
Aquaculture, 169
Areography, 280, 282
Argentina, 401
Argentine, 363
Arizona willow (*Salix arizonica*), 321
Arnold, Matthew, 403
Arthropods (*Cystiphora schmidti*), 214
Asia, 401, 403
Assimilative capacity, 344
Aster, 208
Atya, 317

Atyid, 317
Atyid shrimps, 317
Australia, 31, 129, 214, 215, 287, 289, 352, 353, 355, 356, 357, 401
Australian, 354
Australian Conservation Foundation, 355
Australian mining, 353
Australian print media, 353
Awaous tajasica, 317

Babbitt, Bruce, 24
Babtria tibiale, Acasia appensata (Geometridae), 151
Bacteria, 111, 402
Bailey, Liberty Hyde, 194
Balance of nature, 18, 60, 61, 62, 63, 84, 94, 192, 237, 238, 366, 369, 370, 382, 383
Baltic Sea, 308
Banana plantations, 314
Banff, 292
Barrett, Julianna, 366
Bay checkerspot butterfly (*Euphydryas editha bayensis*), 44, 47, 48, 49, 50, 51, 53, 54, 55, 56, 57–58, 59
  Checkerspot butterflies, 48
  Checkerspot butterfly (*Euphydryas editha*), 287
Bayesian approach, 235
Bayesian models, 234
Bean, Michael, 349, 367
Bear, 14
Beavers (*Castor* spp.), 88, 126, 144, 152, 190, 366
Beecher, Henry Ward, 194
Bees, 285, 286
Beetles, 257
Big Cypress Swamp, 380
Biocentrism, 8
Biogeochemistry, 188, 379, 384, 386
Biogeographers, 216
Biogeographic theory, 201
Biogeography, 188, 190, 274–284, 395
Biological control, 215
Biological oxygen demand (BOD), 143
Bioreserve, 236, 237, 238, 239, 242
Bioreserve design, 239, 242, 244
Birds, 101, 127, 129, 134, 135, 151, 152, 212, 218, 247, 270, 292
  avifauna, 126
  blackbirds, 131, 132
  Britain's birds, 130
  Hawaiian birds, 135

icterids, 125, 126, 131, 132, 134
  orioles, 131, 132
  passerines, 125, 129
  rails, 130
  tricolored blackbird, *Agelaius tricolor*, 134
  warblers, 125, 126, 131, 132, 134
Bison (*Bos bison*), 25, 92
Black bear (*Ursus americanus*), 283
Black crappies, 309
Black rhinoceroses (*Diceros bicornis*), 62
Black Sea, 308
Black-footed Ferret (*Mustela nigripes*), 188
Blackstone River, 309
Bolivia, 40
Botanical gardens, 127, 209
Bradshaw, A.D., 89
Brazil, 40
Breeding Bird Survey (BBS), 131
Brevard County, Florida, 211
Britain, 221, 287, 357
Brown tree snake, 131
Bryophytes, 292
Budapest, Hungary, 308
Budworms, 267
Bureau of Land Management, 321
Burning, 337
Burundi, 40
Bushes, 154
Butterflies, 218, 220, 221
Butterfly *Melitaea cinxia*, 50, 54, 220, 281, 331

Calcium, 257
California Condor, 188
California gnatcatcher (*Polioptila californica californica*), 44, 47, 48, 51, 53, 54, 55, 56
California's Central Valley, 183
Canada, 42, 129, 131, 132, 188
Canadian Rockies, 292
Captive breeding, 202, 394
Captive breeding programs, 210
Captive wild populations, 299
Carbon, 139, 144, 255, 259
  C, 144
  ^{13}C, 144
Carbon dioxide, 27, 178, 401
  CO_2, 143, 144
Carbon sinks, 255
Cardiologist, 341
Caribbean, 254, 317
Caribbean islands, 129

Caribbean National Forest, 317
Carpathian Mountains, 308
Carrying capacity, 339, 370
Cary Conference, 4, 11, 187
Cascade Mountains, 10, 141
Caspian Seas, 308
Catalina Island, 209
Catesby, Mark, 25
Cations, 255
Cattle, 314
Cattle grazing, 314
Cattlemen's Association, 18
Central America, 40
Central America's Caribbean Slope, 315
Central America's forests, 39
Central and Eastern Europe, 305, 308, 309, 396
Central Europe, 308
Cereal crops, 402
Chambers, Robert, 194
Champion International, 141
Chaos, 147, 369
Chaos theory, 364
Chaparral, 172, 177
Charismatic megafauna, 140
Chequamegon, 322
Chestnut, 254
Chickens, 32
Chinook salmon, 141
Christensen, Norm, 364, 366
Christianity, 194
Clam, 129
Clean Water Act of 1972, 138, 305, 309, 310, 349, 396
Clearcutting, 149, 150, 151, 154, 155, 339, 373
Clements, Frederick E., 83, 175
Clementsian ecology, 176, 355
Climate, 325
Climate change, 177, 366, 365
Climatic fluctuations, 402
Clopyralid, 215
Coalescent theory, 211
Coalition to Restore Urban Waters, 312
Coastal sage scrub, 47, 48, 189
Coccinella, 160
Coccinellid beetle, 161
Cockburn, Alexander, 24
Coevolution, 285, 287, 289
  coevolutionary change, 290
  gene-for-gene coevolution, 289
  geographic mosaic theory, 290

Coevolutionary studies, 288
Coevolved interactions, 291
Cohen, Joel, 363, 367
Coho salmon, 141, 142, 143
Colonizations, 99, 220, 221
Columbia River, 189, 339
Colwell, Rob, 363, 366
Comiphera, 62
Commoner, Barry, 354
Competitive Enterprise Institute, 23, 24
Comprehensive Management Plan, 342
Congress of the United States, 11, 24, 25
Connectedness, 299
Connecticut, 251
Connecticut River, 238, 242, 244, 246, 247, 250, 343, 366, 395
Connecticut River Tidelands, 236, 237, 241, 242, 243, 247
Connectivity, 6, 101, 103, 104, 105, 106, 155, 243
Conservation
  environmental education, 311
Conservation genetics, 364
Consultative Group on Biological Diversity, 398
Conte, 250
Coral, 190
Coral reefs, 171, 189, 402
Corridors, 48, 54, 103, 213, 233, 236, 244, 246, 277, 282, 299, 335, 391
Corvus monedula (jackdaw), 152
Costa Rica, 140, 313, 314, 315, 316, 396
  Central Mountain Range, 315
  La Selva Biological Station, 313, 317
  Parque Nacional Braulio Carrillo, 315
  Puerto Viejo, 313, 315
  Puerto Viejo de Sarapiqui, 314
  Sarapiqui, 314, 315
  Sarapiqui province, 313
Cowles, Henry Chandler, 83, 177
Craighead, John, 14
Crayfish, 307
Critically endangered, 321, 333
Crop plants, 170
Cryptogams, 326
Crystal River, 144
Cutthroat trout, 141
Cyanobacteria (bluegreen algae), 111, 190, 270
Cyclic fluctuation, 369
Cyclone, 355
Cypress dome, 188
Czech Republic, 308

Danube River, 308
Darwin, Charles, 193, 194, 285, 286
Data Assessment Statement, 329
Decision support system (DSS), 60
  operational framework, 60
  predictive modeling framework, 60
  system response framework, 60
Decision theory, 299
  dynamic decision theory, 299
  Markov decision theory, 298, 299, 300, 301, 303
  state-dependent dynamic decision theory, 298, 303
  static decision theory, 299
Decomposition, 190
Deep ecology, 357
Deer, 195, 266
Deer hunting, 401
Defenders of Wildlife, 324
Defoliates, 267
Deforestation, 32, 40, 52, 127, 314
  Alaska, 32
Demographers, 391, 400
Demographic stochasticity, 376
Demography, 188, 363
Dendrocopus leucotus (white-backed woodpecker), 151
Denitrification, 139
Department of Environmental Protection, 250
Department of Transportation, 380
Desert tortoise (Gopherus agassizii), 321
Desertification, 109, 117
Desertified, 116, 110
Deserts, 38, 110, 373
Designing nature reserves, 345
Dewberry, Charley, 141
Dioxin, 339, 340
Dirt Bike Magazine, 364, 373
Disease, 63, 233, 289, 321, 329
  anthrax, 63
  epidemic, 100, 289
Dispersal, 6, 46, 56, 87, 99, 100, 104, 106, 119, 153, 157, 158, 159, 160, 161, 165, 213, 222, 244, 280, 281, 335, 365
  dispersal ability, 51, 52, 54, 56, 59
Disturbance, 104
Divine creation, 194
Domestic animals, 402
Douglas fir (Pseudotsuga menziesii), 204, 213
Drinking water quality, 314
Drosophila, 210
Drosophila hydei, 210

Drosophila melanogaster, 205, 208
Drosophila mercatorum, 210
Droughts, 62, 97, 100, 190, 281, 318, 339, 357
Dryocopus martius (black woodpecker), 152
DSS, 60, 71, 77
Ducks Unlimited, 18
Dusky seaside sparrow (Ammodramus maritimus nigrescens), 211, 284
Dynamic patch models, 392

E. S. George Reserve, 161, 162, 163, 165
Earth scientists, 400
Earthquakes, 402
Earthworms, 144
Eastern Europe, 308
Ecocentrism, 8
Ecoforum Peace Conference, 308
Ecological integrity, 239, 246
Ecological Society of America, 167, 263, 30
Ecologism, 357
Economics, 363, 388
Economists, 400
Ecosystem management, 4, 7, 12, 18, 23, 24, 25, 26, 27, 69, 98, 106, 109, 154, 167, 168, 171, 172, 174, 176, 185, 186, 192, 262-273, 320-336, 366, 367, 389, 390
Ecotones, 101
Edge effects, 101, 322, 325, 326, 328, 329
Edge habitat, 320
  321, 345
Education, 29, 40, 41, 42, 186, 195, 304, 313, 314, 321, 345, 369, 401
Educators, 347
Effective population size, 203, 205, 206, 207, 286, 290
Eglin Air Force Base, 263, 264, 265, 266, 269, 270
Ehrenfeld, David, 30
Ehrlich, Anne, 30
El Niño-southern oscillation, 100, 259, 339, 402
Elephants, 62, 63
Elk (Cervus elaphus), 63
Emerging diseases, 191
Endangered, 209, 326, 328, 329, 330
Endangered ecosystems, 189, 333
Endangered species, 14, 17, 25, 26, 43, 123, 184, 188, 189, 192, 202, 203, 204, 207, 208, 209, 210, 211, 228, 229, 235, 237, 243, 266, 321, 329, 330, 333, 339, 344
Endangered Species Act, 11, 17, 19, 25, 26, 44, 47, 52, 330, 349, 374, 389

Endangered species management, 232, 233
Endangered species of sunflowers, 209
Endemic species, 87, 125, 130, 131, 135, 190, 229
Endemism, 125, 131, 133, 134, 135, 188, 364, 393
Environmental Defense Fund, 12, 13, 307
Environmental education, 315
Environmental impact assessment, 67
Environmental Impact Statement (EIS), 342, 343, 347
Environmental law, 29, 41
Environmental outreach program, 315
Environmental Protection Agency (EPA), 11, 341, 347
Environmental stochasticity, 376
Environmentalist, 348
Epiphytes, 255
Epixylic mosses, 151
"Equilibrium Paradigm", 237, 238
Erosion, 260
Essex Land Conservation Trust, 251
Eupitecia groenblomi, 151
Eurasia, 214
Euro-American, 343
Europe, 31, 32, 33, 218, 378, 403
European rabbits (*Oryctolegus cuniculus*), 287
Evans, Edward P., 194
Everglades, 189, 255, 258, 310, 380, 381
Evolution, 188, 193, 194
Ewel, Jack, 365
Exotic generalists, 149, 154
Exotic invaders, 150, 365
Exotic species, 148, 154, 243, 258, 306, 307, 332
Extinct, 53, 129, 130, 207, 307, 374
Extinction, 19, 26, 40, 44, 45, 48, 54, 55, 56, 57, 58, 87, 99, 100, 105, 126, 129, 130, 131, 135, 149, 156, 157, 159, 162, 163, 164, 187, 191, 202, 203, 207, 208, 210, 217, 220, 221, 222, 225, 228, 231, 232, 233, 256, 274, 275, 276, 277, 278, 279, 280, 281, 283, 284, 286, 287, 289, 299, 300, 301, 302, 333, 329, 394, 398
anthropogenic extinction, 398
population extinctions, 290
Extinction probability, 231, 232
Extinction rate, 301, 303
Extirpated populations, 335

Fall cankerworm, 144
Family planning, 29, 41, 42

Farming, 402
Fecal coliforms, 314, 315
Federal Energy Regulation Commission, 312
Federal lands, 323, 336
Federal Water Pollution Control Act of 1972, 309
Fertilization, 389
Fertilizers, 401
50-500 rule, 205
Finland, 220, 223, 281
Fire, 12, 20, 62, 63, 66, 75, 88, 97, 138, 142, 150, 151, 152, 171, 172, 175, 176, 177, 178, 179, 180, 181, 182, 183, 233, 252, 256, 257, 264, 265, 266, 270, 271, 303, 321, 330, 331, 332, 345, 355, 375
fire management, 176, 177, 179, 180
fire suppression, 321, 332
"spot" fires, 181
Firebreaks, 266, 265
Fireweed (*Epilobium augustifolium*), 159, 160, 161
Fish, 90, 129, 130, 191, 214, 254, 307, 309, 312, 317, 318, 319, 402
Fisheries, 42, 126, 378
Fisheries management, 84, 85, 378
Fishing, 373
Fitness, 203, 207
Flax, 289
Flax rust (*Melampsora lini*), 289
Flood, 142, 242, 246
Florida, 321
Florida panther (*Felis concolor coryi*), 283
Flower-insects, 278
Flowers, 286, 285, 287
Flux of nature, 73, 238
Fodder, 320
Forbs, 335
Forest, 22, 26, 32, 36, 38, 39, 40, 42, 98, 144, 149, 150, 152, 154, 162, 164, 170, 171, 180, 181, 184, 254, 255, 257, 275, 317, 318, 343, 354, 356, 357, 401, 402
boreal forest, 146, 152, 154
boreal forests, 149, 150, 365
closed-canopy forests, 259
conifer forests, 172
deciduous forests, 133, 151
dry forests, 127
eastern deciduous forest, 331
eastern deciduous coastal plain forest, 134
eastern hardwood forests, 330
eucalypt forests, 355
hardwood forests, 307

Irian Jaya, 135
late seral stage coniferous forests, 49
late seral stage forests, 50
longleaf pine, 25, 26, 27
longleaf pine savanna, 332
mature forest, 322, 331
mountain ash (*Eucalyptus regnans*, F.
    Muell.) forest, 354, 355
native forests, 355
old growth, 22, 138, 150, 151, 152, 154,
    157, 158, 159, 189, 213, 214, 270, 271,
    324, 328, 333, 402
pine bush, 331
pine/hardwood forest, 197
rainforest, 40, 127, 189, 259, 355
reef, 355
second growth, 331
sequoia mixed conifer forests, 177
spruce forests, 150
taiga, 147, 153
trees, 150
tropical forests, 38, 146, 154
wet forests, 152
young forest, 50
Forest ecosystem, 183
Forest fires, 324
Forest management, 150
Forest plantations, 149
Forest regeneration, 260
Forested ecosystem, 183
Forest-interior species, 320
Forestry, 63, 84, 85, 154, 169, 353, 355
Forestry, timber harvest, 50
Fossil fuels, 401
Fragmentation, 52, 103, 104, 105, 217
Frampton, George, 24
Freshwater, 188
Freshwater fishes, 129
Freshwater mollusks, 130
Freshwater mussels, 129
Freshwater shrimp, 318
Frogs, 134, 156
Frosts, 233
Functional redundancy, 98, 140, 169, 190, 243,
    244, 366, 373
Fungal, 144
Fungi, 326
Fungus, 268
Furbish lousewort (*Pedicularis furbishiae*), 85,
    279
Fynbos, 129, 130

Gambel's oak (*Quercus gambelii*), 281
Game, 327
Game management, 188
Gap analysis, 274, 282, 283, 332
Gap dynamics, 150
Gene flow, 213, 285, 287, 288, 289, 290
Generalist exotics, 153, 155
Generalist mammals, 154
Generalist species, 152
Generalists, 147, 148, 149, 152, 153, 154, 155
Genetic assimilation, 216
Genetic differentiation, 211
Genetic diversity, 170, 202, 203, 204, 205,
    208, 209, 210, 213, 216, 286, 394
Genetic drift, 206, 205, 207, 286, 287, 290
Genetic management, 203, 207
Genetic resources, 203
Genetic risks, 230
Genetic structure, 202-216
Genetic variability, 205
Genetic variation, 202, 206, 209, 211, 213,
    286, 287, 307
Geochemistry, 137
Geochemists, 315
Geographic Information Systems (GIS), 21,
    105, 157, 189, 223, 266, 282, 283, 326,
    327
Geology, 137, 343
Geomorphology, 137
George Mason University, 372
Geothermal processes, 314
Geranium bohemicum, 151
Geranium laniginosum, 151
Gerenuk (*Litocramius walleri*), 62
Germany, 308, 401
Gilbertiodendron, 257
Gilchrest, Wayne, 24
Gingrich, Newt, 338
Giraffes (*Giraffa camelopardis*), 62
Glanville fritillary butterfly (*Melitaea cinxia*),
    220, 223, 225, 281
Gleason, Henry, 84
Gleasonian, 18
Glickman, Gretchen Long, 366
Global climate patterns, 369
Golden Rule, 193, 194
Gopher, 257, 260
Gordon, Doria, 365
Gorillas, 270
Gould, Stephen Jay, 86, 91
Graduate students, 318

Grant's gazelles (*Gazella granti*), 62
Grasses, 150, 152, 255, 257, 321, 335
    annual grasses, 48
    perennial grasses, 255
Grassland, 32, 47, 48, 149, 153, 170, 256, 260,
        343, 402
    temperate grasslands, 146
Gray kangaroos, 357
Grazing, 109, 118, 321, 328, 335, 357
Great Lakes, 214, 283
Green movement, 355
Green sea turtles, 212
Grizzly bear, 14, 159
Groundwater, 314, 380
    contamination, 309
    deterioration, 309
Guam's birds, 131
Guavas, 258
Gulf of Mexico, 211
Gypsy moth, 144

H. Allen Mali Research Fund, 250, 251
Habitat edges, 321
Habitat fragmentation, 6, 43, 44, 45, 46, 50,
        52, 53, 59, 82, 83, 88, 93, 97, 99, 105,
        123, 156, 157, 159, 191, 202, 204, 212,
        213, 217, 218, 219, 221, 286, 287, 322,
        326, 328, 330, 333
    genetic isolation, 328
Habitat islands, 275
Haeckel, Ernest, 83
Hanski, Ilkka, 365
Hardy, Thomas, 194
Hardy-Weinberg equilibrium,, 287
Hares, 152
Harold Sioli, 364
Harper, John, 254
Hawaii, 130, 258
Heartrot fungus, 268
Heavy metals, 308
Hells Canyon CMP [Comprehensive Manage-
        ment Plan] Tracking Group, 342
Hells Canyon National Recreation Area
        (NRA), 342, 343
Herbaria, 326
Herbicides, 215, 401
Heritage Programs, 283
Heterozygosity, 203, 205, 206
Highway Departments, 347
Hippopotamus, 77
Hochbaum, Albert, 195

Holsinger, Kent, 364
Home Builders Association, 18
Homo sapiens, 153, 178
Homozygosity, 210
Honduran, 39
Hot springs, 145
Housatonic Rivers, 251
Hubbard Brook, 188
Hubbard Brook Ecosystem Study, 356
Human population, 31, 367, 391, 394
    Human population growth, 29, 30, 31, 39,
        40, 41, 188, 191, 391, 401
Hunting, 30, 40, 62, 373, 389
    Hunting organizations, 342
Hurricanes, 88, 264, 402
Hybridization, 202, 208, 209, 211
Hydrocarbons, 183
Hydrography, 244
Hydrologist, 341
Hydrology, 75, 76, 137, 174, 180, 185, 189,
        242
    Hydrologic conditions, 247, 258
    Hydrologic cycle, 171
    Hydrologic flows, 171
    Hydrologic maps, 380
    Hydrologic parameters, 242
    Hydrologic processes, 242, 246
Hydropower Reform Coalition, 312

I'iwi, 287
Illinois River, 307
Immigration, 222, 275
Imperator cylindrica, 257, 365
Inbreeding, 205, 97
    Inbreeding depression, 202, 204, 205, 210
Indicator species (ISs), 229, 243, 247, 251,
        325, 327, 328, 329, 330, 333
Indirect effects, 136, 139, 340
Indonesia, 135
Industrial melanism, 191
Insect pests, 149
Insects, 127, 150, 152, 157, 159, 160, 254,
        270, 278, 291, 317
Institute of Ecosystem Studies, 4, 12, 372
Instrumental value, 137
Insular habitat, 275
International Institute of Tropical Forestry, 317
Intrinsic values, 137
Introduced species, 41, 188, 190, 191, 214,
        326, 327
    Introduced trees, 128

Invasions of new species, 291
Invasive species, 215, 247, 249, 251
  Invasive plant species, 214
  Invasive weeds, 243
Invisible present, 366
Ips. (bark-beetles), 151
Island biogeography, 4, 217, 218, 226, 227, 274–284, 299, 322
Island biogeography theory, 95, 105, 298, 299, 395, 396
Israel, 109
  Beer Sheva, 111
  Negev, 109, 111, 114, 116
  Negev Desert, 108, 109, 111
  Sayeret Shaked, 111
  Sayeret Shaked Park, 111
Ivory, 41

Janus, 91
Japan, 31, 401
Jasper, 292
Jensen, Deborah, 349
Jewish National Fund (JNF), 110
John Muir Woods, 401
Journalism, 356
  scientific journalism, 357
Journalists, 353, 355

K- and *r*- selection, 357
Karner Blue Butterfly, 331
Kenya, 60, 61
  Tsavo, 61, 62, 63
  Tsavo National Park, 62
Keystone species, 140, 327, 356
Knowlton, Nancy, 365
Kootenay, 292
Kruger National Park, South Africa, 391

La Selva Biological Station, 316
Laboratory microcosms, 291
Ladybird beetle (*Coccinella septempunctata*), 156, 160, 161
Lake Erie, 214
Lakes, 139, 140, 144, 150, 151, 188, 267, 275, 309, 402
Lakeside daisy (*Hymenoxys acaulis* var. *glabra*), 208
Land ethic, 193–198
Land management, 345
Land snails, 130
Lande, Russ, 365
Land-grant universities, 84

Landsat Thematic Mapper, 283
Landscape mosaic, 224
Larix sibirica (Siberian larch), 151
Latin America, 403
Laws of ecology, 354
Lawton, John, 365
Leaf-cutting ants, 257
Leafy spurge (*Euphorbia esula*), 214
Leopold, Aldo, 84, 91, 101, 193-198, 345, 367
Leopold Committee, 63, 175
Leopold Memorial Reserve, 197
Lesser kudus (*Tragelaphus inbergis*), 62
Lichens, 111, 152, 292
Likens, Gene, 29
Lions (*Panthera leo*), 63
Liquid nitrogen zoos, 191
Living dead, 217, 226
Local extinctions, 275, 278, 286
Logging, 138, 142, 144, 173, 214, 257, 321, 322, 355
Long Island Sound, 242
Long Term Ecological Research, 317
Longleaf pine, 25
Longleaf pine/wiregrass, 330
Loxia curvirostra (red crossbill), 151
Loxia leucoptera (two-barred crossbill), 151
Loxia pityopsittacus (parrot crossbill), 151
Lumbering, 373
Lupine, 331

Macroevolution, 190
Macrophytic patches, 111, 112, 113, 117, 118, 365
Madagascar, 188
Mahogany (*Cercocarpus traskiae*), 209
Mammals, 127, 129, 130, 152, 270, 292
Mangrove swamp, 188
Markov models, 396
Marsh, George P., 194
Marsh restoration, 247
Masked bobwhite (*Colinus virgianus ridgwayi*), 333
Maximum sustainable yield (MSY), 339, 370
Mediterranean Sea, 154, 214
Meffe, Gary, 30
Melampyrum spp. (cow wheat), 151
Melitaea cinxia, 100
Menges, Eric, 85
Mesopredator release, 149
Metapopulation, 19, 45, 47, 52, 54, 57, 58, 90, 93, 99, 100, 106, 155, 156, 157, 160, 161, 162, 164, 166, 189, 217-227, 244, 274,

278, 279, 280, 281, 285, 293, 298, 300, 301, 394, 395, 396
extinction/colonization processes, 301
metapopulation dynamics, 4, 57, 217–227, 280, 301
metapopulation extinction, 300
metapopulation management, 300, 301,302
metapopulation model, 99, 281, 300, 301, 365, 391, 394
metapopulation paradigm, 279
metapopulation persistence, 93, 99, 100, 104
metapopulation structure, 166, 201, 217–227, 278, 287, 288, 289, 290, 291, 293, 327
metapopulation theory, 93, 95, 99, 100, 278, 280, 298, 300, 394
Metapopulation as, 278
Meteorology, 137
Methane, 32, 401, 402
Mexico, 48, 129, 132
  Baja California, 47
Meyer, Judy L., 30, 365
Mice *(Mus musculus)*, 154
Micro-macrophytic patches, 116
Microbes, 140
Microcosm studies, 288
Microphytic soil crust, 111, 112, 113, 114, 115, 116, 117, 118
Microtus agrestis (field vole), 151
Migration, 318
Migratory birds, 254, 256
Mineral cycling, 180
Minimum dynamic area (MDA), 325, 330, 331, 332, 397
Minimum viable metapopulation size, 217, 219
Minimum viable population, 90, 149, 231, 298, 325, 339
Mink *(Mustela vison)*, 339
Mississippi River, 307
Mitochondrial DNA (mtDNA), 211, 212, 284
Models, 330
Montane mammals, 284
Moose *(Alces alces)*, 152
Moses, 193
Mosses, 111, 152
Moths, 291
Motorcycles, 373
Mottled duck *(Anas fulvigula fulvigula)*, 283
Mount St. Helens, 156, 159, 365
Mountain goats *(Oreamnos americanus)*, 282
Muir, John, 194
Muller's ratchet, 207

Multiple-Use, Sustained-Yield Act of 1960, 24
Murphy, Dennis, 365
Mussel, 129
Mustela nivalis (least weasel), 151
Mutations, 205, 207, 208, 289
Mutualism, 291
Mycorrhizae, 152
Myers, Norman, 134
Myxoma virus, 287

National Academy of Sciences, 41, 310, 380
National Biological Service (NBS), 18, 321, 326, 332
National Environmental Policy Act (NEPA), 333, 342
National Park Organic Act, 364
National Park Service, 12, 175, 177
National Research Council, 21
National Science Foundation, 317
National Wetlands Inventory (NWI), 246
Native Americans, 176, 178, 179, 342, 343
Natural resource management, 188, 323
Natural selection, 187, 188, 190, 191, 205, 206, 210, 278, 286, 287, 290
The Nature Conservancy, 248, 249, 250, 251
Nature preserves, 275
NBS, 336, 333
Nematode, 254
New Zealand, 31
Newman, Francis, 194
Newspapers, 348, 353
Nickel, 257
Nitrate, 144, 317
Nitrification, 139
Nitrogen, 144, 255, 259, 307
  Nitrogen cycle, 139, 144
  Nitrogen cycling, 190
  Nitrogen fixation, 190
  Nitrogen fixers, 169
  Nitrogen leakage, 365
  Nitrogen mineralization, 144
  Nitrogen-fixing species, 254
"Nonequilibrium Paradigm", 237, 238
Nongovernmental organizations (NGOs), 372, 374
Nonindigenous earthworms, 258
Nonindigenous species, 191, 258, 259, 274, 281
Nonnative pests, 324
Non-point-source pollutants, 306
Noon, Barry, 365

North America, 129, 135, 149, 218, 292, 305, 307, 308, 378, 403
North American Benthological Society, 312, 313
Northern Rockies ecoregion, 292
Northern spotted owl (*Strix occidentalis caurina*), 22, 44, 47, 49, 51, 53, 54, 55, 56, 57, 59, 158–59
Notophthalmus viridescens, 164
Nutrient cycling, 188, 190, 357

Oceanic islands, 322
Olympic National Park, 282
Oregon, 141, 213
  Knowles Creek, 141
Organic Act of 1916, 175
Organization for Tropical Studies, 314, 316, 317
Oryx, 62
Osborne, Fairfield, 195
Overcutting forests, 322
Overgrazing, 332
Overstocking range lands, 322
Owens Valley, 183
Owl, 22, 56, 50, 157, 159
Oxygen, 139, 317
Ozone depletion, 341
Ozone hole, 366

Pace, Michael L., 29
Pacific islands, 129, 130
Pacific Northwest, 189, 50
Pacific Rivers Council, 141, 312
Parakeets, 254
Parasite transmission rates, 288
Parasite virulence, 285
Parasites, 285, 288, 289, 290
  epidemics, 289
  (infectious) parasites, 290
  virulent genotypes, 290
Parrots, 254
Patch dynamics, 60, 61, 73, 74, 75, 76, 77, 83, 85, 86, 88, 91, 95, 264, 355, 356
Patchiness, 5, 6, 82, 86, 96, 93, 94, 95, 98, 99, 106, 103, 104, 105, 108, 116, 111, 113, 114, 115, 117, 118, 119, 123, 124, 147, 150, 155, 157, 159, 201, 218, 236, 241, 244, 253, 340, 362, 364, 392
  functional patchiness, 148, 153
  functional physical patchiness, 155
  graded patchwork, 5, 82
  metapopulation models, 99

metapopulation persistence, 93, 99, 100, 104
metapopulation theory, 93, 95, 99, 100
patch dynamics, 96, 98, 101, 105
patch theory, 95, 96, 98, 99, 106
patchiness paradigm, 106
patchwork mosaics, 105
physical patchiness, 148, 155
Pathogens, 95, 149, 170, 289, 332, 402
Pels fishing owl, 77
Percolation models, 104, 106
Periphyton, 144
Pest insects, 155, 170
Pesticides, 140, 191, 308, 381, 401
Pests, 95, 332
Pharmaceuticals, 401
Philippines, 135
Phosphate, 317
Phosphorus, 139, 143, 144, 254, 270
Picea abies (Norway spruce), 150, 151
Pickett, Steward, 23, 30, 366
Pigs, 258
Pimm, Stuart, 366, 364
Pine barrens, 335
Pinus silvestris (Scots pine), 150, 151
Piping plovers (*Charadrius melodus*), 247
Plant productivity, 113
Planthopper, 272
Poland, 308, 309
Polish Inspectorate for Environmental Protection, 309
Political parties, 355
Politicians, 400
Pollinate, 291
Pollination, 254
Pollute, 183, 308
Pollution, 42, 77, 188, 246, 247, 249, 306, 309, 317, 324, 339, 373
  non-point-source pollution, 310
Polymerase chain reaction, 212
Polynesians, 129
Ponds, 126, 156, 157, 161, 162, 163, 164, 165
  intermediate ponds, 164
  nonpermanent woodland ponds, 162
  permanent ponds, 162
  temporary, 162, 164
  temporary ponds, 162
Population, 4
Population bottleneck, 206
Population extinction, 281
Population genetics, 201, 202-216
Population growth, 32, 39, 40, 41
Population viability analysis (PVA), 4, 44, 90,

203, 228-235, 277, 278, 299, 330, 338, 364, 365, 391, 394, 395
Populus tremula, 150
Porcupine River, 188
Post-Pleistocene extinctions, 284
Prairie, 177, 335
  Palouse prairie, 321
Prairie dogs, 190
Prairie grasses, 170
Prairie potholes, 310
Prairie soil, 189
Prescribed burning, 335
Prickly lettuce (*Lactuca serriola*), 281
Primary production, 139, 188, 190
Primary productivity, 110, 126, 140, 254, 402
Pringle, Cathy, 364
Prodoxid moth (*Greya politella*), 291
Pseudacris crucifer, 163, 164
Puerto Rico, 317, 396
  Caribbean National Forest (CNF), 317, 319
Puerto Viejo School System, 316
Pulliam, H. Ronald, 22, 30, 365, 366
Pulsatilla vernalis (spring anemone), 151
Purple loosestrife (*Lythrum salicaria*), 215, 243
Pyrite, 139

Quantitative genetics, 4
Quasi-extinction, 228

Ralls, Kathy, 364, 365
RAMSAR convention, 242
Rana lessonae (pool frog), 151
Ranchers, 141
Range management, 84
Rangelands, 214
Rats (*Rattus norvegicus*), 154
Recolonizations, 99
Recovery plans, 333
Recreation, 343
  motorized, 343
  nonmotorized recreation, 343
Red-cockaded woodpeckers, 25, 26, 27, 266, 268, 269
Reforestation, 180
Refuge design, 276
Reich, Robert, 23
Representative ecosystems, 364
Reptile, 292
Rescue effect, 222, 278
Research Natural Areas, 332, 335
Reserve design, 106, 229, 236-251

Resistance alleles, 289
Resource management, 18, 24, 28, 84, 85, 91, 323, 367
Resource managers, 90
Restoration, 6, 7, 27, 48, 74, 88, 89, 90, 108, 109, 124, 136, 180, 195, 196, 197, 198, 243, 248, 249, 250, 252, 253, 261, 312, 333, 334, 335, 337, 384, 388
  ecosystem rehabilitation, 127, 292
Rhinoceros horn, 41
Rhinoceroses, 62
Rio Espiritu Santo, 317, 318
Rio Mameyes, 318
Riparian habitat, 141, 171, 312, 313, 339, 343
Risk analysis, 338, 376, 397
Risk assessment, 311, 337, 338, 340, 341, 344
  ecological risk assessments, 337, 338, 339, 340, 341
  toxics risk assessment, 341
Risk-based decision making, 338, 339
River
  Knowles Creek, 141
River restoration, 141
Rivers, 72, 73, 75, 76, 77, 126, 130, 136, 138, 139, 171, 189, 246, 306, 335, 341, 364
Roads, 328
Robertson, Bill, 363
Rodents, 152
Rogers, Kevin, 367
r-selected, 357
Rush skeletonweed, 214
Rust fungus (*Puccinia chondrillina*), 214
Ruttan, Vernon W., 42
Rwanda, 40

Safe Drinking Water Act, 310
Sage scrub, 47
Salmon, 173, 307, 321
Salmonids, 142
Saltwater marsh, 188
San Francisco, 401
San Juan, 318
Sand County, 196
Sand pine scrub, 182
Savanna, 108, 110, 177, 254
Savannization, 116, 110, 113, 118
Sawgrass marsh, 197
Saxifragaceous host plants, 291
Scandinavia, 150
Schweitzer, Albert, 194
Science and Stewardship Program, 250

Scientific and Technical Advisory Committee, 312
Scops owl (*Otus scops*), 280, 281
Screwworm, 210
Sea otter (*Enhydra lutris*), 190
Sea turtle, 254
Seabirds, 256
Seaside sparrows, 211
Selection coefficient, 207, 208
Self-incompatibility, 207, 208, 209, 211, 216
Sequoia National Park, 183
Serotinous cones, 177
Serveen, Chris, 14
Seston, 140
Sewage, 308, 309
Sewage treatment plant, 308
Sexually selected traits, 270
Shifting mosaic, 98
Shrimps, 317, 318, 319
Shrublands, 180
Shrubs, 111, 112, 255, 257
Sierra Club, 313, 321, 322, 323
Sierra Nevada, 177, 181
Silviculture, 266
Silvio Conte Fish and Wildlife Refuge, 250
Simberloff, Dan, 364, 365
Siuslaw National Forest, 141, 213
Siuslaw River, 141
Sivilculturalists, 171
Skeletonweed, 215
SLOSS, 277
Small mammal, 267
Smallmouth bass, 309
Smith, Adam, 363
Sociology, 137
Sodium-chloride-bicarbonate (Na-Cl-HCO₃) water, 315
Soil, 39, 41, 42, 74, 85, 111, 112, 115, 116, 145, 148, 149, 150, 153, 253, 254, 257, 259, 260, 261, 270, 325, 343, 401
    moraine soil, 149
    sediment soil, 149
    serpentine soil, 48, 49
    Soil erosion, 354
    Soil fertility, 259, 260, 382
    Soil heterotrophs, 260
    Soil nutrient, 259
    Soil organic matter, 260
    Soil quality, 260
Solidago virgaurea (goldenrod), 151
Solute chemistry, 314

Sonoran pronghorn (*Antilocapra americana sonoriensis*), 321, 333
Soulé, Michael, 87
Source-sink dynamics, 19, 22, 47, 95, 157, 166, 279, 395
South Africa, 60, 61, 67, 129
    Kruger National Park, 60, 61, 62, 70, 72, 73, 75, 76
Southeast-Asian, 257
Southern Africa, 129
Soviet Union, 305, 308, 309, 396
Spatial statistics, 266
Spatially explicit population models (SEPMs), 156, 158, 159, 222, 266, 330
Specialist-generalist, 154
Specialists, 147, 148, 149, 154, 155
Sphagnum moss, 190
Spirituality, 357
Spix's macaw, 212
Sport fish, 327
Sport fishing, 214
Spruce budworms, 267
Stabilizing selection, 213
Stable isotope analysis, 144
State natural heritage databases, 326
Static patch models, 392
Steelhead, 141, 307
Stewart, Potter, 23
Strategy of hope, 60, 66
Stream, 139, 140, 142, 171
Streams, 144, 306
Succession, 6, 36, 83, 84, 86, 87, 95, 117, 118, 150, 151, 152, 154, 160, 162, 163, 164, 165, 176, 177, 178, 237, 254, 257, 303, 324, 328, 355
    successional patterns, 328
    successional sequences, 328
    Succession model, 164
    Successional stages, 326, 332, 369
    Successional states, 303
Sugg, Ike, 23, 24
Sulfur, 139, 255
Sulfur reducers, 169
Sumatran rhino, 299
Sustainable development, 363
Swamps, 254, 257
Systematics, 42
Systematist, 212, 216, 291

Taiga, 152
Tall Timbers Research, Inc., 270

Tall-grass prairie, 189, 197
Technology transfer, 60, 65, 68, 70
Television, 352, 353
Terrestrial habitat islands, 322
Thanasimus formicarius, 151
The Nature Conservancy, 242, 246, 248, 249,
 250, 251, 270, 283, 307, 313, 343, 349
The Philippines, 40
Theory of island biogeography, 322
Thermodynamics, 188
Thoreau, Henry, 194
Tidal marshes, 246
Tidal wetlands, 242, 246
Tidal Wetlands Research Grant Program, 250
Tidelands, 343
Tiger, 41
Timber, 320, 321
 Timber harvest, 320, 328
 Timber harvesting, 321, 332, 354, 355
 Timber production, 322
Time, 355
Time Australia, 353
Topography, 325
Total maximum daily load, 339
Tourism, 318
Toxic, 341
Toxic chemicals, 309
Toxicologist, 341
Toxics, 326
Toxins, 143
Trace gases, 401
Tracy, Dick, 364
Transgenic organisms, 191
Tree, 26, 40, 42, 49, 108, 127, 142, 150, 152,
 154, 155, 175, 209, 255, 256, 257, 321,
 326, 365, 402
 broad-leaved trees, 254
 cedars, 141, 142
 conifers, 151, 176, 253
 Douglas fir, 214
 functional patchiness, 147
 functional physical patchiness, 147
 giant sequoia, 177, 181
 giant sequoia-mixed conifer, 181
 incense cedar, 177
 longleaf pine, 263, 266, 270, 271
 longleaf pine ecosystems, 265
 longleaf pine systems, 269
 longleaf pine trees, 268
 old-growth, 271
 physical patchiness, 147

 pine, 150, 177
 spruce, 150, 151, 267
 white fir, 177
 yellow poplar, 257
Tree ring fire, 176
Tribes, 342
Trout, 144, 309
Tucson effect, 366
Turkey, 266
Turner, Monica, 365
Twain, Mark, 86

Umbrella keystone species, 44
Umbrella species, 391
Ungulates, 254, 63
Unionid mussels, 307
United States, 3, 34, 36, 42, 60, 62, 63, 123,
 129, 131, 132, 138, 144, 211, 214, 230,
 254, 283, 292, 305, 307, 309, 310, 311,
 317, 318, 321, 342, 396, 401
 Alaska, 292
 Appalachians, 126, 134
 British Columbia, 292
 California, 44, 47, 48, 49, 50, 126, 134, 189,
  287, 321, 366, 367
  Los Angeles, 47
  San Diego County, 47
  San Francisco Bay, 49
 Colorado, 363
 Connecticut, 250, 251
 Department of the Interior, 17
 Everglades, 20
 Florida, 189, 255, 263, 264, 265, 366, 367,
  380
 Hawaii, 127, 189, 287, 365
 Hawaiian, 127
 Hawaiian island, 128
 Idaho, 177, 292
 Illinois, 209
 Indiana, 164
 Maryland, 24
 Michigan, 156, 157, 161, 163, 164
 Moloka'i, 127, 128
 Montana, 177, 214
 New England, 242, 243, 244
 New Hampshire, 188, 242
 New York City, 36
 New York State, 36
 North Carolina, 27
 North Dakota, 214
 Ohio, 208, 209

Oregon, 22, 51, 177, 204
Puerto Rico, 318, 364
Rhode Island, 309
Texas, 25
U.S. and Polish Academies of Science, 308
U.S. and Romanian National Academies of
    Science, 308
U.S. Army Corps of Engineers, 347
U.S. Congress, 24, 396
    House, 11
U.S. Department of Agriculture, 138, 177
U.S. Department of the Interior, 177
U.S. Endangered Species Act, 3
U.S. Environmental Protection Agency, 310
U.S. Fish and Wildlife Service, 25, 26, 250,
    284, 325
U.S. Forest Service, 14, 177, 180, 321, 322,
    325, 343
U.S. National Park Service, 282
U.S. Postal Service, 14, 177, 180, 321, 322,
    325, 343
U.S. Supreme Court, 349
Virginia, 25
Washington, 22, 51
Washington, D.C., 36
Washington D.C., 34
Worcester, Massachusetts, 309
Wyoming, 214
Wyoming Game and Fish Department, 14
Yellowstone, 12, 63
University of Florida, 270
University of Georgia, 316
    University of Georgia's Institute of Ecology,
        315
University of Puerto Rico, 317
University of Wisconsin, Madison, 324
Urban areas, 312

Vaccinium myrtillus [bilberry], 152
Vines, 255
Virulence, 289, 290
Viruses, 402
Vistula River, 308
Vogt, Bill, 195
Volcanic activity, 314
Volcanos, 402

Wallace, Alfred Russel, 194
Warming, Eugenius, 83

Water chemistry, 364
Water hyacinth, 189
Water quality, 173, 247, 249, 311, 313, 314
Waterfowl, 309
Watershed management, 182
Watershed protection, 314
Weeds, 354
Westminster Review, 194
Wetlands, 41, 137, 242, 246, 248, 307, 309,
    310, 349, 373
Wetlands regulation, 310
Whales, 92, 254
White perch, 309
Whittaker, Robert, 29
Wien, John, 363
Wilcox, Bruce, 87
Wild and Scenic Rivers, 335
Wild and Scenic Rivers Act, 138
Wild flax (Linum marginale), 289
Wildebeest (Connochaetes taurinus), 189, 190
Wilderness parks, 175
Wilderness preserve, 182
Wilderness Society, 313
Wildlife and Wildlands Institute, 14
Wildlife fisheries management, 370
Wilson, E.O., 91
Wisconsin River, 196
Wolf (Canis lupus), 152, 189, 195
Woodpeckers, 26, 27, 144, 152, 155
World Conservation Strategy, 275
World Wildlife Fund, 307

Xenobiotics, 191
Xiphocaris, 317

Yale School of Forestry and Environmental
    Studies, 22
Yellow perch, 309
Yellowstone fires, 179, 182
Yellowstone National Park, 62, 63, 141, 176,
    182
Yoho National Parks, 292
Yucca moths, 291

Zaire, 40
Zebra mussel (Dreissena polymorpha), 214
Zebras (Equus sp.), 62
Zooplankton, 139, 140